T0353461

Mastering in Music

Mastering in Music is a cutting-edge edited collection that offers twenty perspectives on the contexts and process of mastering.

This book collects the perspectives of both academics and professionals to discuss recent developments in the field, such as mastering for VR and high resolution mastering, alongside crucial perspectives on fundamental skills, such as the business of mastering, equipment design, and audio processing.

Including a range of detailed case studies and interviews, *Mastering in Music* offers a comprehensive overview of the foremost hot topics affecting the industry, making it key reading for students and professionals engaged in music production.

John-Paul Braddock is a UK based mastering engineer at Formation Audio and is the co-chair of the Audio Engineering Society's Mastering Group.

Russ Hepworth-Sawyer is a mastering engineer for MOTTOsound, teaches part time at York St John University, UK, and is the co-chair of the Audio Engineering Society's Mastering Group.

Jay Hodgson is a mastering engineer at MOTTOsound and a professor of popular music studies at Western University, Canada.

Matthew Shelvock made the transformation from session musician to mastering scholar, but is currently leading business affairs for Chillhop, the well-known lo-fi hip hop radio and label.

Professor Rob Toulson is a leading scholar in the field of music production and is the CEO of RTSixty Limited, the company behind music based applications such as variPlay, iDrumTune Pro, and Drummer ITP.

Perspectives on Music Production

This series collects detailed and experientially informed considerations of record production from a multitude of perspectives, by authors working in a wide array of academic, creative, and professional contexts. We solicit the perspectives of scholars of every disciplinary stripe, alongside recordists and recording musicians themselves, to provide a fully comprehensive analytic point-of-view on each component stage of music production. Each volume in the series thus focuses directly on a distinct stage of music production, from pre-production through recording (audio engineering), mixing, mastering, to marketing and promotions.

Series Editors
Russ Hepworth-Sawyer, York St John University, UK
Jay Hodgson, Western University, Ontario, Canada
Mark Marrington, York St John University, UK

Titles in the Series

Mixing Music
Edited by Russ Hepworth-Sawyer and Jay Hodgson
Audio Mastering: The Artists
Discussions from Pre-Production to Mastering
Edited by Russ Hepworth-Sawyer and Jay Hodgson
Producing Music
Edited by Russ Hepworth-Sawyer, Jay Hodgson and Mark Marrington
Innovation in Music
Performance, Production, Technology, and Business
Edited by Russ Hepworth-Sawyer, Jay Hodgson, Justin Paterson and Rob Toulson
Pop Music Production
Manufactured Pop and BoyBands of the 1990s
Phil Harding
Edited by Mike Collins
Cloud-Based Music Production
Sampling, Synthesis, and Hip-Hop
Matthew T. Shelvock
Gender in Music Production
Edited by Russ Hepworth-Sawyer, Jay Hodgson, Liesl King and Mark Marrington
Mastering in Music
Edited by John Paul Braddock, Russ Hepworth-Sawyer, Jay Hodgson, Matt Shelvock and Rob Toulson
Innovation in Music
Future Opportunities
Edited by Russ Hepworth-Sawyer, Justin Paterson and Rob Toulson

For more information about this series, please visit: www.routledge.com/Perspectives-on-Music-Production/book-series/POMP

Mastering in Music

Edited by John Paul Braddock,
Russ Hepworth-Sawyer, Jay Hodgson,
Matt Shelvock, and Rob Toulson

LONDON AND NEW YORK

First published 2021
by Routledge
2 Park Square, Milton Park, Abingdon, Oxon OX14 4RN

and by Routledge
52 Vanderbilt Avenue, New York, NY 10017

Routledge is an imprint of the Taylor & Francis Group, an informa business

British Library Cataloguing-in-Publication Data
A catalogue record for this book is available from the British Library

Library of Congress Cataloging-in-Publication Data
Names: Hepworth-Sawyer, Russ, editor. | Hodgson, Jay, editor. | Shelvock, Matthew T., editor. | Braddock, John-Paul, editor. | Toulson, Rob, editor.
Title: Mastering in music / edited by John-Paul Braddock, Russ Hepworth-Sawyer, Jay Hodgson, Matt Shelvock & Rob Toulson.
Description: [1.] | New York : Taylor and Francis, 2021. | Series: Perspectives on music production | Includes bibliographical references and index.
Identifiers: LCCN 2020029330 (print) | LCCN 2020029331 (ebook) | ISBN 9780367227197 (paperback) | ISBN 9780367227319 (hardback) | ISBN 9780429276590 (ebook)
Subjects: LCSH: Mastering (Sound recordings) | Sound recordings--Production and direction.
Classification: LCC ML3790 .M354 2021 (print) | LCC ML3790 (ebook) | DDC 781.49--dc23
LC record available at https://lccn.loc.gov/2020029330
LC ebook record available at https://lccn.loc.gov/2020029331

ISBN: 978-0-367-22731-9 (hbk)
ISBN: 978-0-367-22719-7 (pbk)
ISBN: 978-0-429-27659-0 (ebk)

Typeset in Times New Roman
by MPS Limited, Dehradun

Contents

Preface

The UK chapter of the Audio Engineering Society's Mastering Group held the organisation's first mastering specific conference (AES:UKMC18) on the 22nd and 23rd of September, 2018, in central London (see http://aesmasteringgroup.org/ for related details). The steering committee for AES:UKMC18, comprised of editors John-Paul (JP) Braddock, Rob Toulson and Russ Hepworth-Sawyer, initially planned to publish these papers within a post-conference book. However, after realising the extent to which mastering has become a highly relevant (and sometimes controversial) topic amongst audio and music experts, the editorial team decided to move forward with an open "Call For Papers" to allow participation from those who were not able to attend AES:UKMC18. As a result, we moved this volume to the Perspectives on Music Production series, and included mastering engineer and academic Jay Hodgson and label executive and mastering scholar Matt Shelvock within the editorial committee. Readers should note that a next AES Mastering Conference is expected to take place — later than anticipated — in 2022, due to the Coronavirus pandemic.

As with other Perspectives On Music Production collected volumes, no narrative has been set in advance, and none of the authors have been commissioned for contributions. Instead, the call for contributions was open, the conference call broad, and, as a result, Mastering in Music captures a wide set of today's most relevant discussions by researchers, practitioners, and thought-leaders in the realm of audio mastering.

In the name of exploring relevant controversies surrounding audio mastering, two chapters challenge and investigate some commonly held beliefs of audio mastering engineers. Starting with loudness, Jay Hodgson's opinion piece, which postulates that the often-discussed modern loudness 'constraints' are nothing more than freelancer marketing and sales tactics, provides a fresh perspective from a commercial and creative viewpoint. Using a more empirical approach, Scott Harker investigates some of the rumoured playback differences which are said to exist between dominant platforms for music consumption. His chapter presents one of the first data-driven evaluations of this phenomenon available in print.

Darcy Proper, the first female to win a Grammy for audio mastering, and Thor Legvold provide a deeply insightful piece on mastering in surround sound. Aside from some scant attention at conferences and online journals, this topic thus far has been scarcely discussed within peer-evaluated literature. The editors believe this chapter will serve as a new point of departure for studies in surround mastering.

In mastering, and other creative fields, artificial intelligence has become an area of emergent controversy. Andrew Whelan and Thomas Birtchnell cover the ramifications of this technological 'sea-change' as it were, as do JP Braddock and Russ Hepworth-Sawyer, to a lesser extent, via their recommendations for improving upon audio mastering formatting and delivery workflows. In a tangentially related chapter, Steve Collins, Adrian Renzo, Sarah Keith, and Alex Mesker discuss continuities in automated audio mastering and the disruptive powers which have shaken up the industry.

Exploring the hybrid technical and creative dimensions of audio mastering, Austin Moore elucidates the concept of 'glue' in audio mastering. Moore has already done so regarding the use of the 1176 limiter in mixing and music production, and his expansion of this discussion in relation to mastering is timely (Moore & Wakefield 2017). Neil O'Connor explores the aesthetics of the device-modelling and emulation paradigms within mastering plug-ins. Since the use of the DAW and plug-ins remain dominant in today's production scene, this provides a highly relevant discussion for researchers and practitioners.

Mastering in Music also contains numerous chapters which investigate topics such as (i) connections between the creative act of mastering and personalised (or embodied) approaches to the art, (ii) the spaces and places in which mastering occurs, and (iii) how these elements ultimately impact genre and creativity. The 'Creative Mastering Studio' by Alex Hinksman explores environmental interactions in the studio and the interface between the engineer and the rooms they occupy. Regarding more personalised approaches to audio mastering, Stephen Bruel AESUK:MC2018 paper on remastering the Sunnyboys is replicated as a chapter here. A roundtable discussion considering the success of mid-career mastering engineers also provides a more embodied perspective on current approaches to the art of mastering. The perspectives of more established engineers are shared in the interview with Darcy Proper (AES:UKMC18), and the thought-provoking (and amusing) opinion piece provided by Crispin Herrod-Taylor, of Crookwood Designs on the topic of sound quality. The chapter by artist and researcher Stereo Mike (aka Mike Exarchos) discusses the concept of sound staging as it applies to sampling previously mastered records as well as the creation of new music, via a theoretical investigation of the paradigm of hip-hop production.

Carlo Nardi's illumination on the shifting discourse of audio mastering sets the scene for the future. Nardi, as ever, explores the field and how the practice is dynamically shifting towards a new cultural narrative. His paper answers the questions: when does mastering start, how is it automated and what is the new economy around the field? Automation is discussed as well as concepts of new sound aesthetics in an interesting piece from Holger Lund with respects to vinyl.

Although little in this book deals with the topic of pedagogy, JP Braddock's insights on the role of listening and ear training in audio mastering help to bolster this growing area of discussion. We predict that this type of audio education could become standardised music curriculum in the future.

The volume closes with a piece by JP Braddock and Russ Hepworth-Sawyer dreaming about a day when export format management is not such an arduous task – a day when it becomes impossible to miss one of the numerous different file types we have to export for the same master recording. We hope a positive use of artificial intelligence in audio mastering can one day be applied in this regard.

It is our hope that this book provides a discussion of what's happening now and the conversations that emerged from the 2018 conference. Please look out for details of the next AES Mastering Group conference, hopefully in 2022.

REFERENCE

1. Moore, A. & Wakefield, J. (2017) An Investigation into the Relationship Between the Subjective Descriptor Aggressive and the Universal Audio 1176 FET Compressor. In: Audio Engineering Society 142nd, May 20–23 2017, Berlin.

Part One

Mastering: practice

1

The creative mastering studio

Alexander Hinksman

INTRODUCTION

In September 2014 I began a doctoral study programme to explore audio mastering as a new creative culture of post-production. Between April 2015 and June 2018, I interviewed 20 of the world's leading engineers in the field. Each mastering engineer gave rigorous insight into the cultures that underpin their creative processes and my study took seriously these practitioners, like more conventional 'producers', as creative participants in modern aspects of recorded music production. My thesis recognised and further established that although theoretical studies and debates around mastering have remained largely absent, particularly throughout the academic sphere of Media and Cultural Studies, the aesthetic of commercial and mainstream popular music had come to be a subject of the mastering engineers' creative methods. I explained that mastering engineers continue to imbue the consistent timbre and entire playback time of popular music recordings – single or album – through essential processes that are relative to each engineer's agency when using select combinations of creatively affordant signal processing tools in unique studio environments. For the chapter you read today, I have chosen to draw from aspects of my research that focus specifically on the spaces occupied by mastering engineers. This is because for years, many academics and documentary filmmakers honed in on the culture of spaces where the recording or mixing of 20–21C popular music took place. This rich scholarship worked to bolster the cultural significance of such locations. Discourses around reputable studios, such as those explored by Cogan and Clark [1], actively constructed renowned 'temples of sound', where alchemic creative practices were performed through the tracking and mixing stages of production. The mass media also established recording studios as creatively significant and iconic locations bound to artists or producers with star status – places 'where the magic happens' [2, pp. 64–65] (see also [3,4]).

Similar groups of academics and the mass media also observed how conventional record producers evolved from technicians to creatives in demand for recording and mixing. The recording process clearly shook its 'lab coat' image over the course of the 20C [5, p. 220]. Today, scholars such as Matthew T. Shelvock [6], Russ Hepworth-Sawyer, Jay Hodgson [7] and myself exist as part of a growing collective who recognise how analogous aspects of the mastering industry and its development remain comparatively overlooked – how some of the studios that mastering engineers have used to perform sonic adjustments on hit records are absent from wider industry discourse or theoretical studies and debates around popular music production. As part of my resolve to explore studio spaces used for mastering in this chapter, I will account for why this neglect may be so. Moreover, I will offer nuanced perspectives on the concept of mastering studios being better understood as creatively significant places.

THE CREATIVE MASTERING STUDIO

Irrespective of many reputable mastering engineers having moved office at various stages of their careers, an issue I will unpack later in the chapter, the need for mastering engineers to be aurally attuned with the sonic characteristics of their listening space would prove to be an incontestably popular demand. Many of the mastering engineers I interviewed confirmed how efforts are typically made to adjust the acoustic properties of studios in order to construct, in Geoff Pesche's terms, 'controlled environments for listening', as opposed to recording. Pesche, of *Abbey Road Studios* in London (UK), prompted me to compare the swapping of mastering rooms with divorce, insofar as moving somewhere that sounds totally different, having worked in the same room all the time, would require the engineer to reattune. Adam Gonsalves (*Telegraph Mastering*, Portland, OR) positioned mastering as 'the final critical chance at QC from somebody who does this all day in a room specifically prepared for the task'. Robin Schmidt (*24–96 Mastering,* Germany) twice upheld that operating consistently within the same acoustic environment is vital bedrock to the mastering process; 'you press play and you know exactly what you're listening to', he said. From our interview 2015, I gleaned that Schmidt had previously hired an acoustician to design his room in Karlsruhe. Also speaking in 2015, Greg Calbi confirmed Fran Manzella as the reputable acoustician behind the majority of mastering rooms inside *Sterling Sound*'s former and sole location at 88 10th Ave New York, NY. 'He's a genius', stated Calbi. Following my interview with Calbi, *Sterling Sound* publically announced their impending departure from 88 10th Ave and their appointing of Thomas Jouanjean's *Northward Acoustics* to design their new facilities in Edgewater, NJ and Nashville, TN. By 2015, Jouanjean had designed the main studio at *Stardelta Mastering* – a rural Devonshire (UK) facility owned and operated by Lewis

Hopkin, who I later interviewed in 2016. By 2018, Jouanjean had also been commissioned to redesign the mastering suite at Adam Gonsalves' *Telegraph Mastering*. Hopkin described Jouanjean, his choice acoustician, as 'a fantastically knowledgeable guy'.

Spending time with a cross section of mastering engineers affirmed to me that the conventional goal of any specialist asked to design a listening space or control room would be to construct the 'flattest' and most clinical listening environment possible in accordance with presenting circumstances; even the most sophisticated approach to acoustic design and correction will deviate from a hypothetically or mathematically optimal benchmark when unique structural or spatial limitations are imposed. I also learned that internal fixtures and everyday furnishings could affect the acoustic temperament of spaces used for mastering. Lewis Hopkin explained, 'I knew Thomas [Jouanjean] had designed a pretty much perfect acoustic environment. We looked at plots on a screen and the response was as flat as it was going to be' – *for his particular room*, a repurposed Victorian Baptist church, I add. Thus, whilst efforts can be made to achieve sonically and mathematically optimal benchmarks through artificial acoustic treatment, I suggest that each particular mastering room would likely offer nuance and subjectivity to the listening experience. With this being proposed, it is essential I draw attention to how, as Shelvock [6, p. 201] explained, 'phenomenological evaluation of a record's timbral and dynamic configuration informs every audio mastering session'. Standing by this notion, I affirm that we should consider each creative and critical choice made by mastering engineers as a function of the listening experience afforded by their unique but understudied environment. This idea is further informed by a history of music industry personnel making sense of recording studios as musical instruments in their own right. Susan Schmidt Horning [4, p. 90] cited early tropes that would reinforce this concept – *Columbia*'s '30th Street' came to be regarded as the studio equivalent of a *Stradivarius* violin, for example (see also [8, p. xiii], [9], [10, p. 85]). In part, these sorts of impressions are born out of the view that recording spaces offer desirable and distinctive acoustic reverberances that engineers capture through tracking. The former *Liederkranz Hall* in New York City also garnered a reputation for its acoustics. In Schmidt Horning's [4, p. 87] terms, the facility placed 'new emphasis on the sound of the studio, not just the music being recorded'.

Assimilating all this, I suggest that if spatially and sonically acclimated mastering engineers remain in high demand, then their studio spaces deserve much greater recognition and study as culturally or creatively significant places. My argument becomes more justified when considering how imaginably hundreds of label personnel, engineers, and spaces with unique acoustics could be involved in the pre-production or tracking of any one album; numerous other engineers and spaces may then be involved in mixing. Additionally, fast Internet connectivity and digital multitrack production has enabled patchworked, networked, and digital audio workstation-based approaches to production in all areas of the market (see [11, pp. 186–209], [12,13], [14, pp. 9–13]). Thus, lengthy and costly efforts can be spread out over networks of tracking through to mixing, and at the bottleneck of the process, in a new era of more mobile producers, a sole mastering engineer will insist on reshaping these efforts or performing sonic adjustments as a function of their acoustic environment. Under these circumstances, I suggest that each song, track, take, or overdub that pertains to a patchwork project will share in a common thread that is subject to the physical space used for mastering. Moreover, entire discographies can share in a common geographical relevance through mastering.

Numerous other concepts were informed or brought to the fore through my interpretations of research presented so far. First, I noted that whilst mastering engineers may be prone to construct their role as one that offers creative interjections at the final stages of production, some of the interviews had prompted me to consider how such offerings should only be made in mathematically regulated environments. I suggest that this notion would reinforce popular interpretations of mastering as an amalgam of art and science. Second, by considering mathematically devised rooms as a high requisite for their creative work, this would foster the perceived necessity of hiring a specialist to master recordings at a professionally treated facility. It could be entertained that some of the mastering engineers I interviewed also and inadvertently presented more subtle ways through which the same necessity could be encouraged – Calbi describing Manzella as 'a genius' or Hopkin describing Jouanjean as 'a fantastically knowledgeable guy', for instance. All this being said, I observed how not all 20 of the leading engineers would have been in a position where they could have announced having chosen to hire an internationally renowned specialist to ensure the acoustics of their studio are treated or prepared to a more clinical specification. Thus, while some engineers may choose to promote the mathematically devised room as a high requisite for creative work, others engineers such as Jon Astley (*Close To The Edge,* London, UK) and Simon Heyworth (*Super Audio Mastering,* Devonshire, UK) may bind their creative proficiency to deep-rooted and personal familiarity with the unique acoustic properties of a more organic space.

In September 2015, I noted that the mastering room at Jon Astley's home did not show regular indications of having undergone radical levels of artificial acoustic treatment. 'I know [this room] very, very well', said the engineer, who proceeded to explain that the room's ornamental wooden paneling 'tends to absorb quite a lot'. He added, 'The windows are recessed, so you're getting no zing from the glass and my chimney is a bass trap'. By my interpretation, despite Astley having expressed a clear awareness of undesirable acoustic phenomena and how such phenomena may be prevented, the engineer proceeded to convey an innate familiarity with and preference for the natural aural characteristics of his room. Astley confidently signified his favoured listening spot as an area just behind where I sat. 'I know what's happening [there]', he said. Astley then encouraged me to consider how, once engineers have gotten used to their particular room, it may seem counter-intuitive for them to go about making further artificial acoustic adjustments. Ten months after I interviewed Jon Astley, Simon Heyworth remarked that his own home studio, situated in a granite-walled roundhouse, 'is not an easy room'. Like Astley's room, the unique space did not appear to have received extensive outfittings by an acoustician. I did however observe that Heyworth had fixed some modular acoustic panels to the roundhouse walls. He said, '[The room] sounds great though. I quite like the edginess of having to work hard and having to listen carefully to what's going on, and then be able to say, "this is fantastic", "this is a great listen"'. To this, Heyworth added, 'In doing that, what comes out the other end seems to work on all systems'. The engineer had explained all this after justifying his philosophy behind working within the space, by my interpretation of the discussion. Heyworth had stated, 'You can have a room that is perfect acoustically, but not very interesting to work in and actually what you do doesn't necessarily sound great on all systems. It might sound a bit bland. It might be right, but somehow it's all about the spaces in between for me. How the brain reacts to the feel of the performance, especially with acoustic or electric music. Of course, electronic music might react differently'. Having interviewed both of these engineers in their studios, I sensed that Astley and Heyworth bound their own creative proficiency to deep-rooted and personal familiarity with the unique acoustic properties of, by comparison, lesser treated spaces. My interviews with Astley and Heyworth also helped illuminate the concept of mastering engineers performing creative work in direct response to the acoustic characteristics of a single space with which, as the industry voice crucially implores, they should be deeply accustomed and expert.

From all that I have referenced so far, it is apt to emphasise familiarity as the common thread that connected much of the discussion around spaces used for mastering. Broadly speaking, the mastering engineers I interviewed had verified that they operate only in acoustic environments to which they are accustomed. With this being said, I could identify two historic cases where industry reporters and pro audio manufacturers honed in on instances where mastering engineers had formerly compromised this ethos (see [15–17]). American singer-songwriter Tori Amos was reported to have asked Jon Astley to master 'The Choirgirl Hotel' (1998) off location at *Martian Engineering* in Cornwall, and Mandy Parnell (*Black Saloon Studios,* London, UK) was reported to have assisted Icelandic singer-songwriter Björk (Guðmundsdóttir) with completing her 2011 studio project 'Biophilia' at *Ö&Ö - Addi 800*'s studio in Reykjavik. 17 years after the release of 'The Choirgirl Hotel', Astley entertained that mastering at a different location had been an unusual prospect.

The engineer described occasionally and formerly having performed mastering off location with other artists, such as Jools Holland. 'I'd thought, "this is actually quite an interesting route to take", he said, and added, 'But it doesn't really work that well for me'. My interview with Astley encouraged me to consider how transporting equipment off location may benefit the client in so far as them being able hear the mastering process in an environment to which they are accustomed. In such instances, however, the mastering engineer would need to trust their client and the client's ears, as practitioners such as Astley may not be attuned to the sonics of the space. I felt that in a 2016 interview with *Prism Sound* [18], Mandy Parnell conveyed how 'bigger clients' are often conscious of a link between the expertise of mastering engineers and their sustained relationship with the acoustics of a room. Parnell indicated that such clients 'won't move with you', and that 'it takes a couple of years for you really to get into your [new] room and be in control of it'. Though both engineers had worked off location at previous points in their respective careers, I felt that discussions with Astley and Parnell had upheld that creative mastering work is best performed and demanded to be performed in familiar settings. This had been in spite of pro audio industry media stoking the novelty of more networked or portable approaches to production and post-production in the present day. I suggest that this had also been in spite of popular discourse and textbooks commending all that is possible through fast Internet connection and digital technologies that facilitate low-cost opportunities for amateurs to engage in creative mastering processes (see [11, pp. 186–209], [12], [14, pp. 9–13]). With this being said, it is valid to consider that mastering engineers may defend the concept of working in a known space for the simple reason that it would be in their own best interests of managing such challenges imposed by industry discourse and industry media. I also suggest that if engineers defend the importance of familiar mastering rooms in such a way, then this may heighten perceptions of their work as so-called 'dark artistry' or a valuable and tradable form of creativity. The notion of mastering as a 'dark art' is a construction that I draw further attention to later in the chapter.

Whilst familiarity had surfaced as a common topic that connected much of the discussion around spaces used for mastering, I had also learned that the character or design of spaces used for mastering would vary across different facilities and engineers. No matter the circumstance, I suggest that any engineer may justify that the design of their space will enable them to carry out creative work effectively. Those operating out of mathematically devised rooms could justify their doing so through science or through describing the genius of a specialist acoustician. Engineers operating otherwise could justify their doing so through other means. Moving on to other issues, I will now demonstrate how raising the profile of mastering studios to the point of being more widely understood as culturally and creatively significant places is a feat made difficult through various customs.

The research I have presented so far has already unearthed that established mastering engineers can operate in contestably more modest, residential and, or, less acoustically treated spaces. This is interesting, given that Louise Meintjes [19] actively used the term 'iconicity' to denote the visual appeal or condition of 20C recording studios as fetishably iconic, and costly architectural acoustics would become a key part of forming their mystical image. Other aspects of my research informed my understandings of the extent to which it is also customary for established and famed musicians or artists to not attend their mastering sessions. As a researcher, it took a visit to a highly reputable and regarded studio complex for me to fully understand the significance of this. A number of other visits also made clearer the fact that mythologised spaces are so constructed as a result of their closer and more direct connection to creative work performed by more prominent names or historic figures. To show how I arrived at this understanding, I will now offer a reflection on my experiences as a visiting researcher. Subsequently, I will explore the dynamics of unattended sessions.

*

The date is 11 March 2016 and I arrive for the first time in St. John's Wood, London. Prior to meeting with mastering engineer Miles Showell, I browse a shop set adjacent to the Georgian townhouse face of the broad complex that is *Abbey Road Studios*. Music fans lay flowers near a zebra crossing in tribute to the late Sir George Martin – the so-called 'Fifth Beatle' whose passing occurred only three days prior. The fans also add to the vast amounts of faded graffiti on the whitewash walls that perimeter both properties: 'George forever'. Back in the shop, I observe the proud display of noteworthy instruments and recording equipment owned by or loaned to the Studios. Numerous placards uncover the fundamental histories of 'the most famous *recording* [my emphasis] studios in the world' – a slogan *Abbey Road* project extensively via their displays and merchandise. Copies of Alistair Lawrence's [20] 'Abbey Road: The Best Studio in the World' are stacked in abundance ready for purchase. I choose a hardcover from the top of the pile and a T-shirt.

Showell and I are buzzed through numerous locked doors that separate the public from a studio the engineer shares with Frank Arkwright – an experienced and established practitioner in his own right. We pass 'Studio 1' – a 4,876 square foot space where, I recall, the scores of three 'Lord of the Rings' and four 'Star Wars' films were recorded. We pass 'Studio 2' – half as big, albeit big enough to home a famed staircase and to record both *The Beatles* and *Pink Floyd*. It was as Showell welcomed me inside 'Room 30', the newest of several mastering suites, that I recognised how visiting 'the most famous *recording* studios in the world', my emphasis again, was a necessary step for me to take if I were to fully comprehend what Greg Calbi had said five months prior. Mastering, Calbi said, 'it's really not the sexiest part of the recording process'. Mastering, in its former years, Calbi explained, 'happened in a small room, a little bigger than a closet, where you didn't have the musicians and you didn't have a lot of the fun that goes on in the studio'.

In a place where comparisons of space and design could be made, I also perceived the looks and dimensions of Showell's modern room to be comparatively humble when matched against those of iconic Studios 1 and 2. Room 30 offered enough space to home Showell's equipment, mount artificial acoustic treatment on the walls, and position hefty *PMC* loudspeakers in ways that would ensure equal triangular separation between engineer, left monitor and right monitor. The analogue tape machine, cupboard, and sofa positioned at the rear of this set up left scarce space for manoeuvre, or for musicians to be present, and I would later learn that the lathe Showell had been using for cutting records at half-speed was, at the time, stationed in 'Room 5' – the mastering suite operated by Geoff Pesche. On my second visit to the Studios, Pesche revealed that for many years prior to his own arrival in Room 5, the mastering engineer who occupied this also comparatively humble space was the late Chris Blair – 'Mr. Abbey Road Mastering. Worked here forever…'

In hindsight, I would learn that Blair's occupation of Room 5 was a fact that is missing from the pages of a book I had spotted at the back of Pesche's room – the engineer's own copy of the same hardback I had previously bought from the shop. I later noted that Alistair Lawrence [20, p. 192], author of the book, had cited Blair for having mastered *Radiohead*'s sophomore album 'The Bends' (1995) and then later 'OK Computer' (1997). On page 197, Blair was also named for having mastered *Manic Street Preachers*' 'Everything Must Go' (1996). Blair's final occurrence in the book could be found on page 282, where I spotted him pictured and noted for having enjoyed a '35-year career at the Studios'. Prior to Lawrence's book being published, the author had also spoken with one current mastering engineer at *Abbey Road Studios*, Sean Macgee, who offered brief explanations of disc cutting technologies on pages 59–60. Despite mastering having a small degree of presence in the book, Alistair Lawrence neglected to research or comment on the history of rooms and spaces that mastering engineers had occupied in order to carry out work that is clearly respected by their peers at *Abbey Road*. The outsider remains unaware of when such spaces were built, changed or assigned to different engineers. At best, readers learn that 'the TG12310 transfer console […] was installed in six of the mastering suites in the 70s to optimize transfer of audio signals to vinyl' (p. 261). On my third and final visit to the Studios, current 'Head of Mastering' Lucy Launder offered fond memories of Chris Blair. 'He was a star', she said. 'One of the top mastering engineers at the time'.

I suggest that the comparative lack of profile Alistair Lawrence awarded to rooms used for mastering and disc cutting at *Abbey Road Studios* is demonstrative of the fact that these processes are often understood as 'bridges' [21, p. 11] or 'gateways' [22] between production and manufacture. It would have long been considered unfitting, to a certain degree, for manufacturing facilities concerned with procedural disc duplication or even vinyl pressing to be historicised in details commensurate with the sorts of mythical 'temples of sound' used for recording in the mid-20C and onward. I also suggest that the comparative lack of profile Lawrence awarded to mastering speaks of how authors and wider discourse will tend to focus on places that are more closely bound to the classically fabled processes of penning, recording or 'producing' popular music. I suggest that authors and wider industry discourse will focus on these places due to their closer and more direct connection to creative work performed by more prominent names or historic figures such as *The Beatles* and Sir George Martin. Mastering, in its former years, as Greg Calbi had affirmed, 'happened in a small room, a little bigger than a closet, where you didn't have the musicians and you didn't have a lot of the fun that goes on in the studio'. Despite this, and for reasons I have begun to establish so far, I maintain that authors who attempt to document the rich history of recorded music production should dispel this stigma and now offer increased focus on mastering as a creative interjection made at critical stages of the production process – one that it is subject to the room used for carrying out the work.

I would later experience a further and somewhat validating epiphany in respect to what Calbi had told me, and this occurred in the July that followed my final visit to *Abbey Road Studios* in June 2016. Simon Heyworth sat me in the so-called 'sweet spot' of his mastering studio in rural Dartmoor – 'not an easy room', he'd said. The engineer permitted me to hear one of my all time favourite songs played out of the very Dunlavy speakers through which, and in the very room in which it had been mastered. I remember being captivated by the fact I was able to hear Imogen Heap sing 'Wait It Out' in this way. The song had appeared on her 2009 album 'Ellipse', which won 'Best Engineered Album, Non-Classical at the 52nd Annual GRAMMY Awards. Admiring Heyworth's contribution to her recording, I asked the engineer to describe how Heap felt and how she reacted when hearing her finished album while sat in the very spot where I had been placed. She had not attended the session, I learned.

*

From my interview with Adam Gonsalves in 2015, I gleaned that approximately 30–40% of mastering clients had been interested in attending their mastering session at *Telegraph Mastering*. Similarly, from my later interview with Mandy Parnell in 2016, I gleaned this fraction to be around one third at *Black Saloon Studios*. 'Not too often', revealed Jon Astley, when I asked about how regularly he would conduct attended sessions. 'Because [the studio] is in my house and I don't want to do it day in, day out, but I do offer this maybe once or twice a week'. Similarly, Gonsalves explained that he began limiting himself to performing two attended sessions per week and that he had been considering reducing this number. From my interview with Gonsalves, I also gleaned that it is often a specific type of client who will request to attend a mastering session. Gonsalves described prospective attendees as artists for whom he would often be conducting a 'first record' mastering session – 'they've never seen [mastering] before'. Jon Astley associated the prospect of attending a mastering session at his *Close To The Edge* as more important for unsigned artists, and I suggest that this would correspond with what Gonsalves had previously said. Having interviewed Astley, I deduced that the preponderance of work carried out for signed artists would be done so unattended at *Close To The Edge*, and Scott Hull (*Masterdisk,* USA) offered further insight into the nature of attended sessions. My interview with Hull had prompted me to theorise how, after a few records, the trust relationship established between facility and client could in itself eliminate the necessity for attending sessions.

Gonsalves, one of the engineers whose interview had prompted me to draw a link between attended sessions and the unsigned artist, associated his proposed reduction of attended sessions with crossing a certain threshold of busyness. I qualify 'busyness' to denote success in attaining regular work from labels or similar, and also the need to keep up with scheduling. Upholding this, Robin Schmidt had alluded to how attaining more regular streams of work from clients such as record labels could perhaps lessen the necessity of operating in an accessible part of a major city, or as part of an established studio complex. Schmidt identified a modern tendency for established engineers to depart larger inner-city mastering facilities, such as *Sterling Sound* in the USA, move out into the country and convert spaces such as garages into studios. 'You get the files online', he said, being the first engineer to hint at vast changes in digital technology and network infrastructure that have been embraced by music industries of the 21C. At a time prior to each interview, the Internet would inform me that both Schmidt and Gonsalves, two of a younger albeit established generation, owned and operated *24–96 Mastering* and *Telegraph Mastering* respectively from rooms within family-sized homes on residential streets. They served regular and respected clientele – as did a more senior Scott Hull, who chose to move *Masterdisk* from New York City and into the Peekskill suburb of the northern New York metropolitan area, where he spoke for an interview in 2016. From Peekskill, Hull will cut vinyl records for the likes of *Dave Matthews Band* – an artist recongised to have amassed the largest number of concert ticket sales in the 2000s globally. 'You can set up your studio anywhere that has high-speed Internet', he stated, and I further deduced from our later discussions that the trend is for mastering engineers to now be working somewhat unattended. Whilst Mandy Parnell's *Black Saloon Studios* is conveniently situated just a five-minute walk away from the accessible *Walthamstow Central* London tube station, it nonetheless operates from a residential and suburban street. From my interview with Parnell, I grasped that most of the Studios' clients enquire from overseas and also that demand for her expertise is largely based on previous work or from recommendation. Though Parnell understood that established engineers have moved out to the country and still get work, the engineer expressed satisfaction in living her London lifestyle.

Exploring how frequently artists will attend mastering sessions offered further insight into why mastering rooms are comparatively absent from wider discourses that have addressed or mythologised spaces where creativity is performed as part of the whole production process. I have argued that discourses celebrating the penning, recording and 'production' of popular music have strong tendencies for privileging spaces occupied by established artists and, in a more conventional sense, 'producers'. I further suggest that if mastering engineers are now more physically isolated from these people, operating from contestably smaller, rural and, or, residential locations, then this will detract from understanding their rooms as sites where artistic endeavours are fulfilled and creative methods of working are performed. Furthermore, I contend that technological affordances, ever-developing infrastructures of the digital age and also rising costs of inner city real estate, will all together promote the isolation of mastering studios from star figures who can assign their work to any of the engineers I have interviewed. Likewise, it will encourage the growing trend of engineers departing from established mastering facilities. When demand for specialist disc cutting and stereo mastering involved transferring audio from larger physical mediums of storage, such as magnetic tape, I suggest that clients would have enjoyed the convenience of inner city locations or the greeting of a more commercial state of affairs at facilities such as *Sterling Sound* and *Masterdisk* in New York City, or *Abbey Road Studios* and *Metropolis Studios* in London. But with less or no reliance on freight, established engineers are now in a better position to join the growing list of peers who have jumped ship, moved office, and set up shop at a residential location of choice or perhaps a place of economic convenience.

Having established all this, it is necessary for me to divert our attention back to the matter of familiarity. Regardless of circumstances and incentives that may inspire an engineer to depart from an established studio, or an established studio to depart from an inner city location, it could be acknowledged that the very act of doing so would contravene an industry voice that has implored for mastering engineers to operate only in acoustic environments with which they are deeply accustomed. I have already cited Mandy Parnell, who conveyed in a 2016 interview with *Prism Sound* [18] that 'bigger clients' are often conscious of a link between the expertise of mastering engineers and their sustained relationship with the acoustics of a room. Parnell also indicated that such clients 'won't move with you', and that 'it takes a couple of years for you really to get in to your [new] room and be in control of it'. It would seem, however, that not all engineers might have agreed with my interpretations of what Parnell had said, if she had been referring to the prospect of engineers aurally and sonically adjusting to newer studio spaces. For example, though Geoff Pesche encouraged me to compare the swapping of mastering rooms with divorce, he had also qualified that it took about ten days for him to adjust to the sound of Room 5 at *Abbey Road Studios*. I learned that Pesche had helped to refurbish Room 5 after the passing of Chris Blair and his own departure from *Townhouse Studios*, then owned by *Virgin Records*. Similarly, Lewis Hopkin moved *Stardelta Mastering* and began operating in its current premises, outfitted by Thomas Jouanjean, as of January 2015. Hopkin had experienced, quote, 'instant' satisfaction with performing creative work in the newer space. 'There was no adjustment process', said the engineer. 'The room was spot on'. Hopkin later explained that he mastered a number one record just one day after he moved to the new studio. Hopkin also conveyed that the mathematically devised room should afford engineers with the ability to carry out creative work effectively. 'If mastering engineers do one thing', Hopkin argued, in a hypothetical sense, 'then they say, "we have listened to your music on the finest available monitoring and in the finest available environment. We've made adjustments based on what we heard there and we hope that they translate into the wider world quite well"'. It was just prior to this stage of our interview when Lewis Hopkin had explained, 'I knew Thomas [Jouanjean] had designed a pretty much perfect acoustic environment. We looked at plots on a screen and the response was as flat as it was going to be'.

Geoff Pesche and Lewis Hopkin were two engineers whose interviews had encouraged me to broaden my grasp on the concept of profound familiarity with a space as vital bedrock for carrying out effective mastering work. Moreover, the insights of theirs I have shared had prompted me to take a step back and critically examine this prospect as a moot bargaining chip that could be fostered from within the industry and that could serve to promote or construct the creative proficiency of the engineer, unless circumstances are different and, or, other assets take precedence – operating in a space that is professed to be near-mathematically perfect, for instance. Immediately following our discussions of room acoustics, but in no regard to my own reflections I have just outlined, Lewis Hopkin had noted that 'quite a lot of dark science has been banded around to build mystique'. To this he added, 'There's no witchcraft in [the studio.] Just equalisers and compressors'. And whilst the broader cross section of the 20 mastering engineers I spoke with had also been keen to dismiss any clear-cut notions of mastering as a dark art, I do suggest that such images could be fostered through some of the more underlying ways that a mastering engineer may construct him or herself as a creative contributor to the process of recorded music production. In reflecting on my interviews with Pesche and Hopkin, I had taught myself to question whether assertions of having developed profound familiarity with a particular studio space would be examples of one such mechanism.

After I had interviewed Pesche and Hopkin, my discussions with David Mitson inspired me to question the necessity of operating in familiar, mathematically devised, or somewhat palatial spaces altogether. Mitson formerly worked as a mastering engineer at *Sony Music* (previously *CBS Records*) in Los Angeles, and he was one of the last engineers I interviewed. In July 2016, at the time of us speaking, Mitson was mastering on a freelance basis and under the name of *Mitsonian Institute* in the West Midlands, UK. Though Mitson explained that he owns and uses a pair of loudspeakers to which he is accustomed, he prompted me to be objective and entertain the concept of mastering through headphones – despite how such a concept would cut hard against the grain of how the wider majority of mastering engineers agree to work. By July 2016, I had learned that many mastering engineers would insist practitioners in their field could check on headphones, tell if something is wrong on headphones, but they would refrain from mastering entirely with headphones. After speaking with Mitson, I identified the most outspoken outlier and one of the very few proponents of actually using headphones outside the group I interviewed as mastering engineer Glenn Schick. Schick outlined his philosophy on headphones in a very public and thorough manner via his website. Before I had considered headphones with David Mitson, *Sony*'s former engineer had begun alluding to, quote, 'dogs and ponies, "The Emperor's New Clothes", smoke and mirrors'. With more time spent talking to Mitson, I perceived that dogs, ponies, smoke, mirrors, and 'The Emperor's New Clothes' may have altogether encapsulated the engineer's reasoning for having me question some essential mastering studio dogma. 'The Emperor's New Clothes' (Hans Christian Andersen, 1837) is often used as metaphor to illustrate the commitment people have to concepts that are socially accepted as logical or true, when it is against the social norm to question their validity. 'I don't want to sound too strident', Mitson had said. 'There is definitely room for [established engineer] and [their setup]. If Taylor Swift comes in with an entourage, then they need all the facilities and somebody's got to cater to them'. And the engineer stated that none of his clients attend mastering sessions at *Mitsonian Institute*. 'It's as simple as that', he settled.

And so, in the process of reflecting on Mitson's interview and my resulting interpretations, I had started to broaden my grasp on the entire concept of needing a large 'studio', in the classic sense of the word, to carry out effective mastering work. I had been reminded of Robin Schmidt, who when discussing how trust is built between studio and client, had alluded to a philosophy that underpinned his own approach to interior design at *24–96 Mastering* in Karlsruhe. Through experience, Schmidt had learned that working in a studio deficient of, quote, 'super polished' façades could make gaining the trust of prospective clients quite difficult. 'I wanted [my mastering studio] to look in such a way that when people walk in [or see it online, I add,] they know someone's put thought into it', said the engineer. 'It's not to mesmerise or blind people; it's just about gaining that initial trust'. For me, this would later raise questions concerning some of the more strategic and not so creative benefits of operating in established or generously furnished spaces. But whilst David Mitson's words and Glenn Schick's operations had prompted me to maintain a fully objective perspective on the notion of dedicated or plush studio space as a requisite for good mastering work, I still recognised that Schick's approaches to mastering would conflict with those adopted by the cross section of engineers I interviewed and beyond. Moreover, Glenn Schick's more recent approaches to work had arrived after him having used dedicated studio space to master a wide variety of records that were recorded by household names. Therefore, I also considered it valid to question the extent to which demand for any specific engineer's creative input is largely subject to their discography, notwithstanding the realities of their current studio setup. With all this being said, I will conclude my exploring of studio spaces used for mastering by stating that the preponderance of engineers would strongly suggest headphones be used for 'checking' (at best!), and that loudspeakers allowing music to propagate through air should be used as a primary reference setup for listening. In simple terms, there is strength in numbers and time spent in the field has led me to assert that the convention of mastering with a studio monitoring setup is far too entrenched, at present, for headphones to be considered as a pervasive solution to concerns of 'bigger clients' who 'won't move with you', or having to reattune to a new studio.

CONCLUSION

Over the course of this chapter I have explored the creative world of leading mastering engineers by drawing on aspects of my research that consider the spaces they occupy. Although I have questioned the construction of profound familiarity with a space as vital bedrock for effective mastering work, and although I have also questioned the necessity of operating in a mathematically devised space, I do suggest that one particular idea stands to reason. I propose that we have been somewhat taken in by the grandiose nature of studios such as *Abbey Road* or *AIR*, both of which have been understood as culturally and creatively significant through artist narratives and historic sessions of tracking or mixing. Through becoming more cognisant of the range of ways in which a mastering engineer can work with their room as they perform creative interjections to a mixdown recording, we can also begin to consider the creative and cultural significance of the mastering studio. Whether the engineer's room is treated or untreated, I have upheld that the acoustics of a particular studio will offer nuance and subjectivity to the listening experience and the creative mastering process. As I progressed through the chapter, I began outlining various customs that could present as obstacles to our making sense of mastering studios in this way. One custom is that of established mastering engineers operating out of contestably more modest, residential and, or, less acoustically treated spaces. I suggest that from the perspectives of those working outside of the mastering and wider recorded music industries, the aesthetic of some setups may be thought of as banal or less relevant when compared against some of the long celebrated and grandiose studios that have permeated discourse, artist biography and myth concerning the tracking or mixing of popular music. In this chapter, I also explored the issue of established mastering studios remaining and becoming increasingly more isolated, physically and socially, from creative processes undertaken by key figures or prominent names in popular music. This is subject to developments in digital technology and network infrastructure. Through my interpretations of research presented in the latter stages of this chapter, I entertained that there could be more strategic and not so creative benefits to operating in established or generously furnished spaces. I also questioned the extent to which demand for any specific engineer's creative input could be subject to their discography, notwithstanding the realities of their current studio setup.

Contributed in loving memory of Patrick Bryan Hinksman and with gratitude to the mastering engineers who were so generous with their time.

REFERENCES

1. Cogan, J. and Clark, W. Temples of Sound: Inside the Great Recording Studios. Chronicle, San Francisco, (2003).
2. Anderton, C., Dubber, A. and James, M. Understanding the Music Industries. Sage, London, (2012).
3. Massey, H. The Great British Recording Studios. Hal Leonard Corporation, Milwaukee, WI, (2015).
4. Schmidt Horning, S. Chasing Sound: Technology, Culture & the Art of Studio Recording from Edison to the LP. The John Hopkins University Press, Baltimore, MD, (2013).
5. Hull, G., Hutchinson, T. and Strasser, R. The Music Business and Recording Industry. 3rd edn. Routledge, New York & London, (2011).
6. Shelvock, M. T. Audio Mastering as a Musical Competency [Ph. D. Thesis]. The University of Western Ontario, (2017).
7. Hepworth Sawyer, R. and Hodgson, J. Audio Mastering: The Artists (Perspectives on Music Production). Routledge, London, (2018).
8. Moorefield, V. The Producer as Composer: Shaping the Sounds of Popular Music. MIT Press, Cambridge, MA, (2010).
9. Eno, B. PRO SESSION - The Studio As Compositional Tool. Typed and supplied by Bass, D. (1979). Accessed December 2019 from http://music.hyperreal.org/artists/brian_eno/interviews/downbeat79.htm.

10. Marrington, M. Composing with the Digital Audio Workstation. In: Williams, J. and Williams, K., editors. The Singer-Songwriter Handbook. London, Bloomsbury, (2017).
11. Bregitzer, L. Secrets of Recording: Professional Tips, Tools & Techniques. 1st edn. Focal Press, Oxford, (2009).
12. Hawkins, E. Studio-in-a-box: The New Era of Computer Recording Technology. EM Books, Vallejo, CA, (2002).
13. Théberge, P. 'The Network Studio: Historical and Technological Paths to a New Ideal in Music Making'. Social Studies of Science, Vol. *34*, No. 5, (2004), pp. 759–781. Available at: http://dx.doi.org/10.1177/0306312704047615.
14. Wyner, J. Audio Mastering: Essential Practices. Berklee Press, Boston, MA, (2013).
15. Inglis, S. Mandy Parnell: Mastering Björk's Biophilia. Accessed December 2019 from https://www.soundonsound.com/people/mandy-parnell-mastering-bjorks-biophilia.
16. Miller, J. Tori Amos: Tales from the Studio. Audio Media. (2003). Accessed September 2015 from http://www.yessaid.com/interviews/03–10audiomedia.html.
17. SADiE. Mandy Parnell Uses Her Prism Sound 'Box of Tricks' to pull Björk's Biophilia together. Accessed December 2019 from http://www.sadie.com/sadie_news.php?story=0093.
18. Prism Sound. Mandy Parnell Interviewed in Her Studio Black Saloon. (2015). Accessed December 2019 from https://www.youtube.com/watch?v=bGiYzlNC3NE.
19. Meintjes, L. The Recording Studio as Fetish. In: Sterne, J., editor. The Sound Studies Reader. London, Routledge, (2012), pp. 265–282
20. Lawrence, A. Abbey Road: The Best Studio in the World. 1st edn. Bloomsbury Publishing, London, (2012).
21. Katz, B. Mastering Audio: The Art and the Science. 1st edn. Focal Press, Oxford, (2002).
22. Nardi, C. 'Gateway of Sound: Reassessing the Role of Audio Mastering in the Art of Production'. Dancecult: Journal of Electronic Dance Music Culture, Vol. *6*, No. 1, (2014), pp. 8–25.

2

Surround + immersive mastering

Darcy Proper and Thor Legvold

INTRODUCTION

Immersive and surround sound is literally all around us, as part of our daily lives and experiences as we move through the world. Cinema-goers today can expect an immersive or surround sonic experience as part of the film. Since television moved to HDTV (High Definition Television) nearly two decades ago, 5.1 surround sound has been required for broadcast [1]. For computer gaming enthusiasts, nearly all of the sounds in the game are designed for surround or immersive audio, as are other forms of AR (Augmented Reality)/VR (Virtual Reality) entertainment. Any visitor to an amusement park will recognise that an immersive soundscape is an integral part of the experience, particularly for rides and attractions, while there are sound artists such as Bill Fontana, David Miles Huber, and Ozark Henry, among others, who design soundscapes that envelop and surround the listener adding a compelling auditory component to their music, art installations, museum exhibits, and special events. Work is being done in the medical field exploring the use of immersive audio and virtual reality to help speed patients' recoveries from strokes and other neural damage. And let's not forget that real life itself is immersive – our ears are receiving audio cues from all directions, all the time.

Mastering for surround and immersive audio follows the same general principles as for traditional stereo, after all, the goal is the same. It is the last stage of adjustment and the final polish, serving as a bridge between the artist's/producer's musical intention and the end listener, taking into consideration requirements of the final delivery format and any relevant technical standards as well as the artistic aspects of the project itself. However, whereas the practice of mastering for stereo is well established and its tools have gone through many decades of development and improvement in the hands of talented industry professionals, the areas of surround and, more recently, immersive audio represent a relatively new and less well-charted domain.

In this chapter we'll attempt to explore the realities of working in immersive audio and share the experience and advice of pioneers in this developing field with those who want to learn more. We'll focus specifically, although not exclusively, on mastering music intended for playback in discrete channel-based and object-based formats, and touch on working with Ambisonics and other alternative formats where possible. Ambisonics has found a niche in academia and research, as well as gaming, especially with AR/VR, but hasn't (yet) proven viable as a consumer music format, although that seems to be changing of late with adoption by Sennheiser, Qualcomm, and others. As such we won't spend as much time discussing mastering for Ambisonic delivery formats but encourage interested readers to watch for further developments in music production on this front.

A BRIEF HISTORY OF SURROUND AND IMMERSIVE AUDIO

Before looking at specific concerns and practices for surround and immersive audio, it is perhaps valuable to first define some terms.

Surround sound is defined as single layer audio with more than two channels of information – 'layers' indicating speaker placement on a horizontal plane, 'channels of information' indicating discrete, non-correlated audio streams or channels. Stereo, for example is two channels in one layer, which can be comprised of Left/Right or Mid/Side.

Surround sound began in earnest around 1970 with quadrophonic, now often referred to as '4.0'. It was four channels of audio reproduced by four speakers arranged in a square around the listener – left & right front, and left & right rear at 90-degree angles. Some creative mixing was done for 'Quad' but while it often sounded great in the studio, getting it to the consumer was difficult. Vinyl formats required special playback and decoder equipment and offered less than ideal channel separation due to the physical limitations of the vinyl itself, while quadrophonic 'reel-to-reel' and '8-track tapes' were cumbersome and lacked convenience. Hence quadrophonics was deemed a commercial failure and phased out by the late 1970s.

Modern Surround sound is probably most well known in cinemas and home theatres in the form of 5.1 and 7.1 surround (Left, Centre, Right speakers across the front with Left and Right speakers in the back for 5.1, while 7.1 adds two additional speakers as left and right side fills.) A more complete overview is available on Wikipedia [2].

Music productions in 5.1 surround first reached the home in the mid-1990s via DVD-Video. (DVD stands for Digital Versatile Disc.) DVD-Video was developed by the DVD Consortium (later called the DVD Forum) which was a large group of major electronics and media companies. The focus of the format was, as one might suspect, on video, but 5.1 audio could be encoded on the disks in a compressed format via Dolby Digital or DTS 96/24.

In 1999, Sony and Phillips introduced the Super Audio Compact Disc (SACD), intended as a replacement for the aging Compact Disc format. It offered high resolution audio using Direct-Stream-Digital encoded 1-bit audio with greater dynamic range (120 dB), extended frequency response (20 Hz–50 kHz), 110 minutes playback time in stereo per disc and up to six channel surround sound. Mastering Engineer Simon Heyworth was there at the beginning, having been asked by Sony to help remaster legacy albums for the new format, including the now iconic Tubular Bells album. Heyworth worked directly from analogue tape safeties of the original 16 track masters and he recalls, 'it was just like being in the studio when we first mixed it, it was pristine, and sounded fabulous!' [3]

Shortly thereafter, the DVD Forum launched an alternative format, DVD-Audio. It offered many of the same advantages as SACD – high resolution multi-channel audio, supporting up to six channels at 96 kHz 24 bit (and 192 kHz 24 bit in stereo or mono) plus long playing time and planned compatibility with the large number of already installed DVD-Video players in home systems. Audio was encoded on the disc either as linear PCM (Pulse Code Modulation) data or MLP (Meridian Lossless Packing).

In spite of their excellent sound quality, there are a variety of reasons why SACD and DVD-Audio failed to gain traction in the market and ultimately floundered. Despite the current limited attraction of the formats, a small number of releases are still being produced, especially for SACD/DSD. There are lessons to be learned by the failure of either format, from the format war and marketing campaigns that confused consumers, to the additional playback equipment required, and the difficulty in creating adequate and convenient references for client approvals.

In 2006, the Blu-Ray Disc was officially introduced. With its significantly increased storage capacity and playback bandwidth, it was a natural successor to SACD and DVD-A. Blu-Ray could accommodate 6 or more channels of 96 kHz 24 bit audio (in some cases much higher sample rates) on one disc in uncompressed linear PCM format. Household data transmission and storage capabilities were not yet up to a specification that would easily allow large high-resolution audio files to be streamed or downloaded, so a physical medium was of interest (and still is in many parts of the world.) Stefan Bock and the team of engineers at msm-studios in Munich developed the underlying Javascript code for PureAudio Blu-Ray – fully compatible with standard Blu-Ray players and focused on audio-only releases (or those with very limited video content). Pure Audio Blu-Ray allows the consumer to easily operate the disc without the need for a video screen; playing back with the simplicity of a CD and switching between multiple audio streams at the touch of a button, thus making it a truly convenient universal consumer playback format.

Naturally, as technology was expanding, progress wasn't going to stop at mere surround sound on the sonic front. In the early 2010s, the spirit of 'onwards and upwards' compelled audio engineers to start exploring with height for true three-dimensional reproduction of sound, and Immersive audio was born. Immersive audio builds on the surround formats by adding additional layers above or below the main horizontal plane, most often in order to represent height. As with many emerging and new technologies, there are a plethora of competing formats and approaches, which can be broken down into three main types: discrete (channel-based), object-based, and scene-based. These will be covered in more depth as you read further in this chapter.

With the ubiquity streaming today, there are a number of formats for delivering surround and immersive content electronically. Prime among them is Fraunhofer's MPEG-H which accommodates encoding of discrete audio channels, audio objects, or higher order Ambisonics (HOA), with decoding implemented at playback. Primarily aimed at the broadcast, streaming and post-production markets, it employs lossy compression on the audio to reduce bandwidth requirements – similar to MP3 and AAC encoding [4]. There are also a number of online sites such as HDTracks, AcousticSounds, and record label-based sites selling high-resolution downloads which include surround and immersive content, making it available as FLAC, ALAC, WAV, AIFF, and DSD files.

WHY IMMERSIVE AUDIO?

Immersive audio allows creative people to break free from the constraints of the 'stereo straightjacket' that has existed since the 1950s, allowing much greater artistic freedom. Immersive audio provides exponentially expanded opportunities for an artist to express him/herself on an encompassing stage, and greater freedom to create a compelling emotional connection and riveting listener experience. It represents an entire new playground on which artists can explore and create.

We live and operate in an immersive world, our auditory perceptual system registering and decoding sound waves all around us in real time as we move through the world. Stereo represents a convenient compromise but neglects the fact that sounds from an instrument or scene activate a space and interact with it. Our brains perceive stereo as an emulation of reality, and process accordingly to recreate an illusion. With immersive audio, no such processing is necessary, there is no illusion, rather a complete sound field ready to be experienced directly, which creates a much more powerful connection to the listener.

Immersive productions generally require far less signal processing than stereo productions, since there is no need to try to stuff a

360-degree experience into two channels. The result is that the music sounds more natural, lifelike, and realistic. This enhances the realism and relationship with the music and what the artist is trying to convey. Working in immersive can be more forgiving in that you can hear all elements much more clearly and how they work together musically without fighting the other elements for space as they would in a typical stereo production. At the same time, it's much more revealing of poor technique and mistakes on both the artist's and engineer's side. Daniel Shores explains his philosophy when mastering immersive audio similarly, 'I want to be able to not hear anything that isn't music. No speakers, artefacts, etc., just full transparent 3D, fully immersed and connected with the music' [5].

Immersive is quickly replacing stereo as the de-facto format for television, and has done so long ago in the film world. HDTV mandates 5.1 surround sound as a minimum for broadcast, UHD offers additional object-based formats such as Dolby Atmos alongside Ambisonics. Sports has been one broadcast area to rapidly adopt new technology and multichannel formats in order to increase viewer excitement and engagement. MPEG-H 3D audio streams are used by several national broadcasters and offer enhanced viewer involvement; a listener can decide which objects or mix version he wants to hear, e.g. one with the dialog 3 dB louder, or the crowd at a sporting match attenuated. Cinema moved to surround decades ago, and most major titles today are released in immersive Dolby Atmos, DTS-X, or Auro-3D/AuroMax. More and more car audio systems offer surround or immersive audio by default, actually upmixing stereo content to suit.

All of these aspects serve to elevate the argument for working in immersive audio, most of all the ability to provide a uniquely engaging and emotionally moving experience far beyond what we experience with stereo material. As Stefan Bock of msm-studios notes, 'I've seen artists with tears in their eyes as they heard their songs in immersive for the first time. It's a powerful experience' [6].

AUDIO FORMATS

As previously mentioned, there are three main categories of formats currently being used in music production for immersive audio: Discrete (Channel-based), Object-based, and Scene Based (Ambisonics). The latter two are designed to separate the production playback configuration from the listener's and adapt to whatever playback system the consumer has available. All of these formats may be optionally delivered in headphones, with several options for encoded headphone reproduction. The more advanced headphone systems include some sort of HRTF (head-related transfer function) profile which characterises the effect of the head and structure of the ear on the perception of sound, creating a better listening experience in headphones. Some companies like Smyth Research are selling units that measure the user's own head in an actual room for a completely personalised measurement, while Sony currently uses a cellphone picture of the user's ears to find an appropriate match in their database.

In Discrete immersive audio, the speaker layout is predefined and standardised (see Figure 2.1). The main formats of this type are NHK 22.2 [7] and Auro-3D [8], which offers 9.1, 10.1, 11.1, and 13.1 variations. Auro-3D offers three distinct horizontal layers, the main layer where traditional stereo and surround speakers would be placed, at ear level, a height layer further up and surrounding the listener, and a 'voice of God' speaker top centre. 22.2 was developed by NHK (Japan Broadcasting) initially as an experimental format to complement their new Super HD video format (now standardised as UHD TV [9–11]) and later found broader acceptance within the audio and post-production community. It too offers three layers, a main ear-height layer, an upper layer, and a lower layer below the main one (Figure 2.2).

Object-based audio can be challenging to comprehend, and object-based formats differ considerably in their approach from discrete

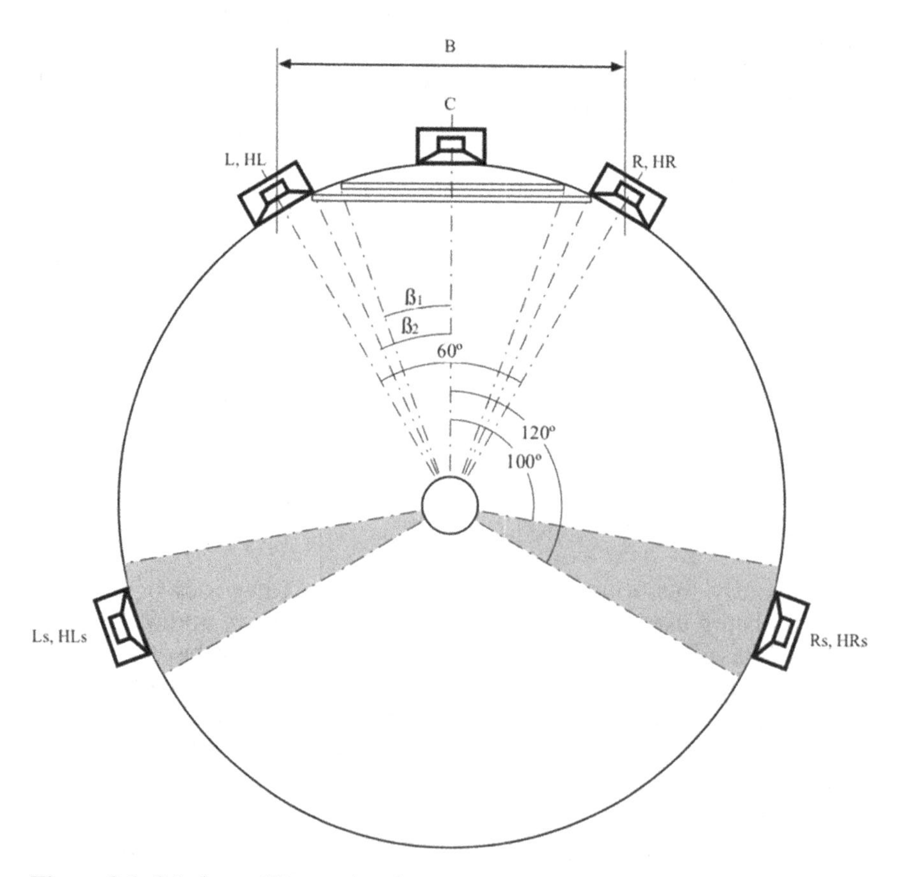

Figure 2.1 9.1 Auro-3D speaker layout

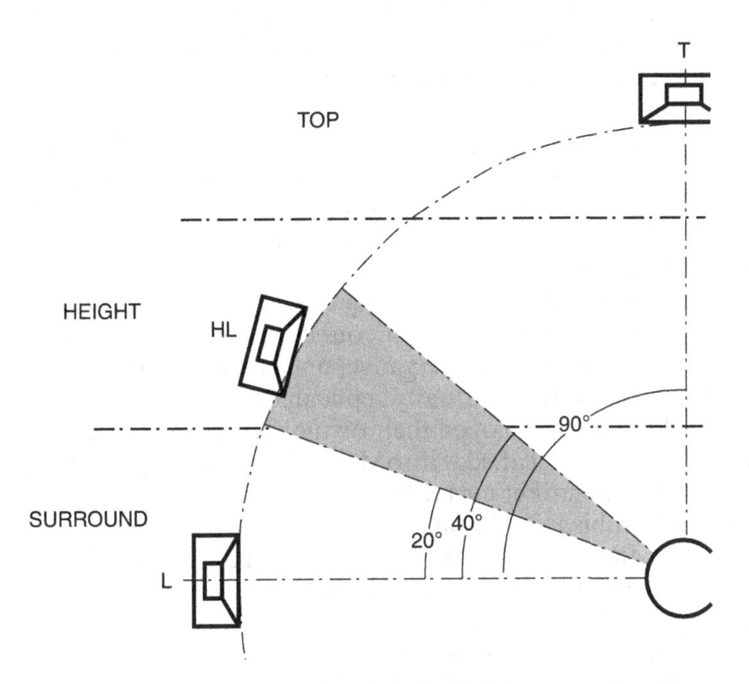

Figure 2.2 Auro-3D speaker elevation, showing optional top speaker

ones by separating the production and playback environments so that the speaker configuration of the playback system may cover a wide range of variations. Put simply, object-based audio is audio that contains metadata that describes it, including gain, placement in a soundfield (be it mono, stereo, surround, or immersive), and more. The major difference between object-based and discrete formats is when, where, and how assignment to channels occurs. In discrete it's done in the mix room (or mastering room) and locked into specific channels at delivery. In object-based it's defined in the mixing or mastering room through metadata, but not actually performed until played back on the listeners' system.

As with discrete formats, there can be a discrete 'bed layer' of channels, usually 5.1 or 7.1, and in Atmos offering two height channels (7.1.2). In addition to this, one works within a virtual audio sound field where objects (independent audio elements with accompanying metadata) are placed and manipulated within the sound field. While the 'bed' channels are static and fixed to speaker positions, objects may be manipulated individually across the entirety of the sonic space. On playback, a decoder calculates how to recreate the master based on the encoded metadata and given physical speaker configuration. This means that the engineer can never be completely certain how things will translate on any given system, but in practice it generally seems to work well. The two dominant object-based formats today are Dolby Atmos and DTS:X, which are installed in the majority of movie theaters in the USA and Europe as well as built-in to home audio receivers around the world. At the time of this publication, Dolby Atmos has begun to expand its focus to include music production on a large scale, partnering with UMG and WMG labels and with streaming services Amazon Music HD and Tidal, while DTS:X has chosen to focus mainly on the commercial cinema market and television post-production, and therefore is not commonly found as a consumer music release format (Figures 2.3 and 2.4). Sony's 360 Reality Audio is a newly-released object-based format based on 13 speakers (5 in a surround configuration on the main level, 5 corresponding speakers above, and three in front below the main level) which are then virtually replicated in headphones for the end listener. Auro has developed their own object-based format in collaboration with Barco, called AuroMax, that is also primarily focused on the cinema market currently.

The concept of ambisonics began with the late Michael Gerzon, who built on Alan Blumlein's work developing stereo, and has been developed and refined beyond his initial research. Ambisonics takes a very different approach than both discrete and object based formats, as Julien Robilliard, product manager at Fraunhöfer IIS explains, 'Ambisonic signals are scene-based audio elements that describe not individual sources (like channel-based or object-based formats), but rather the sound field as a whole from one point in space' [12].

Currently the most widely used ambisonics format is the basic four-channel first order B-format. Somewhat oversimplified, the four

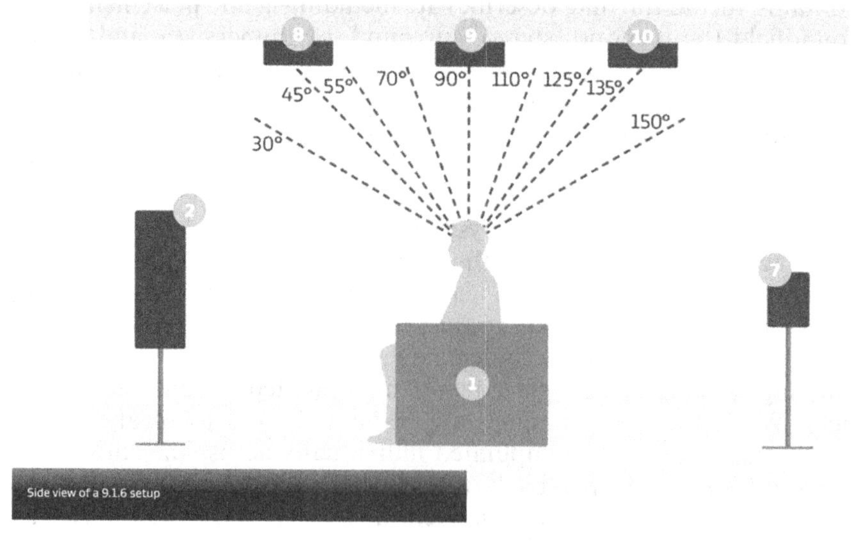

Figure 2.3 The authors' interpretation of Dolby Atmos 9.1.6 – plan view of main layer speaker layout

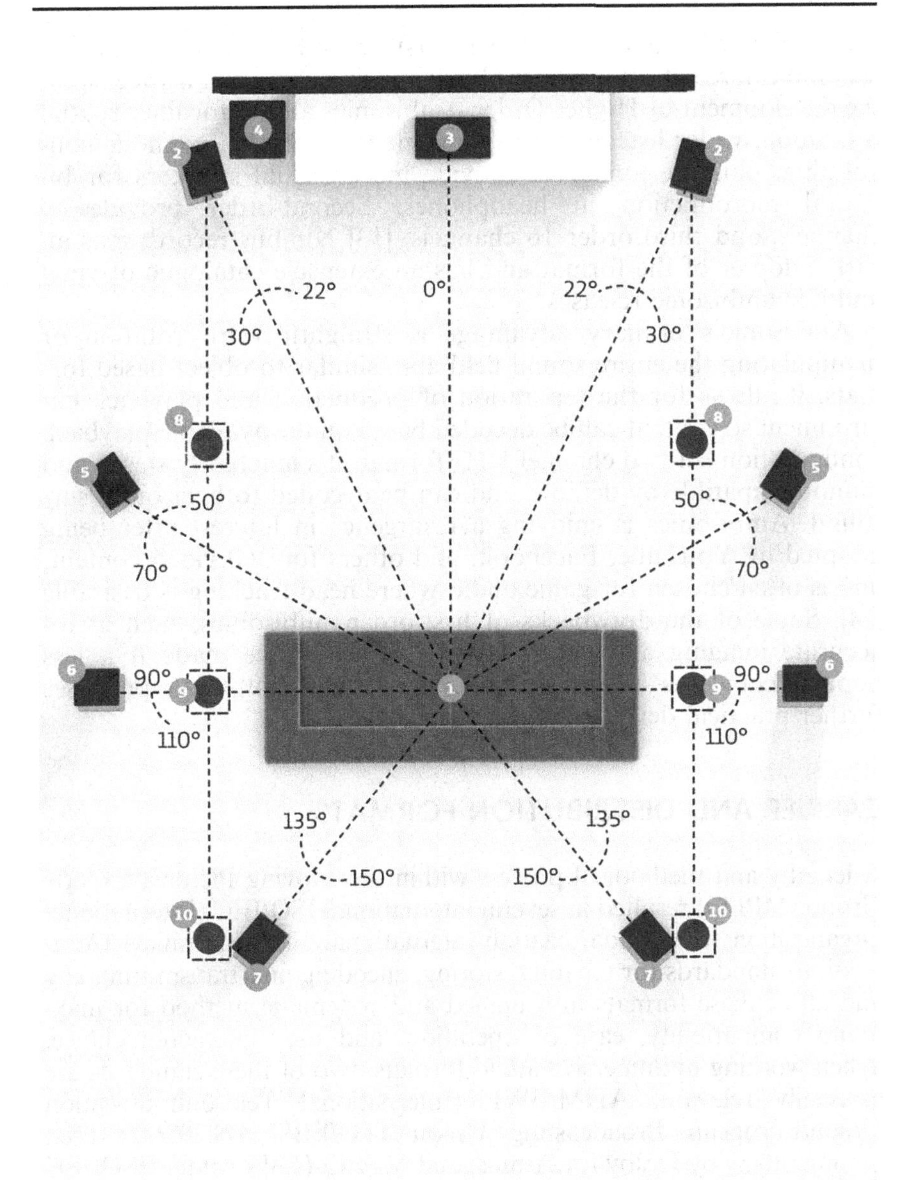

Figure 2.4 The authors' interpretation of Dolby Atmos, elevation view of height channels. Note that height channels may have front, middle and rear, and that the angle between L and R pairs and the listening position should also be approximately 45°. Notice the difference in angle relative to the listener compared to Auro-3D, which specifies 30°

components (W, X, Y, and Z) can be said to represent a directionality in a 360-degree sphere – centre, front/back, left/right, up/down. By manipulating differential gain and phase relationships, these channels can be combined to create a three-dimensional representation of sound. While spatial resolution in first-order B-format is relatively low and can result some rather blurry imaging and a small sweet spot, the development of Higher Order Ambisonics allows for finer spatial resolution, wider listener sweet spot, more flexible microphone convoluting, plus the ability to provide more virtual speakers for binaural reproduction in headphones. Second-order provides 9 channels, and third-order 16 channels. [13] Nimbus records was an early adopter of the format and has an extensive catalogue of vinyl and CD ambisonic releases.

Ambisonic's primary advantage is straightforward rotation or manipulating the entire sound field and, similar to object-based formats, it allows for the separation of production and playback environment so content can be decoded based on the available playback configuration. In two channel UHJ format, it's inherently stereo and mono compatible by default, and can be decoded to horizontal surround. Ambisonics is enjoying a resurgence in interest after being adopted by YouTube, Facebook, and others for 360 video content, and is often chosen for game audio where head-tracking is desirable [14]. Some of the drawbacks of first-order ambisonics, such as inaccurate imaging and comb filtering effects, have made it a less popular choice as a music production format, but HOA promises further practical development in that field.

CARRIER AND DISTRIBUTION FORMATS

A lengthy and methodical process within the Moving Pictures Experts Group (MPEG) resulted in several international ISO/IEC (International Organization for Standardization/International Electrotechnical Commission) standards for creating, storing, encoding and transmitting any and all of these formats in a unified and systematic method for maximum compatibility, ease of operation, and user interaction/choice. When working in immersive audio formats, two of these standards are especially relevant, ADM/BWF (International Telecommunication Union/European Broadcasting Union ITU/EBU BS.2076-2) used among others by Dolby for Atmos, and MPEG-H 3D Audio (ISO/IEC 23008-3), a lossy encoded 48 kHz streaming file format used by recent newcomer Sony 360 Reality along with a number of other companies [15–17]. 360 Reality was unveiled at the CES in New York City autumn 2019 and has support from several major labels as well as AMAZON, TIDAL, Nugs.net, and Deezer [18]. Readers are encouraged to read further on the benefits and features these standards afford content creators and artists for creative possibilities beyond traditional stereo reproduction.

Physical media is alive and well, despite rumours to the contrary, and is still the dominant format in markets where internet infrastructure may not allow high speed broadband. Blu-Ray Disc (BD), often in the form of PureAudio Blu-Ray, is the most common physical format for immersive audio for music productions, with a small number of SACDs still being produced.

DIGITAL AUDIO WORKSTATIONS

In order to work in any of these formats, it may be necessary to choose a Digital Audio Workstation (DAW) that supports the desired production format. Some common choices are Avid ProTools, Merging Pyramix, Steinberg Nuendo, Cockos Reaper, and Fairlight (built into Davinci Resolve), although there are other options available. Be aware that some production plugins (e.g. Dolby Atmos) only support a select few of these DAWs, currently ProTools, Nuendo, Logic, and Ableton, so check to ensure any plugins or other processing required for your desired format is supported, as well as the required bus-width.

More detailed information about these and other formats is available online for free. For the duration of this chapter, we will refer to both surround and immersive as simply immersive audio, as both envelope the listener and offer the potential to create a soundscape around the listener, unlike stereo.

ROOM SETUP AND CALIBRATION

Any audio mastering needs to take place within a calibrated and tuned monitoring environment. The acoustics involved when working in immersive are complex and should ideally be referred to a specialist's attention.

For surround work, a 5.1 or 7.1 monitoring setup should be chosen with identical full range loudspeakers at each channel position. This is covered in more depth in NARAS and ITU specifications available online at: https://www.grammy.com/sites/com/files/surroundrecommendations.pdf (P&E Wing Recommendations for Surround Sound Production) and www.itu.int (BS.775-3). Note that the placement of the rear speakers differs between the two standards. The ITU standard recommends the speakers be placed quite wide at 110-degrees from the centre, while the P&E Wing recommendation states a range of 110-degrees to 150-degrees with 135–150-degrees being considered optimal, placing the speakers closer together behind the listener. It is generally viewed as a matter of taste, determined by the preference of the listener and, as often as not, the physical layout of the room itself. The wider ITU set-up is often preferred for content where the rear speakers contain mostly audience or ambient information, while the narrower 135–150-degree configuration allows for better phantom imaging in the rear.

For immersive mastering, additional height speakers need to be installed and calibrated to match the other transducers according to

Dolby, Auro or Sony 360 specs, also available online at each respective manufacturer's website. Auro has typically embraced a 5.1.4 configuration for music production, now expanding to 7.1.4. Sony's 360 Reality Audio specifications indicate that production work should be done on a 13 channel speaker set-up reflecting the virtual speaker configuration used for their headphone-only consumer playback. This is similar to the Auro-3D 9.1 configuration with the addition of an upper centre speaker and three lower speakers – L, C, R – below the main ear-level set, and without a subwoofer. Dolby considers 7.1.4 the minimum configuration for a home Atmos mix room, and for music 9.1.6 is considered optimal [19,20].

While the number of speakers involved may be the same in some cases, there are differences in speaker layout between Atmos and Auro so it isn't a matter of simply setting up more or less identical speaker setups and switching formats in the box, although with some adjustment in speaker placement one can transition between the two with a minimum of effort. The primary difference is the placement and layout of the height channels, as both use the same standard for the main layer (5.1 or 7.1).

Feedback from a number of mastering engineers suggests that Auro tends to be a bit better at separating instruments and space vertically compared to Atmos. Presumably, this is due largely to the height speakers being placed at a 30° angle above the listening position and directly above the main layer speakers for Auro, rather than the 45°-angle overhead position recommended by Atmos. Ralph Kessler described a project that had a church choir up in the balcony by the organ – in Auro the sense of height and natural placement choir relative to the orchestra was preserved, whereas in Atmos one could risk having the choir singing from overhead (which may be a very desirable effect in other cases) [21]. For overhead effects, Auro does offer the option of a centre height speaker directly above the listening position, sometimes called the 'Voice of God' speaker.

In any case, we would currently recommend a minimum seven channel main layer and four height, plus LFE as a minimum for professional mastering work.

When working in ambisonics, traditional surround or immersive setups are suboptimal, a regular polygon or solid is preferred, depending on order, but rarely practical. For first order ambisonics a square or cube is ideal, for second or third a hexagon or octagon (octahedron or dodecahedron in 3D, respectively). I.e. one generally needs a number of speakers equal to the number of virtual speakers each order provides, and placement is significantly different than traditional surround and immersive formats. There are methods of decoding ambisonic material for 5.1, Atmos and Auro-3D, but all offer significant compromises, due in part to the non-symmetrical layout of these more common playback formats [22–24].

Both object-based and ambisonic formats are commonly reproduced on headphones, so a good pair of cans and a quality headphone amplifier are essential studio items for checking playback translation. Most engineers will find that the immersive headphone listening experience is greatly enhanced when listening through a custom HRTF profile rather than standard presets or none at all [25,26].

And for yet another perspective, extensive research performed by the NHK came to the conclusion that 22.2 was the ideal balance between reproduction accuracy and fewest number of loudspeakers for immersive audio reproduction, although it's hard to consider 24 speakers 'few'. One major finding of the research was that a 120° spread across the front speakers is a minimum for accurate imaging, something all other formats lack – 5.1, 7.1, Dolby, and Auro all opting for 60° coverage [27–29]. That being said, there's a long list of incredibly compelling and beautifully crafted productions in everything from 5.1 to Auro to Atmos, so the more standard 60-degree front spread standard seems to currently be the more commercially viable approach.

After all this confusion about how many speakers and where to put them, It's worth stating again, having the correct number of speakers in all the right places will be useless if the system is not properly calibrated. Precise guidelines are available from the various format developers as to how they assume the room to be calibrated for accurate listening in their format.

MASTERING

Signal processing audio in surround and immersive mastering is much the same as in stereo, with most of the heavy lifting done through a combination of EQ for tonal shaping and timbre, and compression for overall density, rhythm, and timbre. Peak limiting to avoid clipping and overloading DA converters is the same in both formats, although, when working in immersive, one doesn't typically find the same blind pursuit of loudness that is so common in stereo. There are more channels, more room for instruments and space to exist, and generally an expectation from the listener of a realistic and engaging presentation, something that over-compression works against. Additional signal processing may include a number of additional tools, such as noise reduction and cleanup, harmonic distortion to add texture, and reverb to help solidify a sense of continual enveloping space. That being said, all of the engineers interviewed in connection with this chapter agreed that when working with surround and immersive material, a lighter hand is generally preferred with regard to signal processing compared to working in stereo.

Whereas in stereo one needs to check for mono compatibility, in immersive one needs to check both folddown to lower channel count formats (e.g. from Atmos or Auro-3D to 7.1 or 5.1) as well as L-C-R, stereo, and mono. Additionally, each format has predefined folddown coefficients for moving from one format to another, although these may not provide the best sounding results that properly convey the artists creative intent. It's important to check and decide whether to accept the standard methods or to prepare separate masters for each format – time and budget permitting.

When working in immersive formats there are a few surprising processing tools rarely found when working in stereo, chiefly the use of reverb. While it's generally rare to add reverb when mastering stereo material, the sense of a wholistic unifying and enveloping space tends to be much more critical for suspension of disbelief when working in immersive audio formats. Many immersive productions may sound amazing when sent off to mastering, but 'don't sound like they're under one roof' as engineer Stefan Bock of msm-studios notes [6], and a careful touch of quality reverb 'helps to connect the elements in the mix, as a common issue is that mixing multiple signals [in immersive] doesn't always sound cohesive'.

Due in part to maintaining the integrity of the soundfield in immersive, there are times when up-mixing will be used to help move material from stereo (e.g. where stems or the original tracks are unavailable) to an immersive format to integrate with other material. This is especially relevant in reissues or remastering of catalogue content for immersive. Audiotech Digital Limited's Penteo seems to be the dominant tool for this today, although Nugen's Halo Upmix has also a growing popularity among professional users.

Noise reduction isn't uncommon in stereo mastering but can be more critical when working in immersive as there is less masking going on due to the expanded soundstage, so problems are much more apparent and potentially distracting, particularly the 'under water' warbling effect of heavy-handed broadband noise reduction. Again, this should primarily be an issue in the case of re-issuing catalogue material repurposed for immersive release. Newer recordings made with immersive in mind would presumably be made in a manner to keep the noise floor to a minimum.

Getting the low end 'just right' is an important part of the art of mastering, and is perhaps one of the hardest things to perfect. It requires a monitoring environment capable of reproducing the full frequency spectrum, which is challenging in all but the best rooms. In addition to the usual aesthetic considerations for getting that perfect bottom end, it's important to be aware that, when working in immersive, bass management comes into play. Bass management is a system that utilises the subwoofer to reproduce frequencies below the low-end limitations of a consumer's main monitors, augmenting the system in order to provide full-range playback. So, in addition to any existing material in the LFE channel, bass management adds frequencies from all other channels below a defined cut-off frequency to the subwoofer. Conversely, in systems without a subwoofer, some bass management systems will send the LFE content to all main channels. Naturally, it is then important to ensure that the phase and timing relationship between the LFE signal and the low-frequency content of the other channels is compatible to eliminate the potential for comb-filtering or artificial build-up in the low end which will result if the timing and phase are not correct.

Mastering for discrete formats

When working in discrete (channel-based) formats, the assumption is that the listener will have the same speaker configuration as the mastering studio, just as in stereo. This simplifies the workflow considerably, as each speaker maps to one audio track from start to finished master.

Other than the number of channels, mastering for discrete channel based immersive audio is very similar to traditional mastering in stereo. The approach is largely the same, with the same kind of adjustments possible on individual tracks and channels – filtering/EQ, compression, limiting, etc. The seemingly ever-expanding number of channels has led to many engineers working entirely 'in the box' on the workstation of their choice, but others have found it worthwhile to expand their traditional analogue toolset to accommodate more channels, allowing them to achieve their sonic goals in immersive with the same familiar tools they have enjoyed using in their stereo work.

Despite the trend for streaming playlists, rather than albums, many artists still prefer to release albums intended to be heard in a particular sequence, taking the listener on a musical journey from beginning to end. The traditional aspects of working in discrete immersive allows for the same 'album-friendly' workflow that has been an important part of stereo mastering for decades – beginning with the first track, building the album track-by-track from there, creating well-timed pauses and perhaps interesting transitions throughout – all on a toolset (be it analogue, workstation-based, or most likely a combination of the two) that is comfortable and familiar. Albums may eventually be rendered out song-by-song for delivery, of course, but a coherent album master is often the top priority, be it for download or physical release.

Keeping the intended sound field intact is a crucial aspect when working in immersive and, with so many more channels involved, it is critical to preserve the inter-channel timing and phase relationships carefully created by the recording and mix engineer. Pristine clocking and correct delay compensation are essential to keeping the sound field stable.

Many engineers find it helpful to approach an immersive audio track in smaller groups of tracks, allowing them to hear what is happening in particular channels without the distraction of the others. Taking the channels in sensible pairs or individually can help to get a handle on what is happening where. For example, one might first get a sense of the front left and right, then the centre, and next the left and right rears. After that, the upper channels might get a listen and a bit of EQ-ing, etc. However, as any good mix engineer will tell you, getting the 'perfect sound' in any particular channel or pair of channels does not necessarily result in the perfect sound when all of the channels are played back together. Ultimately, any adjustments you make will have to work to create the big picture, even if a particular channel sounds less than ideal when soloed. It's the overall balance and timbre that are of interest in the end.

Linking of any dynamic processing also usually adheres to groups of channels, but occasionally applies globally to all channels. The key is to preserve the stability of the soundfield. With the vast number of channels involved in immersive productions, it can be a challenge to find compressors and limiters capable of linking in all of the various ways that might be useful to achieve the desired effect. If working in groups of channels rather than globally, be sure to listen for any undesired instability created by the dynamic processing. As always, the focus in mastering is on enhancing the artistic intent and connection with the listener while preserving the continuity and integrity of the overall sound field. Since sheer loudness is generally of less concern in immersive, a lighter hand may be called for.

Mastering for object-based formats

Object-based mastering can be performed largely the same as discrete by using beds, which are channel-based fixed placement of audio according to speaker placement, with a caveat. In Atmos these beds, or fixed positions, correspond to either a 5.1 or 7.1 main layer, but have only two channels for height (L/R). For most immersive productions, the front and rear height channels will carry distinct information based on their relative position to the sound source and listener. In Atmos this necessitates the use of additional objects. Objects are simply audio tracks that contain or have associated descriptive metadata that defines where in the sound field the audio belongs, rather than hard coding it 'in the mix' as with discrete formats.

Because up until recently, Dolby's focus with Atmos has been primarily on film and television post-production, music production workflows are relatively new. While there are a few pioneers developing tools and techniques for hybrid workflows, most of the production work being done in Atmos is 'in the box' – the currently supported workstations being ProTools, Nuendo, Logic, and Ableton. For mixing, the engineer can work in a more-or-less traditional fashion, but setting up the session to be mapped out as 'static' bed tracks and 'dynamic' objects. There are 128 objects available so an engineer choosing to use the maximum 'bed' of 7.1.2 would potentially have 118 additional objects at his/her disposal. Three-dimensional panning is implemented in the workstations themselves. The outputs of the workstation are fed into the Renderer (a.k.a. RMU, Production Suite, or Mastering Suite). The engineer can configure the desired monitor set-up in the Renderer, and listening through this device, can audition the results, including downmixes, and adjust as necessary. The engineer then records the bed channels and objects into the Renderer and, from there, a proprietary ADM file can be created. The ADM file contains the bed and object audio files, object panning meta-data, object output assignments, and bed & object group meta-data. For mastering, this ADM file can be reopened in the workstation and adjusted as desired, then exported again in the necessary configurations for the final deliverables.

Sony's 360 Reality Audio format currently utilises a workflow whereby an engineer creates stereo stems by whatever means he/she normally mixes. These stems are then pulled into the Architect software as up to 128 objects, where they are assigned as static or dynamic. Once in Architect, only their spatial orientation and gain can be adjusted and automated. When the mix is finished, Architect exports a set of master files, rendering out the number of channels corresponding to the desired resolution/bitrate. Those renders are then loaded into Sony's proprietary MPEG-H encoder to generate the deliverable files [30].

For both Dolby Atmos and 360 Reality Audio, a 96 kHz workflow is fully supported, but final deliverables are at 48 kHz for distribution to the consumer.

It's worth noting that currently, these object-based workflows generally focus on a song-by-song approach rather than a consolidated album approach. With a bit of ingenuity, one can theoretically generate a set of songs at appropriate relative levels and with the necessary post-gaps to create an album flow when those files are played back-to-back, but this is not yet as intuitive as working in a more traditional channel-based workflow. Furthermore, some streaming services may not yet support the capability to play these immersive formats back to the listener 'gapless-ly'. As object-based technology adapts to accommodate music production, as music engineers get more comfortable and creative in their use of the technology, and as streaming services develop, we can likely expect this to become easier.

Mastering for ambisonic formats

Ambisonic or scene-based mastering is similar to object-based in the sense that the production playback configuration is independent of the consumer listening setup, with the decoder responsible for correctly recreating the master as it was finalised in the studio.

Since the vast majority of music productions are not recorded using ambisonic microphone techniques, we have found that most engineers working on music for ambisonic distribution work in a discrete channel-based set-up which allows them to use traditional production techniques. Then they listen through the ambisonics encode-decode process for the necessary delivery format, making any necessary adjustments to get a good representation of the material. Because of this and the number of competing systems and formats, we can only offer generic advice consistent with mastering in general – make it sound as compelling as possible, and work within the constraints of whichever ambisonic system is being used. Most modern ambisonic toolkits offer the ability to use EQ, dynamic range control and even reverb within their framework.

Mastering for multiple formats

These days, music productions are often targeted for release in multiple formats in order to reach the widest possible audience – from stereo to a variety of immersive release formats, rather than committing to just one.

This being the case, many engineers feel it best to begin with the discrete stereo master.

First of all, working on the stereo version first allows the mastering engineer to get familiar with the musical material when there are only two channels to consider. Decisions such as the sequence of the songs, the timing between them, and relative levels can all be determined while working in this 'easy to manage' form. Also, if one starts off in immersive, even a stellar performance and mix can sound slightly disappointing when 'downsizing' from immersive back to stereo listening. 'Disappointed' isn't generally the most productive frame of mind for a mastering engineer anxious to deliver his/her best work to the client.

Once the stereo is done, most engineers we've spoken to at this early stage in the game have indicated that they are able to embrace their most musical and comfortable workflow by creating the immersive master first in a discrete, channel-based format. From that, the object-based or ambisonics versions will be a kind of 'translation' or remapping of the discrete immersive master making the necessary adjustments for the additional release formats.

DELIVERABLES

Depending on the desired delivery format, there may be a number of different deliverables prepared from a mastering session.

Deliverables for discrete formats

Working in a channel-based format is arguably the easiest as far as deliverables go. Each channel of audio maps to one loudspeaker, and a master deliverable will normally consist of a set of WAV files, one per channel. File lengths should be identical and sample accurate when lined up, with each channel clearly labelled according to the labelling convention for the given format.

In some cases, such as for Auro3D, the 5.1.4 stream may be played back as a downmixed 5.1 stream to accommodate the playback system, so you may be asked to provide downmix coefficients for the channels. The discrete audio stream is then encoded into a rendered Auro3D stream which losslessly decreases the bandwidth to allow it to fit on a Blu-Ray or be broadcast. This encoding process may take place at the mastering house using an available Plug-in Suite, at the authoring facility, or at Auro's own facility.

If the material will be released on physical media such as Blu-Ray, it's important to work in communication with your authoring facility to ensure they get what they need in order to guarantee that your master makes it to the carrier format unchanged. Many encoders, for example, require a certain amount of dithered 'silence' (generally around three seconds) as pre-roll before the audio begins to allow the processor to ramp up, and of course, when working with picture, it's important to deliver at the appropriate time base for synchronisation.

Deliverables for object-based formats

Deliverables for an Atmos master may consist of one of two formats. DAMF (Dolby Atmos Master File) was the original format developed by Dolby and includes three files – a table of contents, all audio interleaved, and all metadata. DAMF is not as common today and has been largely superseded by ADM BWF.

ADM BWAV (Audio Definition Model Broadcast Wav) is a single BWAV file with embedded metadata following the ITU BS.2076 recommendation developed by Dave Marston of the BBC [15]. It's a common format for delivery to streaming services and is used by Netflix, Apple, Disney, and others, and includes all session information encoded into the BWAV itself as metadata. It has become the de facto standard in today's production and the dominant immersive delivery format. An ADM BWAV loaded back into ProTools or Nuendo will actually recreate the stems from the session, and can function as a session archive, and is for all purposes a lossless BWF audio file at whatever sample and bit rate is desired, with object position and panning metadata either in the header chunk or as a separate file. The embedded decoders and renderers take care of the rest when played back on consumer devices.

For MPEG-H, Fraunhöfer's MPEG-H 3D Audio authoring plug-in will allow exporting both MPEG-H as well as the previously mentioned ADM/BWF and supports a number of audio workstations [31]. Blackmagicdesign's Davinci Resolve will render MPEG-H natively with support built-in. Sony's 360 Reality Audio format is ultimately delivered as proprietary MPEG-H files, generated from the set of pre-renders created in the Architect mixing/mastering session. These pre-renders correspond to the various resolutions/bitrates of the resulting MPEG-H encodes which are then distributed and decoded at playback for the consumer via the affiliated streaming services and personalised using Sony's smartphone headphone app [32].

Deliverables for ambisonic formats

Deliverables for ambisonic format need to be encoded for subsequent decoding on the listener's system. There are two systems, AmbiX and FuMa, which vary not only in channel order (AmbiX uses WYZX ordering, while FuMa uses WXYZ), but also in normalisation weighting [33]. Encoders are available from a range of vendors. Converting from one to the other is a mathematically straightforward process. A first order deliverable will normally consist of an interleaved four channel WAV file, while second & third order will have 9 and 16 interleaved channels, respectively. Ambisonics in first order has traditionally been delivered as UHJ, a matrixed two channel delivery file that maintains stereo and mono compatibility but can be decoded on the listener side for horizontal surround [34].

Due to the rapidly evolving nature of ambisonics and plethora of tools and playback platforms available, the reader should refer to ambisonic specific references.

QUALITY CONTROL AND CLIENT APPROVAL

When working in discrete formats, it would be ideal to be able to assume that the client playback system will be reasonably close to the mastering studio, assuming the system has been set up and configured correctly, which is no different than with stereo. Unfortunately, some people choose to place the left speaker beside the couch and the right in the kitchen! For those that take the time and care to set up their listening system correctly, they can be assured that what they hear will be close to what the mastering engineer intended, at least as far as the equipment will allow. However, this isn't always the case, and mastering engineers need to be prepared for some clients to QC their work in stereo, without actually having heard the immersive version!

With object-based formats we are largely at the mercy of the listener's configuration and the renderer, which interact to recreate as closely as possible what was heard in the studio and is designed specifically to adapt to a wide range of potential setups. Obviously the greater the number of speakers and channels available, the closer to the original one can come. What one hears in a 9.1.6 Atmos room may or may not be acceptable on a 3.1 consumer soundbar, both due to the fewer number of speakers and the downmix coefficients used. One key point to remember is that the algorithms will include *all* channels in the original master in the downmix.

The Sony Reality 360 format helps to eliminate at least some of these variables as it is always intended for headphone playback. Consumers will choose their own headphones, of course, which allows for a great deal of variation in quality, but the thirteen 'virtual speaker' configuration will at least remain constant.

As mastering engineers, it's part of our job to ensure consistent playback results across a wide range of playback systems. Many professional engineers interviewed for this book suggest creating alternate masters for several potential playback formats, i.e. 5.1, 7.1, Atmos, Auro-3D, but this is generally not time/budget-friendly and it still doesn't eliminate the potential for a listener to select the Atmos stream (thinking for example that it's 'better' because it has more channels) for playback on a mid-range 3.1 or 5.1 soundbar. It would be ideal to have a few playback systems available for checking how the master will translate in real life to consumers – small speakers, soundbar, Amazon Echo Studio, headphones – assuming one's pocketbook allows.

In ambisonics one either needs a room configured for the format, or a headphone setup that recreates the sound field via virtual speakers encoded for binaural playback. It should ideally be QC'ed in the VR environment it's destined for (Facebook, Google, Unity, Unreal, etc). There are simply too many variations to adequately cover here.

Getting a reference to a client for approval when working in immersive can be challenging, and becomes more so when delivering for multiple formats. Artists and producers can't be expected to have multiple high-end playback systems, and attended sessions are less and less frequent. Some engineers rely on a binaural render [35], some create a down mix or fold down to a more readily available format, like 5.1 or even stereo, but that leaves the artist or producer in the dark regarding how the full system playback will sound. The headphone-based formats alleviate this issue since clients will presumably have headphones available, but at the moment, many clients are forced to take a 'leap of faith' and trust that what the mastering engineer has delivered will convey their music to its best advantage in immersive.

GETTING INTO IMMERSIVE AUDIO

For engineers who may be interested in working in immersive formats, there are a few pointers some of the more experienced players have to offer.

First, as with stereo mastering, it's important to understand and gain experience in each stage of production, recording, editing, mixing, and really understand the whole production chain, how each part affects the next.

Second, gain as much experience as possible with the format specific requirements and limitations, codecs, home playback systems, etc. From Amazon Echo Studio to a high-end home theatre and a calibrated Atmos cinema. Really understand how things translate.

Third, understand by beginning in surround, get a deep understanding of how up and down mixing works, and the effects of comb filtering and delay issues. A master can sound great in the room until it's down mixed, potentially causing major problems. Since most consumers will end up hearing some sort of down mix anyway, it's vital for the engineer to understand how things will be affected, and to realise that the more immersive things become, the more combinations there are for things to go wrong.

It's also important not to be intimidated by the high-end rooms out there, as Daniel Shores points out [5]. By all means if you want to learn, try it out as you're able, even if it means getting an inexpensive '5.1 in a box' system to start with – experiment, learn as you go, figure out what works well, what breaks and how to fix it. Listen a lot to what other engineers and producers have done, get past the fascination and newness and instead think about it from an artistic point of view. Flying instruments through the air and spiralling sound effects through space can be lots of fun, but first and foremost, it needs to serve the music.

FINAL THOUGHTS

When done right, immersive audio is an even greater improvement over stereo than stereo was over mono, and it's our contention that immersive is the future of audio, despite the music industry being rather late to the party. It's clear that we're still in the pioneering phase of music production in immersive, and we look forward to the further adaptation of the primarily post-production/film-focused formats to better suit a more musical workflow, as well as the development of more powerful and intuitive tools for the recording, mixing, and mastering engineers working in these formats.

Mastering has traditionally been the art of translating material from studio formats to consumer distribution formats. Setting metadata, downmix coefficients, and other aspects often left to authoring might be more appropriately addressed in the mastering room as part of the creative decision making process. Stereo is still the most popular format, and we will always be at the mercy of how people set up their speakers and the quality of their headphones, but we're looking forward to more compositions imagined and created in immersive audio – artistically compelling and moving content that will inspire listeners to make immersive audio an integral part of the way they enjoy music.

WITH GRATITUDE

The authors would like to extend their sincere thanks to all of you who were so willing to share your vast knowledge and expertise with us: Ed Abbot, Stefan Bock, Mark Drews, Ozark Henry, Simon Heyworth, David Miles Huber, Ralph Kessler, Hiro Komuro, Jeff Levison, Tom McAndrew, Sven Mevissen, Mary Plummer, Ronald Prent, Steve Rance, Michael Romanowski, Daniel Shores, Gus Skinas, Ceri Thomas, Tom Van Achte, Reynaud Venter, and the countless engineers and artists who have inspired, mentored and guided us.

REFERENCES

1. Advanced Television Systems Commitee, *A/53: ATSC Digital Television Standard, Part 5*, 1st edn. Advanced Television Systems Commitee, Washington DC, (2007).
2. "Wikipedia", Wikimedia Foundation, 29 December 2019. [Online]. Available at: https://en.wikipedia.org/wiki/Surround_sound. [Accessed 03 January 2020].
3. Heyworth, S. *Interviewee, Surround Mastering. [Interview].* 13 November 2019.
4. EBU/ITSI, *ETSI Technical Specification 101 154, Sophia Antipolis Cedex: European Telecommunications Standards Institute/European Broadcasting Union*, 2017.
5. Shores, D. *Interviewee. Surround & Immersive Production and Mastering. [Interview].* 8 November 2019.
6. Bock, S. *Interviewee. Surround & Immersive Audio Mastering & Authoring. [Interview].* 23 October 2019.
7. Hamasaki, K., Hiyama, K. and Okumura, R. '*The 22.2 Multichannel Sound System and it's Application',* In: *Audio Engineering Society Convention Paper 6406*, Barcelona, (2005).
8. Van Daele, B. and Van Baelen, W. 'Auro-3d', 28 02 2012. [Online]. Available at: https://www.auro-3d.com/professional/industries. [Accessed 10 October 2019].
9. Hamasaki, K. '*22.2 Multichannel Audio Format Standardization Activity.' Broadcast Technology, No.45, Summer*, pp. 14–19, (2011).
10. International Telecommunications Union, *ITU-R BT.2020-2 Parameter Values for Ultra-high Definition Television Systems for Production and International Programme Exchange*. International Telecommunications Union, Geneva, (2015).
11. International Telecommunication Union, *ITU-R BS.2051-2 Advanced Sound System for Programme Production.* International Telecommunication Union, Geneva, (2018).
12. Carter, J. 'www.techradar.com', 30 03 2018. [Online]. Available at: https://www.techradar.com/news/the-immersive-audio-youve-never-heard-that-could-revolutionize-virtual-reality. [Accessed 02 October 2019].
13. Olivieri, F., Peters, N. and Sen, D. *'A Technology Overview and Application to Next-Generation Audio, VR and 360° Video.'* In: *Scene Based Audio and Higher Order Ambiosonics*. European Broadcasting Union, Technology & Innovation, Geneva, (2019).
14. Rumsey, F. 'Spatial Audio - Channels, Objects, or Ambisonics?' *Journal of the Audio Engineering Society*, Vol. *66*, No. 11, (2018), pp. 987–992.
15. International Telecommunications Union/European Broadcasting Union, *BS.2076-2 Audio Definition Model.* International Telecommunications Union/European Broadcasting Union, Geneva, (2019).
16. International Standards Organization/International Electrotechnical Commission, *Information technology – High Efficiency Coding and Media Delivery in Heterogeneous Environments – Part 3: 3D Audio*, Geneva:

International Standards Organization/International Electrotechnical Commission, (2019).
17. Herre, J., Hilpert, J., Kuntz, A. and Plogsties, J., 'MPEG-H Audio–The New Standard for Universal Spatial/3D Audio Coding.' *Journal of the Audio Engineering Society*, Vol. *62*, No. 12, (2014), pp. 821–830.
18. Byford, S., 'www.theverge.com.' Vox Media, LLC, 15 10 2019. [Online]. Available at: https://www.theverge.com/2019/10/15/20915250/sony-360-reality-audio-release-date-amazon-partners. [Accessed 20 October 2019].
19. McAndrew, T. *Interviewee, Dolby Atmos for Music Production.* [Interview]. 19 November 2019.
20. Netflix, 'Netflix Sound Mix Specifications & Best Practices v1.1.' *Netflix*, 04 04 2019. [Online]. Available at: https://partnerhelp.netflixstudios.com/hc/en-us/articles/360001794307-Netflix-Sound-Mix-Specifications-Best-Practices-v1-1. [Accessed 28 November 2019].
21. Kessler, R. *Interviewee. Surround & Immersive Mastering.* [Interview]. 07 November 2019.
22. Neukom, M. *'Decoding Second Order Ambisonics to 5.1 Surround Systems.'* In: *Audio Engineering Society Convention Paper 6980*, San Francisco, (2006).
23. Lee, R. and Heller, A. *Ambisonic Localisation – Part 2.* In: *International Conference on Sound & Vibration*, Cairns, (2007).
24. Heller, A., Benjamin, E. and Lee, R. *Design of Ambisonic Decoders for Irregular Arrays of Loudspeakers by Non-Linear Optimization.* In: *Audio Engineering Society Convention Paper*, San Francisco, (2010).
25. Paukner, P., Rothbucher, M. and Diepold, K. *Sound Localization Performance Comparison of Different HRTF-Individualization Methods*, Munich: Institute for Data Processing, Munich Technical University, (2014).
26. Nowak, P., Zimpfer, V. and Zölzer, U. *3D Virtual Audio with Headphones: A Literature Review of the Last Ten Years.* In: *Fortschritte der Akustik – DAGA München*, Munich, (2018).
27. Hamasaki, K., Hayama, K., Nishiguchi, T. and Ono, K. *Advanced Multichannel Audio Systems with Superior Impression of Presence and Reality.* In: *Audio Engineering Society Convention Paper 6053*, Berlin, (2004).
28. Howie, W., King, R. and Martin, D. 'Listener Discrimination between Common Channel-Based 3D Audio Reproduction Formats.' *Journal of the Audio Engineering Society*, Vol. *65*, No. 10, (2017), pp. 796–805.
29. Neuendorf, M., Plogsties, J., Meltzer, S. and Bleidt, R. *Immersive Audio with MPEG 3D Audio – Status and Outlook.* In: *NAB BEC Proceedings*, Las Vegas, (2014).
30. Komuro, H. *Interviewee. Sony 360 Reality Audio.* [Interview]. 27 November 2019.
31. Fraunhöfer, I. I. S. 'Fraunhöfer Institute for Integrated Circuits', 29 04 2019. [Online]. Available at: https://www.iis.fraunhofer.de/en/ff/amm/dl/software/mhapi.html. [Accessed 06 November 2019].

32. Sony Home Entertainment & Sound Products. 'Sony Corporation', *Sony, 11* 2019. [Online]. Available at: https://www.sony.com/electronics/360-reality-audio. [Accessed 15 November 2019].
33. C. Nachbar, F. Zotter, E. Deleflie and A. Sontacchi, *AmbiX - A Suggested Ambisonics Format.* In *Ambisonics Symposium*, Lexington, (2011).
34. Leese, M. J. 'Ambisonic Surround Sound FAQ', *Martin J. Leese*, 21 01 1998. [Online]. Available at: http://members.tripod.com/martin_leese/Ambisonic/faq_section07.html. [Accessed 12 November 2019].
35. Romanowski, M. *Interviewee. Surround & Immersive Mastering.* [Interview]. 16 December 2019.

3

Towards a definition of compression glue in mastering

Austin Moore

INTRODUCTION

The mastering engineer has many processing devices at their disposal. Typical mastering processors include EQs (passive, parametric, graphic), de-essers, stereo enhancers, limiters, and dynamic range compressors. While so-called 'hyper-compression' (heavy compression or limiting to achieve loudness), and its unwanted perceptual artefacts, have been much discussed in the literature, little work has been done to investigate other aspects of the mastering process. Particularly, the beneficial sound qualities it imparts onto program material. The work presented in this chapter aims to expand current academic research by addressing the following questions. Firstly, what does the subjective descriptor glue/gel (called 'glue' from hereonin) mean at an objective level? The primary reason for answering this question is to devise an objective definition of this descriptor, which can be used in succeeding academic and professional studies. Secondly, what are the common compressor types used during the mastering process, and what are the typical reasons, other than loudness, for mastering engineers to implement compression in their work? These additional questions are important to consider as they will help shape the definition of glue and give it context. A mixed-method of discourse analysis, grounded theory, text mining, and the Delphi Method was employed to answer the research questions.

BACKGROUND AND RELATED WORK

Literature relating to compression and music production was reviewed to ascertain the current state of the art in the area. It became apparent that previous studies had focused on the effects of hyper-compression, often concentrating on whether its artefacts harm the perceived quality of the audio material. Taylor and Martens [1] claim that achieving loudness is a significant motivation to use compression in mastering, so one can argue this is why hyper-compression is well researched.

Ronan et al. [2] investigated the audibility of compression artefacts among professional mastering engineers. For the study, 20 mastering engineers undertook an ABX listening experiment to determine, whether they could detect artefacts created by limiting. Two songs were processed using the Massey L2007 digital limiter to achieve –4 dB, –8 dB, and –12 dB of gain reduction. The masters (including the uncompressed versions) were then presented to the listeners using the ABX method. The results showed that the mastering engineers found it challenging to discern differences between a number of the audio tracks, particularly those with –8 dB of gain reduction and the unprocessed reference. The same experiment carried out by Ronan et al. [3] used 49 untrained listeners and showed that they were unable to detect up to 12 dB of limiting. The results of these two studies are not surprising to the author of this chapter. Research carried out by Moore and Wakefield [4] into the descriptor 'aggressive' and its association with the 1176 compressor, showed that listeners were unable to discern differences between various combinations of attack and release settings when applied to vocal tracks in three different music productions. Thus, it seems that the subtleties of compression are challenging to discern under double-blind testing conditions.

Campbell et al. [5] conducted a study with a large sample size of 130 listeners. The participants rated mixes with compression which had been introduced at various points in the signal chain, namely on tracks, subgroups, and the master buss. Their results showed that listeners preferred mixes where compression had been applied to individual tracks rather than to groups or on the mix buss. However, their test used identical attack and release settings on all stimuli and made use of the same compressor (Pro Tools Compressor/Limiter), which may have played a role in the results. Adjusting the compressor settings, so they were more appropriate for mix buss processing, and using a compressor with a suitable character for buss processing could have yelled a different outcome.

Wendl and Lee [6] looked into the effect of perceived loudness when using compression on pink noise split into octave bands. For this study, the authors wanted to observe if playback level and crest factor (a measurement of peak to RMS) affected perceived loudness after varying amounts of limiting had been applied. The results showed there was a non-linear relationship between the octave bands and changes to the crest factor. Of interest to professionals is the result which shows that modifications to the crest factor in a band centred around 125 Hz does not correlate with perceived loudness at moderate playback levels. Moreover, the perceived loudness could be louder than expected. The authors recommended that mastering engineers should be cautious of compression activities which affect the low end.

Some interesting work was conducted by Ronan et al. [7] which sits outside of the hyper compression studies reviewed so far. The authors of this paper investigated the lexicon of words used to describe analogue compression. Ronan et al. conducted a discourse analysis on 51 reviews of analogue compressors to look for common descriptors used in the texts and created inductive categories to group the words. The categories they created included signal distortion, transient shaping, special dimensions and most pertinent to this study, a category called glue. The descriptors Roan et al. grouped into this category were 'glue, cohesiveness, gel, well-integrated/homogenous, blend, separation and knitting'. However, the authors of the study did not define glue, other than noting it was perceived as a positive attribute.

Other loosely connected, yet robust academic work, has been carried out by scholars involved with the Semantic Audio Labs and the Semantic Audio Feature Extraction (SAFE) project. SAFE aims to understand the audio features associated with semantic descriptors, which can then be used to create intelligent mixing tools. Stables et al. [8], investigated terms used to describe compression on the mix and conducted hierarchical clustering to look for similarity in the words. They identified glue as one of the terms, but the clustering process did not find any direct similarity with other terms. Glue was loosely associated with the words master, hard, gentle punch, and soft in the upper branches of their dendrogram.

As can be seen, apart from a small body of work, there is a gap in the literature relating to the positive effects of compression, particularly during the mastering process. Anecdotal evidence suggests that many audio engineers and academics are interested in the character compression imparts onto program material; therefore, the lack of work in this area is surprising. Moreover, it is common for modern software tools to be designed with graphical user interfaces which use semantic descriptors. The Waves CLA plugins have controls over parameters such as spank, bite, and growl. Therefore, the work presented in this chapter aims to fill the void in academic studies on the positive effects of DRC, particularly relating to its use in mastering and the semantic descriptor glue.

QUALITATIVE METHODS

The following chapters discuss the work undertaken to answer the research questions. As a starting point, the results from a literature review will be discussed to provide an initial definition of glue. Secondly, the findings from a discourse analysis will be presented to demonstrate typical reasons for mastering engineers to use compression, ways in which they describe glue and the type of compressors they utilise to impart this sonic ingredient. Finally, the findings from a Delphi Study will be discussed to illustrate the method used to generate the definition of glue showcased in the current study.

Defining gel/glue-initial stages

Analysis of non-academic sources was conducted to build an initial definition of glue. This involved searching mastering textbooks [9–11] for discussions of glue, in which the author proposed an objective description of the effect. The results revealed little detail in the texts. However, one source by Cousins and Hepworth-Sawyer stood out as it provided some information regarding how a compressor imparts glue. They stated [12]:

'The entity of glue is just a by-product of gain control, making the track sound like a whole entity rather than its individual parts. From a technical perspective, this could be explained by the fact that all the instruments in the mix are being subjected to the same gain reduction – in effect, the movements of the compressor in and out of gain reduction gives the instruments a common identity. However, discerning how and why a compressor can glue a mix together is an important part of understanding mastering. Put simply, it's often the case that a compressor is used as much for its glue-like tendencies as it is to control the dynamic range of the input, or indeed, to add loudness'. (p. 74)

One can posit that a part of the glue effect, is due to the compressor, imparting it's time constant curve and gain reduction collectively to all elements in the mix. Cousins and Hepworth-Sawyer used the phrase 'common identity', which could be another way of saying that the sound sources have fused. Interestingly, this idea of sonic fusion has been discussed in the literature relating to hearing aid compression. Stone and Moore [13] found that hearing aid compression, specifically fast-acting compression (reported by the authors as being 1–2 ms attack and 360 ms release), can reduce the temporal contrast of sounds, resulting in the brain hearing multiple sounds like one.

As a starting point, Cousins and Hepworth-Sawyer's position on glue is a good one, but more work needs to be done to quantify it at a lower level with a more precise definition. Is it merely the process of giving the elements of a mix 'common identity', or is there more to it?

Discourse analysis

As a further investigation of glue, variations of discourse analysis were accomplished as a two-part study. Firstly, a textual analysis was carried out on two key textbooks, which collated interviews with mastering engineers. The books were The Mastering Engineers Handbook by Bob Owiniski [14] and Audio Mastering: The Artists by Hepworth-Sawyer and Hodgson [15]. This analysis was performed to gather substantive data on the use of compression in mastering, which would feed into the design of surveys used later. The textbooks were searched for content relating to compression and data was recorded using the software package NiVivo. Codes were created to categorise discussions on the use compression and specific compressor styles (note-compressor style relates to the gain reduction method utilised by the compressor design i.e. Valve/Tube, Opto, VCA, FET, etc.). As there is little previous work in this area, the method adopted was based on grounded theory [16]. The theory which developed from the coding was grounded in the data used for analysis. During the initial coding, there was no hypothesis to test; rather, the theory emerged out of the coding.

Reasons to use compression and compressor types

Initially, the interviews were coded to create categories on the general use of mastering compression. For example, if the discussion referred to using a compressor to soften transients, a category called transient design was created and all subsequent discussions of this type were coded into this category. New groups were created when a fresh theme emerged, and the final categories are shown in Figure 3.1.

As can be seen, there are five categories, tonal shaping, transient design, glue, parallel compression, and general dynamic range control. Tonal shaping includes discussions where the mastering engineer stated they used the compressor to alter the tone of program

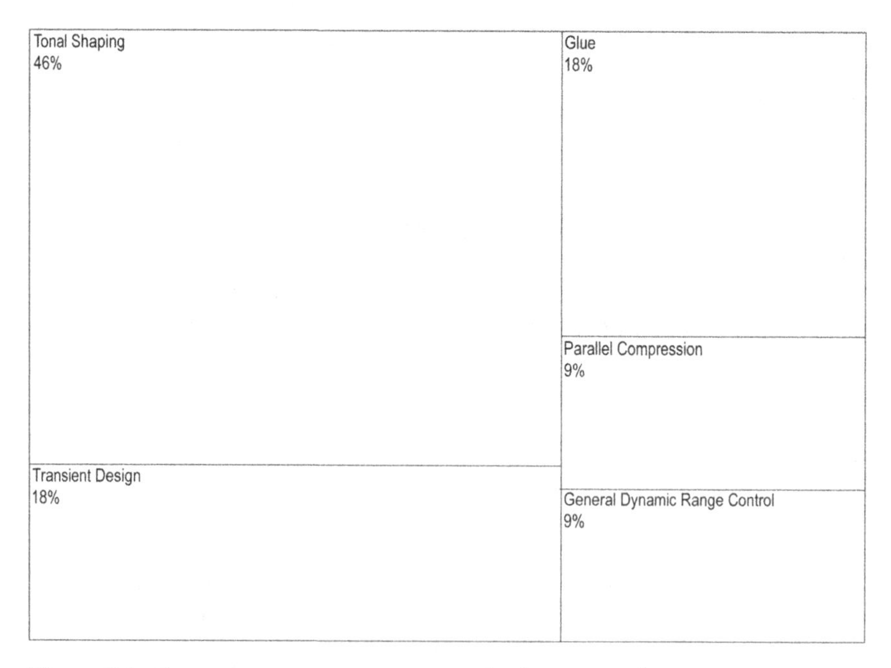

Figure 3.1 General reasons to compress during mastering

material, with a focus often being on tightening up the lower mid-range and bottom end. Passages which used the word glue or gel were coded into the glue category, but unfortunately, the discussions did not give the word an objective meaning. The other categories are self-explanatory and require no further explanation. Baring parallel compression, these codes provided a starting point for the design of the first survey issued as part of the Delphi Study.

Secondly, compressors mentioned by name were coded into categories relating to the style of gain reduction. The motivation behind this was to look for any connection between compressor styles and glue. The results are shown in Figure 3.2, which reveals that Valve compressors were mentioned most frequently, followed by VCA and Opto compressors. A small number of interviews referred to a compressor brand but did not state the model, and these are coded into the unknown category. An interesting result from this study is that FET compressors, which are commonly used in audio mixing, were not mentioned at all. One possible reason for this is that the fast-acting attack times generally associated with these devices is not as desirable for mastering. The study did not show any clear link with compressor style and glue. Still, it did verify anecdotal evidence that valve, VCA, and Opto compressors are the most common styles of compressor in mastering.

While this study revealed some interesting results, the data sample was small. The main reason for this was that the books looked at

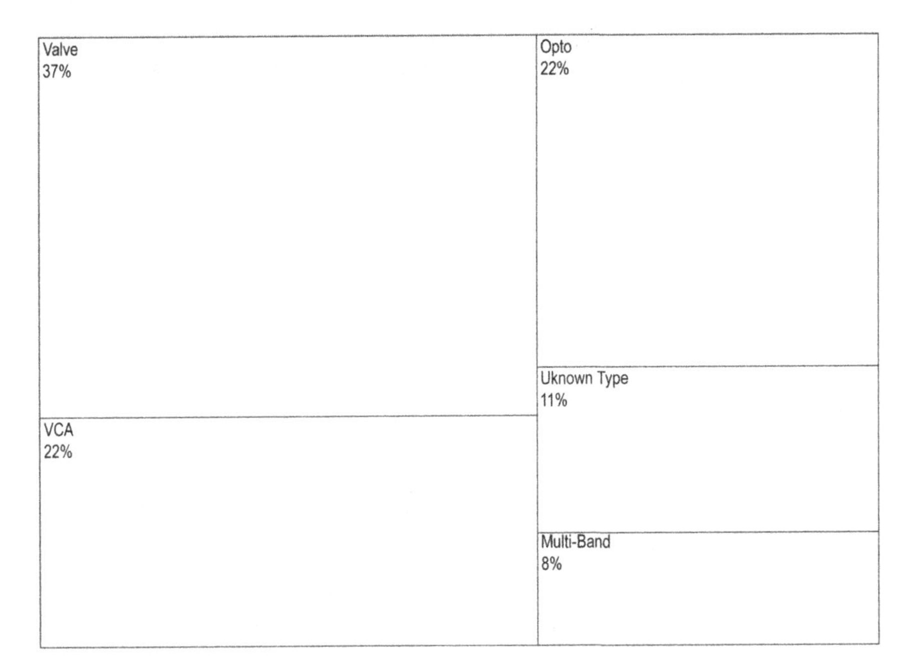

Figure 3.2 Common compressor types in mastering

many areas relating to mastering, meaning that compression was not their main focus.

Web scraping

To extend the previous research, textual analysis was conducted on the mastering sub-form from the online discussion board Gearslutz. All threads ranging from the sub-forum's first thread in 21/09/02 until 12/07/19 (dates in UK format) were extracted using a web scraping function written for the statistical software program R and using the package rvest. All discussions with the word glue or gel in the title were extracted as text and input into NiVivo for analysis.

Figure 3.3 shows the compressor types mentioned in the threads about glue. Again, the results show valve, VCA, and opto as the most common styles of compressors, but the important result to consider here is that these discussions focused specifically on glue, showing that valve and VCA compressor styles are the most frequently used when glue is the desired sonic signature. This is not to say other compressor styles are not implemented, but the data in this sample suggests a bias towards valve and VCA compressors. Again, FET compressors were not mentioned in the discussions.

Grounded theory analysis of the threads was carried out on how the posters described the sound of glue. As with the previous grounded theory study, codes were developed as themes emerged in

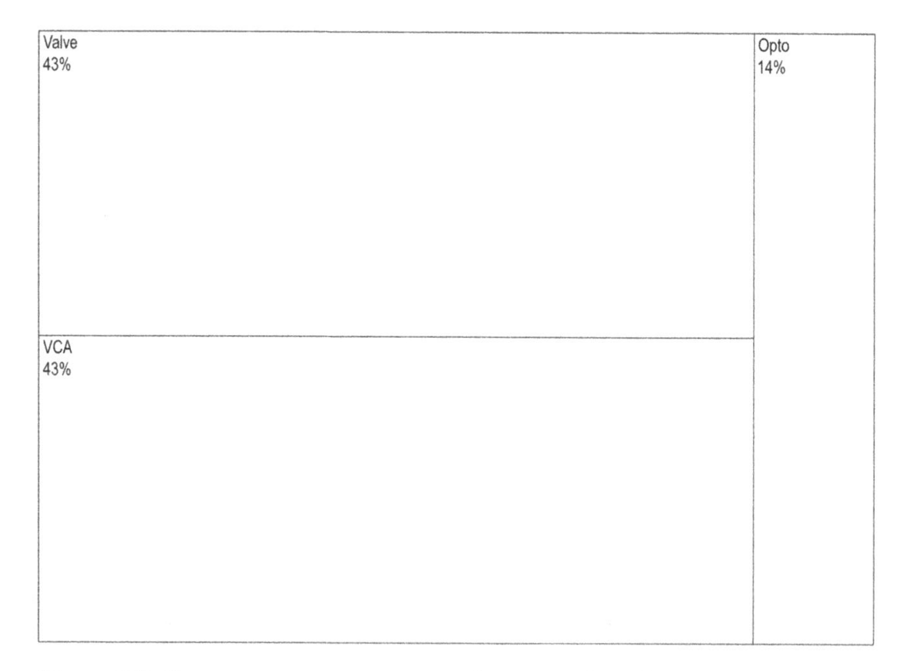

Figure 3.3 Common compressor types for mastering glue

the text. Figure 3.4 shows the main themes. Unsurprisingly dynamic cohesion makes up the majority, but codes relating to reshaping transients, rhythmic movement, manipulation of ambience and frequency related colouration emerged. Analysis of the text relating to cohesion shows that the commenters note glue makes the mix sound more together, uniform and 'tucked in' without sounding overly compressed. The non-cohesion codes are interesting as they reveal more about the perceptual effects of glue. Colouration and movement are the largest of the four codes, with discussions noting how glue can tighten up the low end and make the music swing with the audio material's rhythm.

Finally, the glue threads were studied for discussions relating to compression settings. It was found that all posters suggested the use of slow to medium attack and release settings and a small amount of gain reduction, no more than 2–3 dB. One interesting discussion explicitly related to the Manley Vari-Mu compressor. Posters in this thread recommended, as a starting point, that users set the Manley with its threshold at minimum, release in the middle and attack at the slowest position. From here the input should be increased until the compressor is showing ½ dB of gain reduction and then the threshold reduced to achieve the rest of the desired attenuation. Presumably, the reason for this approach is to drive the input stage into saturation, which posters note is a vital ingredient of this compressors glue effect.

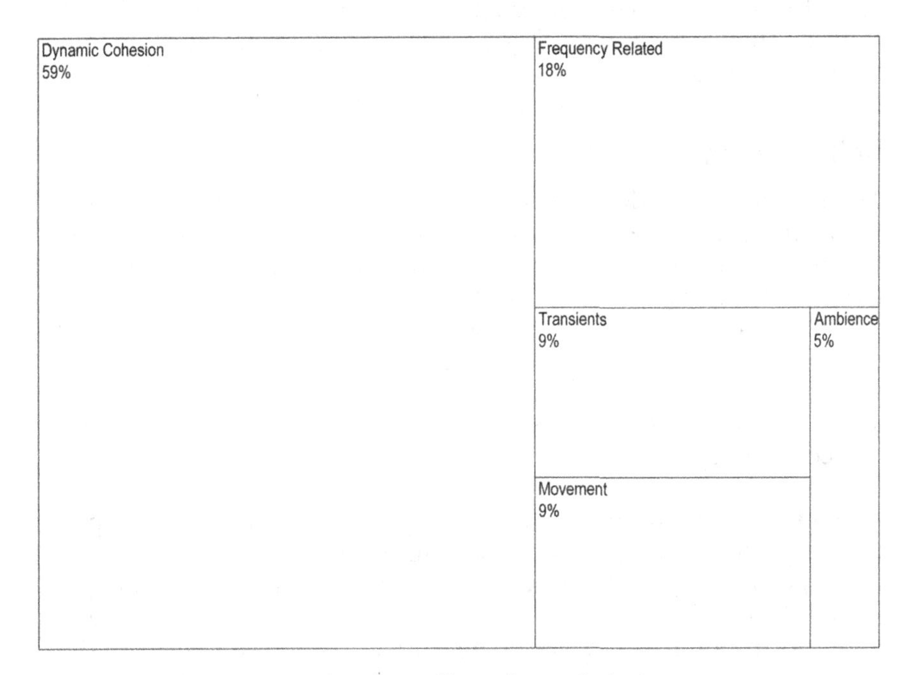

Figure 3.4 Themes emerging from discussions of glue

DELPHI METHOD

The discourse analysis and grounded theory revealed thought-provoking detail on glue and yielded information to be used in subsequent surveys. However, the studies did not fully answer the primary research question, which was, what does glue mean at an objective level? Thus, to address this issue, a further study, making use of the Delphi Method was conducted.

The Delphi Method, is a research technique used to collect the opinions of multiple experts, often via a series of questionnaires and focused discussion. An important aspect of the Delphi Method, is that the participants are anonymous and not aware of each other. This makes the Delphi an attractive method as it ensures participants are not biased by each other's responses, particularly during the initial stages. The process can be used to achieve a consensus opinion on specific issues, often with the goal of setting policies. As noted by Hsu and Sandford [17], the Delphi Method 'has been applied in various fields such as program planning, needs assessment, policy determination, and resource utilization'. The Delphi Method was used in the current study to survey mastering engineers via two online questionnaires, followed by a series of follow up emails. Initially 80 mastering engineers were contacted to take part in the study. They were told that the purpose of the study was to gather their opinions on a subjective descriptor relating to compression in mastering, and that multiple participants would take part. From the initial contact, ten mastering engineers replied to confirm they would participate and were used in the remaining stages of the current study.

Delphi survey stage one

The first stage was designed to develop a list of reasons to use mastering compression, which would then be used in the second stage. During the discourse analysis, a number of reasons for compression use were noted and used in the stage one survey. The primary purpose of this survey was to test whether professional mastering engineers agreed with these reasons and to check for any redundancy. The participants were presented with the 22 reasons shown in Table 3.1. Close inspection will reveal there are 11 main areas, which are consistency, harmonic distortion, movement, ambience levels, transients, general control over dynamics, changes to frequency content, manipulation of stereo image, depth, sustain, and peak limiting. These areas have been split into sub-categories to allow for more precise responses, which has resulted in the 22 reasons presented below. To complete the survey, the mastering engineers were asked to respond with a yes or no answer to indicate whether they agreed with the reasons. The percentage agreement amongst the ten participants is also shown in Table 3.1.

Table 3.1 Agreement on reasons to use mastering compression

Reason to Use Compression During Mastering	Percentage Agreement
Dynamic consistency in the low end	30%
Dynamic consistency in mid-range	40%
Dynamic consistency in the top end	20%
Harmonic Distortion	60%
Movement/Groove	80%
Increased ambience and or reverb levels	50%
Increased transients	30%
Decreased transient	60%
General control over dynamics	100%
Increase in low end frequency content	50%
Decrease in low end frequency content	0%
Increase in midrange frequency content	10%
Decrease in midrange frequency content	20%
Increase in high end frequency content	0%
Decrease in high end frequency content	20%
Increased stereo image	30%
Decreased stereo image	30%
Increased depth	30%
Reduced depth	40%
Increase sustain and low-level details	80%
Decrease sustain and low-level details	10%

It is curious to note that harmonic distortion, movement/groove, decreased transients, general control over dynamics and increase sustain and low-level details are the only reasons with greater than 50% agreement.

Delphi survey stage two

Once the first stage had been completed, any reasons with a response rate of 10% or less (meaning only one participant or no participants agreed) were removed to leave 18 reasons. Thus, the redundant reasons were, decrease in low-frequency content, increase in mid-range frequency content, increase in high-end frequency content and decrease sustain and low-level details.

The second stage, entailed the participants, responding using a 5-point Likert scale to show how likely they felt the remaining 18 criteria were ingredients in the creation of the sound quality glue. Participants used a scale which included the following five points, highly unlikely, unlikely, neither likely or unlikely, likely, and highly likely. The results are presented in Table 3.2, which show the percentage of responses rated as likely or highly likely, plus the mean and the median scores for each criterion.

As can be seen, criteria relating to consistency and general control over dynamics were rated as being the most likely ingredients by the participants, followed by harmonic distortion and movement/groove. However, the median result for harmonic distortion and movement/groove is three, which suggests a weaker agreement amongst

Table 3.2 Potential glue ingredients

Criteria	Percentage Likely or Highly Likely	Mean	Median
Dynamic consistency in the low end	77.78%	3.4	4
Dynamic consistency in mid-range	66.67%	3.3	4
Dynamic consistency in the top end	55.56%	3.1	3.5
Harmonic Distortion	44.44%	3.3	3
Movement/Groove	44.44%	3	3
Increased ambience and or reverb levels	0.00%	2.5	3
Increased transients	22.22%	2.4	2
Decreased transients	33.33%	2.8	3
General control over dynamics	88.89%	3.9	4
Increase in low end frequency content	22.22%	2.4	3
Decrease in midrange frequency content	0.00%	2.1	2.5
Decrease in high end frequency content	0.00%	2.1	3
Increased stereo image	11.11%	2.1	2
Decreased stereo image	11.11%	2.3	2.5
Increased depth	33.33%	2.7	3
Reduced depth	22.22%	2.6	3
Increase sustain and low-level details	22.22%	2.9	3
Peak limiting	11.11%	2,2	2.5

participants. After the participants had rated the possible ingredients using the Likert scale, they were asked to reflect on their scores and come up with a definition of glue. These comments were read to ascertain if they correlated with the scores.

The comments made by the participants were:

Participant 1: 'It is totally dependent on the material supplied. Any of the criteria could have an impact on the "glue/gel" factor'.

Participant 2: 'I consider gel/glue to be a criteria i don't really consider at all when working. Just treating the recording with a specific tool might have an effect of glue/gel, since all elements of the mix get the same treatment. Having said that, my definition of gel/glue would be consistency in dynamic form, mostly groove/movement related. "Applying a (sometimes also almost invisible) dynamic footprint to all elements"'.

Participant 3: 'Usually the imparting a sonic characteristic on the audio which makes the elements less distinct within the mix'.

Participant 4: 'Being honest, I've never been much of a "glue compression" type. I tend to use compression in a more expansion type way (longer attacks, etc). Most music I receive already has compression, and needs more life than less. Put another way, glue tends to make things depend on one another. I look to impart separation with compression, when possible'.

Participant 5: 'A sense of disparate parts becoming part of a single entity. Broadband compression can create an interaction between parts, for example a loud low frequency sound causing a reduction across the spectrum (could assist with groove)'.

Participant 6: 'A process used to help contain material to fit into a box, whether a technical box or a stylistic box. Can also be used to enhance or help focus certain sonic relationships that exist within the mix'.

Participant 7: 'A bit of dynamic compression and uniform THD that adds a sonic cohesion to all parts'.

Participant 8: 'Cohesion in the music's delivery'.

Participant 9: 'The glue meaning, I would define is how the different elements in music material start to share a common bond sonically. I personally believe the majority of this comes best from either a piece of analogue gear or an analogue emulation plugin. Mostly from the harmonic imprint that a piece of gear adds. If one sound in the mix goes thru Gear A then only that sound has that imprint, but if all the sounds (a full mix) goes thru Gear A then the imprint (or harmonic distortion) applies to all of the sounds together. The same thing happens with Analog tape. Similarly, when all of the individual sounds in a mix are mixed on all console channels this effect adds collectively and in addition the stereo bus adds a similar element as the same as any individual piece of gear used in mastering does. The glue also comes from the dynamic envelope moving all of the sound elements together as one. The Attack and Release time constants of course play a big part in the compressor controlling all of the elements uniformly. And lastly the overall frequency response of the gear adds to that imprint. Of course, with level change (or the level pushed thru the gear-which is not always nominal) can also add to help or hinder that effect'.

Participant 10: No response

As can be seen, the participants' comments aligned with the numerical results, suggesting that glue relates to cohesion, consistency, movement, colouration and distortion. To test this theory, an initial definition of glue was devised, which was emailed individually to all participants for their approval and comments. Again, it should be noted the participants still did not know who else was taking part.

Delphi survey stage three

The participants were sent the following definition with the instruction to reply stating whether they approved or disapproved, and if any further refinements should be made. The initial definition of glue was:

> General control over dynamics, with harmonic distortion and a sense of movement and groove, which imparts a cohesiveness to program material.

All participants replied and the definition was met with approval. However, some suggestions were made by the participants, which were used to further shape the definition:

Participant 1: 'Yes, that pretty much sums it up. So that would be an AGREE with the definition'.

Participant 2: 'I generally agree with your definition of glue/gel. Harmonic distortion is not an essential part of it though, in my opinion, your definition might put too much emphasis on it. I think it is an optional part of it and sure also helps to achieve cohesiveness. I achieve a sense of glue and gel also with a very clean analogue compressor (transformer-free), i might even say it is achievable with a digital compressor as well'.

Participant 3: 'Yes, I think that's a balanced description which should leave everyone happy'.

Participant 4: 'I don't disagree with that - harmonic distortion is the only part I wouldn't necessarily agree with personally. that's not an essential part of glue for me. Sometimes you want things held together but cleanly. But overall, yeah. That's a decent summary!'

Participant 5: 'That sounds about right, although I wouldn't necessarily include groove as a defining feature of glue/gel, but more of a by-product...Yes, I mean groove/movement is not a primary feature of glue/gel, and I personally wouldn't specifically reference it in such a concise statement'.

Participant 6: 'This participant did not reply but replied to subsequent commination to approve the final update'.

Participant 7: 'General control over dynamics, with *uniform* harmonic distortion overlaid *to help* impart* a cohesiveness to program material'.

Participant 8: 'Yes but 'harmonic distortion' is not a requirement but could be a part of the outcome - depend on tool selection...

I mean you could have glue/gel without or with harmonic distortion. It's not intrinsic. If you use a tool that saturates/ distorts the sound, i.e., a valve compressor it will had a tad of harmonic distortion but if you use a clean full band compressor such as a TC6000 comp or Weiss there's no 'additive'. All full band compressors can all glue/gel to the mix if applied well. The 'tool type' changes the additional outcomes'.

Participant 9: 'I would agree with the statement. But I do believe the tonal shape of the sound is changed in frequency response and harmonic distortion together from a piece of gear or emulated plugin that is also giving what I would consider that cohesiveness. I feel it's based not only from the transformers or tubes or main contributor of harmonic distortion but also any components or op-amps, etc. in the path. So, I guess my personal feeling is it's not just harmonic content that is contributing as the 'glue', the overall tonal shape in frequency response direction imparted on the sound also plays some part'.

Participant 10: 'To me the sentence without 'with harmonic distortion and a sense of movement and groove' makes more sense in terms of English...Then my only suggestion would be to change movement to swing. So that the audio has a sense of swing and groove'.

It can be garnered from the comments, that although the participants agreed with the definition, some additional clarification was required. The main issue was primarily due to the feeling that harmonic distortion and movement/groove was dependent on the style of compressor and that some compressors might impart glue without either of these two components.

Delphi survey final definition

The comments made in the previous stage were reflected upon, and the definition was updated to increase clarity. The update was then sent to all participants for their input. The participants all responded to say they were satisfied with the amendments, and that they agreed with the definition. Therefore, the final working definition of glue is as follows:

> Gel/glue as it pertains to dynamic range compression is a uniform control over dynamics, which imparts a cohesiveness to program material. Depending on the style of the compressor, this may include the introduction of subtle uniform harmonic distortion, colouration and a sense of rhythmic movement and groove.

The final result from this study provides more detail to the anecdotal definitions of glue which have been used previously. While, the working definition still focuses on cohesion, it includes lower-level attributes, which may play a role in the creation of this cohesion. The author proposes that additional work needs to be done to test this definition and to establish how much the stipulated lower-level details influence the perception of glue. Future work by the author will investigate this area by conducting a series of perceptual listening tests and the results from these experiments will be used to update the glue definition accordingly.

CONCLUSION

The present study set out to determine what constitutes compression glue during the mastering process. The findings suggest that glue creates a cohesiveness to program material, which depending upon the style of the compressor may impart subtle distortion, colouration and rhythmic movement. It appears the most commonly used styles of compressor to create glue are valve designs followed by VCA and opto styles. The results of this study will be of use to academics, professionals, software designers, and hardware manufactures. It provides a definition of glue to use in further research into sonic glue, not only in compression and mastering but in other areas of audio and music production. The study has gone some way towards enhancing our understanding of glue, but additional work needs to be done to achieve a more robust understanding of its perceptual attributes. Ultimately, a perceptual model of glue can be created, which can be used to measure the amount of glue in music mixes. This can then be used to devise a real-time glue meter, which can be integrated into the production process similar to the punch model proposed by Fenton and Lee [18].

The author would like to thank JP Braddock, Joe Caithness Josh Clark, Nick Cooke, Barry Grint, Jay Hodgson, Kevin Lively, Bob Macciochi, Kassian Troyer, and Tim Turan for taking part in the research and providing invaluable input.

REFERENCES

1. Taylor, R. W. and Martens, W. L. 'Hyper-Compression in Music Production: Listener Preferences on Dynamic Range Reduction'. In: *Audio Engineering Society Convention 136*. Audio Engineering Society (2014).
2. Ronan, M., Ward, N., Sazdov, R. and Lee, H. 'The Perception of Hyper-Compression by Mastering Engineers'. *Journal of the Audio Engineering Society*, Vol. *65*, (2017), pp. 613–621.
3. Ronan, M., Ward, N. and Sazdov, R.: *The Perception of Hyper-Compression by Untrained Listeners*. In: *Audio Engineering Society Conference: 60th International Conference: DREAMS (Dereverberation and Reverberation of Audio, Music, and Speech)*, (2016).

4. Moore, A. and Wakefield, J.: *An Investigation into the Relationship Between the Subjective Descriptor Aggressive and the Universal Audio of the 1176 FET Compressor*. In: *Audio Engineering Society Convention 142*, (2017).
5. Campbell, W., Paterson, J. and van der Linde, I. 'Listener Preferences for Alternative Dynamic-Range-Compressed Audio Configurations'. *Journal of the Audio Engineering Society*, Vol. *65*, (2017), pp. 540–551.
6. Wendl, M. and Lee, H.: *The Effect of Dynamic Range Compression on Perceived Loudness for Octave Bands of Pink Noise in Relation to Crest Factor*. In: *Audio Engineering Society Convention 138*, (2015).
7. Ronan, M., Ward, N. and Sazdov, R. *Investigating the Sound Quality Lexicon of Analogue Compression Using Category Analysis*. In: *Audio Engineering Society Convention 138*, (2015).
8. Stables, R., De Man, B., Enderby, S., Reiss, J. D., Fazekas, G. and Wilmering, T.: *Semantic Description of Timbral Transformations in Music Production*. In: *Proceedings of the 24th ACM International Conference on Multimedia*. pp. 337–341. ACM (2016).
9. Katz, B. *Mastering Audio: The Art and the Science*. Butterworth-Heinemann, (2003).
10. Wyner, J. *Audio Mastering-Essential Practices*. Berklee Press, (2013).
11. Waddell, G. *Complete Audio Mastering: Practical Techniques*. McGraw Hill Professional, (2013).
12. Cousins, M. and Hepworth-Sawyer, R. *Practical Mastering: A Guide to Mastering in the Modern Studio*. Routledge Ltd, (2013).
13. Stone, M. A. and Moore, B. C. 'Effect of the Speed of a Single-channel Dynamic Range Compressor on Intelligibility in a Competing Speech Task'. *The Journal of the Acoustical Society of America*, Vol. *114*, (2003), pp. 1023–1034.
14. Owsinski, B. *The Mastering Engineer's Handbook: The Audio Mastering Handbook*. 2nd edn. (2007).
15. Hepworth-Sawyer, R. and Hodgson, J. *Audio Mastering: The Artists: Discussions From Pre-Production to Mastering*. (2019).
16. Strauss, A. and Corbin, J. M. *Grounded Theory in Practice*. Sage, (1997).
17. Hsu, C.-C. and Sandford, B. A. 'The Delphi Technique: Making Sense of Consensus'. *Practical Assessment Research and Evaluation*, Vol. *12*, (2007), pp. 1–8.
18. Fenton, S. and Lee, H. 'A Perceptual Model of "Punch" Based on Weighted Transient Loudness'. *Journal of the Audio Engineering Society*, Vol. *67*, (2019), pp. 429–439.

4

Engineering authenticity

THE AESTHETICS OF DSP MODELLING IN MASTERING PLUGINS

Neil O' Connor

INTRODUCTION

Analogue audio processing and music production equipment and its associated workflow has not only survived numerous evolutions, it has also regained a somewhat mythological status. Over the last decade, as computer processing speed and memory space have become faster and cheaper to implement, this status has been turned into a very directed, and profitable selling point by brands such as Wave's, Universal Audio and SoundToys. In particular, what can only be termed as 'boutique' processes and effects (noise and tape bias and emulation) are becoming more and more incorporated into plugins suites and such developments and trends are, as Toffler [1–5] suggests, 'informed, supplemented, and aided by pro audio marketing, where both the professional and consumer merge as 'the prosumer'.

Cole [6] defines the prosumers status, field position, economic class, and habitus within the recording field as 'a vigorous battle between the so-called prosumer by considering a new socio-historic subject (the prosumer) and space of production (the project studio)'. He comments:

> The project studio and prosumption reconfigure some, but not all, of the dominant relations within the wider social or economic field. Thus, we can see that although prosumers are not economically determined, their relative autonomy is itself directly tied to changes in the structure of social production and consumption.
>
> (*Ibid.*, p. 458)

Almost 20 years ago, McGregor (2002) raised the point that 'prosumers are now more than ever relying on emulations, ones where part of the real system is replaced by a model' while more recently, Pras, Guastavino, and Lavoie (2013) suggests that the 'new technological paradigm of digital technology in music production has changed the role of professionals and their traditional business models'. Through this, the mastering process, as a business model, has become democratised (via the prosumer) in that the lending of an experienced and objective ear is jeopardised. On the other hand, Wishart [7] points out that 'technology offers us tremendous opportunity, it also carries risk, a risk that proposes that there is no inherent virtue in doing everything'. This kind of approach is core to a number of 'all-in-one' mastering plugin suites now available including the Waves Abbey Road TG Mastering Chain.

Technology offers us a tremendous opportunity in its extension of physical hardware capabilities within the digital domain. In 'All Buttons In', Moore (2012) discusses the Urei/Universal Audio 1176 dynamic compressor, a machine crucial in both music production and mastering histories. One process that Moore discusses (as the title of his paper suggests) is the 'all buttons in' mode in which, by pressing in 'all ratio options in', simultaneously, the unit reacts beyond the inventor Bill Putman's intended use. Fifty-three years later, Universal Audio has reimaged the 1176 as a virtual instrument (Figure 4.1), with the same physical 'all buttons in' effect. What is interesting here is that the transitional effects from the physical to the digital are apparent: the imposed technological limitations of 1966 and unwilding freedoms of digital domain processing of 2019 and in such, these considerations are core to the conceptual framework of this chapter.

Frith [8] acknowledged that one of 'the twentieth-century threats to musical autonomy is not the rise of mass music ... but the development of recording technology' and through this development, everything that occurs during the record-making process is subject to mastering. From the tuning of a room, to appropriate monitoring,

Figure 4.1 Universal Audio 1167 plugin.

sequencing, fading, editing, and format optimisation, mastering is more than just the quality control of audio before it is released and consumed by the wider public. Yet as more and more mastering options and possibilities become available, this technology alters the relationship between both the performer, producer/engineer and mastering engineer, in an age where the consumer can now, more so than ever, act as all three.

BACKGROUND

As suggested by Toffler [1], the prosumer sees the 'merging of the professional and the consumer' and this links this new mixture of statuses and/or roles to macro level economic changes that 'heal the historic breach between producer and consumer' (1980, p. 11). In order to frame the 'prosumer' within this discussion, one framework worth considering is that of notions of different forms of 'capital', as proposed by Pierre Bourdieu in *Distinction: A Social Critique of the Judgement of Taste* [9,10]. Bourdieu defines 'capital' in terms of economic, cultural, social, and symbolic relationships, referring to the fact that they are all utilised by members of the social field in order to attempt dominance for their judgements. Bourdieu also points towards what he referred to as 'social assets' (in arts and education) that allow for social mobility beyond economic means. These, it seems, are most likely to determine what constitutes taste in cultural fashion. This, in turn, allows it to become a form of capital:

> There is an economy of cultural goods, but it has a specific logic. Sociology endeavours to establish the conditions in which the consumers of cultural goods, and their taste for them, are produced, and at the same time to describe the different ways of appropriating such of these objects as are regarded at a particular moment as works of art, and the social conditions of the constitution of the mode of appropriation that is considered legitimate. But one cannot fully understand cultural practices unless 'culture', in the restricted, normative sense of ordinary usage, is brought back into 'culture' in the anthropological sense, and the elaborated taste for the most refined objects is reconnected with the elementary taste for the flavours of food.
>
> (Bouriedu, p. 1)

Zagorski-Thomas (2014) expands further on Bourdieu's notion of 'capital' by linking it with a person's 'position', commenting:

> On an economic level, the spending power of an audience and its ability to buy the output and thereby confer value on it is one fundamental use of capital. On the other hand, cultural capital

> relates to the tacit and explicit knowledge that confers power to an individual; social capital relates to power that stems from a person's position within some social grouping and symbolic power relates to ideas such as prestige and honor.
>
> (Zagorski-Thomas, p. 131)

In relation to the mastering process, cultural capital flows from wealth, knowledge, social position, and prestige. Zagorski-Thomas suggests that mastering processes are informed by those who have cultural capital (well known producers for example), which in turn 'informs the practices of a cohort of mastering engineers who admire those in question or the individuals work, mediated by the social field, which affects a part of the cultural domain' (*Ibid,* p. 131). An examination of such frameworks help to highlight the social and cultural implications associated with using equipment like Pultec and Manley, brands far beyond the reach of most bedroom producers. In referring to Moore's (2012) discussion of the 1167, using these devices within the mastering chain, Zagorski-Thomas suggests, lends itself to a certain sense of prestige. Thus, hardware used within the mastering process (compressors, limiters), can become the symbolic indicators of social and cultural capital. Further to this, Kaiser [11] attempts to define the relationship between capital the users relationship with technology and argues that 'there are many aspects of analogue software that cannot be emulated by software plugins', suggesting that:

> The credibility gap of software emulation in music production comprises tentative, olfactory and gustatory sensations, process-oriented aspects of workflow, as well as aspects of a hardware's physical characteristics, and time-dependent aspects.
>
> (Kaiser, p. 3)

Kaiser touches on key factors that influence the notions of 'engineering authenticity' in relation to physical hardware; the physicality of turning knobs to make sonic adjustments and procedural processes (changing tape, alignment, waiting for valves to heat up). This aids towards defining some of the aesthetic differences between using hardware and software devices within the mastering process where adjustments are often made on a finite level. One further consideration is that of the role of ergonomics in that it helps clarify the relationship between the worker and the job at hand and focuses on the design of work areas or work tasks to improve job performance. For the purpose of this chapter, it is proposed here that the ergonomics of each process (physical and virtual) can result in different approaches, broadly defined as follows:

- Physical Processing: the tactile and hands on process of using faders and knobs to adjust parameters within the mastering process

- Virtual Processing: the non tactile process of using a mouse or a touch screen to adjust parameters within the mastering process.

While it is evident that virtual processes can make workflow more streamlined (via automation and digital recall), most mastering engineers generally have an ergonomically hybrid approach; via the use of both physical and virtual sound processing tools. As previously suggested, as the access to virtual processing is now more commonplace to both the prosumer and professional music producer, there is now the very real possibility of traditional mastering processes becoming radically disrupted. The experiential processes that take place often involves the user (the mastering engineer) having a personal level of engagement within the mastering process and as Klett and Gerber [12] argue this involves 'engagement with human socio-cultural notions of noise as desirable or undesirable depending on the genre and taste of the listener'.

Through this process, technology becomes a collaborator and outside the technical tasks that a mastering engineer is responsible for, a reinvention of the role of the mastering engineer is currently taking place via the prosumer; in that the prosumers role is shifting the landscape of mastering. Such considerations, it is believed, are shaping the mastering and the plugins of today and the formalisation of future plugins. As discussed later, AI (Artificial Intelligence) may very well shape the progress of DSP modelling going forward in that the emergence of dedicated products and services in artificial intelligence-driven audio mastering and poses profound questions for the future of both the music and mastering industry, an industry that has already faced significant challenges due to streaming and music's digitalisation of music over the past decade.

ENGINEERING AUTHENTICITY: WAVES ABBEY ROAD TG MASTERING CHAIN

In attempting to achieve 'engineering authenticity' in DSP modelled plugins, reverse engineering and emulation feature within the development of Waves Abbey Road TG Mastering Chain. Chikofsky and Cross [13] describe reverse engineering as 'the process of developing a set of specifications for a complex hardware system by an orderly examination of specimens of that system'. It also allows for the process of analysing a subject system (EMI TG 12410 Transfer Console in this case) to identify the system's components and their interrelationships and create representations of the system in another form (its digital emulation). Emulation, on the other hand, is a common practice used to combat obsolescence or what could be termed 'digital preservation'. When Steinberg launched their VST (Virtual Studio Technology) plugin interface specification in 1996 (released with Steinberg's Cubase 3.02), only a handful of audio processing tools were available. Today, VST plugins are developed in line with the demand for sonic characteristics that both mix and mastering engineers are more commonly asked to provide for clients. In an attempt to emulate the sonic characteristics of hardware, manufacturers play a tit or tat game in trying to replicate characteristic features of hardware processors. Ultimately, emulation relies on components or codes that are able to bring the outcomes of these models closer to reality. Plugins developed with no hardware counterparts on the other hand, take less time to develop from start to finish. However, some may include greater technological challenges that require longer periods of time and effort. Software engineers and DSP programmers who develop audio plugins have to firstly identify a piece of hardware component to emulate and generally follow a number of typical steps, broadly summarised as follows:

- Sourcing the original documents of the hardware that is being modelled (schematics, operating manual)
- Test recordings to evaluate its characteristics
- Prototype rough builds of the plugins
- Compare the plugin to the original hardware via A-B testing
- Beta Testing: evaluating the overall user experience
- Design aesthetics and functionality

In an evaluation of the characteristics of a piece of audio equipment, the exact internal sample rate of the plug-in is also a contributing factor in how the filters interact with each other. Through this process, the sample rate conversions between DAW's and the internal one had to be implemented in a way to properly mimic aspects of the hardware and its analog circuitries including analog filters. Such processes are core to the development of the Wave's Abbey Road Mastering Chain plugin. Modelled on the EMI TG 12410 Transfer Console (Figure 4.2), the console was used in all of Abbey Road's mastering suites since the early 1970s and is still being used today to some degree.

Mirek Stiles, Abbey Road Studios Head of Audio Products, discusses the consoles contribution to the mastering process, commenting:

> I remember seeing the various TG Mastering modules used frequently in the studios and they still are to this day, most recently for mixing Brockhampton's *Iridescence* album which debuted at #1 on the Billboard Top 200. I wasn't so familiar with the mastering rooms' workflow, but from spending time with the engineers it was important to learn that the TG modules are used on almost every project; the only exception being the compressor, which can be a little too harsh for some stereo masters. With this in mind, the Waves and Abbey Road teams went about tweaking the original TG topology. Low level information is more prominent and the VCA curve and attack release circuits were modified to create something more suitable for today's mastering requirements. The end result was signed

Figure 4.2 EMI TG 12410 transfer console.

> off by Abbey Road's mastering engineers and we are all very proud of this plugin. It's an extremely versatile and beautiful sounding tool for both mixing and mastering.
>
> (Stiles, 2019)

As Stiles suggests, the limitation of the original EMI TG 12410 Transfer Console was the machine itself in that the components used may have been of low quality but had a complimentary effect on tone. Such factors are often part of the manufacturers cost effectiveness agenda and now developers and programmers now have the ability to bypass and supplement such issues by providing high fidelity over low DSP loads (TG Mastering Suite currently supports up to 192 kHz). In referencing Bourdieu's notions of capital, Wave's has attempted to use many forms of cultural capital in selling the product, as its marketing campaign points towards, suggesting:

> The distinct solid-state transistor-based sound of the TG12410 has proven itself time and again over many decades. Whether it's used for mastering as a complete console, or for mixing with only select modules, the TG brings nothing less than magic — the same magic heard on albums like Pink Floyd's *The Dark Side of the Moon*, Nirvana's *In Utero*, Radiohead's *OK Computer* and Ed Sheeran's "+". Thanks to this Waves/Abbey Road collaboration, the TG12410 is available outside of Abbey Road Studios and you can now bring the very same magic to your own productions
>
> (Waves, 2018).

The plugin itself (Figure 4.3) comprises a modular series of processors (or 'cassettes') that can be arranged in any order or individually bypassed. Its output can be independently switched between regular Stereo, Duo and Mid-Side processing. In the Expanded View, the module blows it up to fill the whole window, granting access to the Link/Unlink button and separate controls for Left/Right or Mid/Side.

The function of each processor (or cassette) is detailed as follows:

- TG12411 Input: Input processing including phase and L/R Balance (Stereo/Duo and M/S (Mid-Side) Mode)
- TG12412 Tone: 4 Band EQ, 5 filters per band (Stereo/Duo and M/S Mode): Each band centre/corner frequencies (32–128 Hz, 181–724 Hz, 1.02–3.25 kHz, and 4.1–16 kHz)
- TG12413 Limiter: Includes three compressor/limiter types:
    1. Original: models original Zener Diode limiter

Figure 4.3 Waves Abbey Road TG Mastering suite plugin.

2. Modern: models a Waves and Abbey Road design algorithm
3. Limit: models behaviours and harmonic distortion of the original Zener Diode

- TG12414 Filter: high and low pass filters with accompanying presence bell
- TG12416 Output: with stereo spread feature enabling 'tilting' the mid- and side- signals by up to ±5 dB (Waves, 2018).

Both Waves and Abbey Road propose that the TG Mastering Chain, rather than aiming for transparency and precision, is more suited to shaping the 'character' of a mix/output. In contradiction to the mastering process, which in many ways relies heavily on achieving character by being transparent and precise, yet with fixed-frequency equalisation and filters, the tone control offered by the TG Mastering Suite does not suit processes more commonly associated in mastering. One of the more difficult aspects of 'engineering authenticity' within the TG Mastering Chain, or that of any plugin, is that of tape emulation. The cassette, TG12411 (Figure 4.4), includes a tape equaliser function that allows for the filtering or flattening out the frequency spectrum, with the options of NAB-to-IEC (U.S. and European EQ standardisation curves for tape) at 5.7 and 15 IPS (inches per second) and IEC-to-NAB at 15 IPS and 5.7 IPS.

Waves describe the TG12411 Input Cassette functionality in relation to mixing within the digital domain, commenting:

> Originally, in the hardware domain, this was done when playing back from NAB to IEC tape machines either at 7.5 IPS or 15 IPS, in order to provide the equalization required for playing NAB-equalized tape on an IEC machine and vice versa. In your digital mixes, you will not need to use this control for that

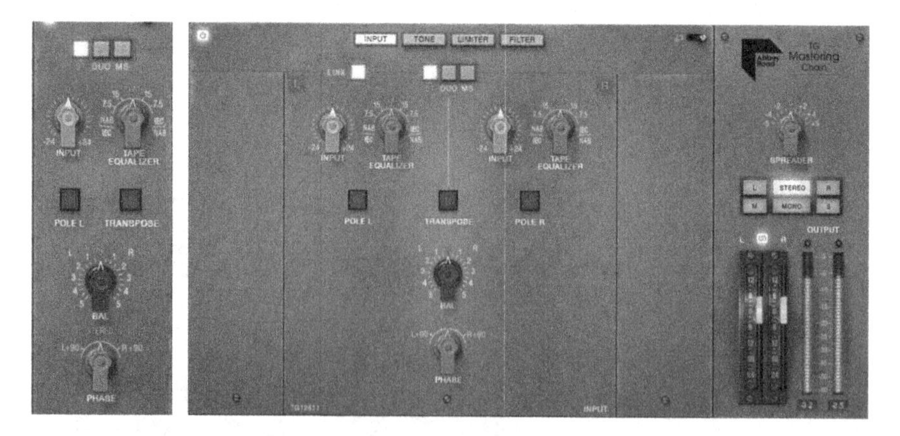

Figure 4.4 Waves TG 12411 Input Cassette.

historical purpose. We have decided to keep this control, however, due to its great sonic character you may find that it gives your tracks interesting characteristics.
(Waves Abbey Road TG Mastering Chain User Manual, p. 11)

Reviews for the Waves Abbey Road TG Mastering Chain have been balanced while some point towards its lack of transparency and precision. Indeed, what seems to be universal is its suitability to genres other than rock and pop music and what seems to be the most beneficial contribution is its fixed frequency equalisers (although this is always the apt equalisation option). Some users complained about the harshness of high end frequencies (8 kHz+) while others saw the benefit of using the plugin as a channel strip on both drum and guitar channels. As ‘prosumer’ reviews, they offer valuable insight into the plugin’s functionality and usability. One observation is apparent from the reviews; this sense of limitation, with fixed filters in the TG 12412 Tone Cassette (Figure 4.5) is perhaps, for some, the most rewarding part of the TG Mastering Chain.

The obvious disadvantage of this is that some of the features can sound a little generic but in many cases of DSP modelling, sometimes less is more. Ultimately, allowing the user to go between both ‘original’ and ‘modern’ modes, the TG 12413 Limiter Cassette (Figure 4.6) can bring about genre specific characteristics that are ultimately deemed appropriate by the user or guided by the tonal qualities of a particular preset.

ENGINEERING AUTHENTICITY: TECHNOLOGY DETERMINISM AND HUMAN AGENCY

The digital emulation of the EMI TG 121410 Transfer Console by Waves opens up an array of questions and considerations in relation

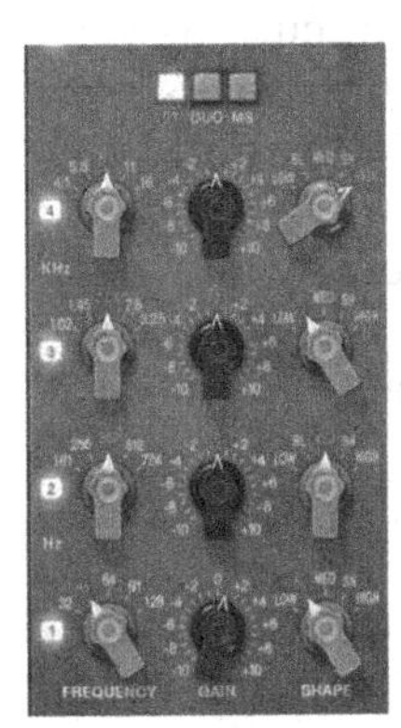

Figure 4.5 Waves TG 12412 Tone Cassette.

Figure 4.6 Waves TG 12413 Limiter Cassette.

to our experience with technology for mastering engineers. As previously discussed, technological capital is not just about simply buying or accumulating technology, it's about people having access to technology, or moreso processes that would usually be unavailable to them. Free multitrack DAWs like Reaper, that offer a suite of free VST composed in a very smartly designed program, have had a huge and profound impact on the music production industry. In order to differentiate themselves, companies like Waves and Abbey Road lean heavily on taglines like 'authentic', 'warm', 'vintage' to help market and sell their products. One of the possible contributing factors to this is the collective consideration that these analogue technologies and processes of the past provide a more distinctive sound. It is certainly possible that, as the mastering chain is the final adjustment before consumption by the general public, that this is more relevant today than ever as the bid to stand out 'sonically' from the crowd becomes more and more important. In many ways, the development of such technology poses an interesting observation in current DSP modelling culture, that of music productions addiction to its own past, something Simon Reynolds discusses in Retromania (2012):

> Is nostalgia stopping our culture's ability to surge forward, or are we nostalgic precisely because our culture has stopped moving forward and so we inevitably look back to more momentous and dynamic times.
>
> ([14,15], p. 22)

Considering the past begs the question if we have now become 'culturally conditioned' in that, if the tools of mass marketing in music production has informed us, through nostalgic approaches, that these emulations are far superior to traditional plugins. In many ways, the accessibility of pirated software has largely informed this process. Reynolds continues:

> Not only has there never been a society so obsessed with the cultural artifacts of its immediate past, but there has never before been a society that is able to access the immediate past so easily and so copiously.
>
> ([14], p. 56)

In this sense of abundance of possibilities (further facilitated through the use of presets), the user is faced with a magnitude of options, each tailored to each project and its needs. Coupled with referentialism towards the machines of the past, our relationship with technology and its influence must be addressed.

We have the freedom to choose whatever mastering tools we feel are appropriate, however, one major consideration is the proposition that the technology that we use ultimately determines this influence. In an attempt to address this and to define 'engineering authenticity' further, two frameworks, technological determinism (TD) and human agency are considered. TD seeks to understand how technology influences human action and thought. Much of its approach is based on the historical observation that new technologies are often released without much thought given towards their impact on both society and culture. Dafoe [16] goes as far to suggest that 'technology transforms societies and cultures in significant ways'. He continues:

> A central issue in the study of technology is the question of agency. To what extent do we have control over the tools we use – and hence also our systems of production, social relations, and worldview? To what extent are our technologies thrust upon us –by controlling elites, by path-dependent decisions from the past, or by some internal technological logic?
>
> ([16], p. 1048)

Herbert Marcuse's One-Dimensional Man (1964) discusses freedom in relation to the commoditisation of technology and proposed the following: is technology independent of social influence or if it is determined by human will alone? Marcuse suggests that:

> Technological rationality, which impoverishes all aspects of contemporary life, has developed the material bases of human freedom, but continues to serve the interests of suppression. There is logic of domination in technological progress under present conditions: not quantitative accumulation, but a qualitative leap is necessary to transform this apparatus of destruction into an apparatus of life.
>
> ([17], p. 11)

In a more contemporary viewpoint, Mayr [18] prescribes human agency to the claim that we do in fact make decisions and enact upon them in the world, commenting:

> Our self-understanding as human agents includes commitment to three crucial claims about human agency: that agents must be active, that actions are part of the natural order and that intentional actions can be explained by the agent's reasons for acting.
>
> ([18–20], p. 1051)

For the purpose of this chapter, it is suggested here that presets, used within the mastering process, can determine or limit an agent and their decision making abilities. Further to this, such determinations, mediated through technology, is largely conditioned by technological change, or as Mayr suggests:

> The defining characteristic of technology is its functionality, not its specific materiality. Technology, thus, (1) denotes those entities–artifacts, techniques, institutions, systems–that are or were functional and (2) emphasizes the functional dimension of those entities.
>
> (*Ibid*, p. 1053)

The discussion of Technological Determinism and Human Agency has attempted to highlight some of the conceptual and philosophical issues associated within the shaping of technology, a consideration valuable in the current marketing of emulated VST plugins. In many ways, technical or technological development can be seen as the key mover in historic and social changes and that this purports towards how man and machine converse; an ever shifting dialogue within both music technology and the mastering process and moreso, its future directions and possible uses.

CONCLUSION

> Mastering your own album is like marrying your first cousin. You never know how the children will turn out, or maybe you do! [21].

In an attempt to define 'engineering authenticity' both the social and cultural considerations of technological development were considered through Bourdieu's theory of 'cultural capital' theory. This uncovered the role of the prosumer and its position within current music production software marketing strategies. In the examination of the Waves Abbey Road TG Mastering Chain plugin, it was highlighted that its modelling and emulation, attempts to contribute sonic character to a mix or a mastering process (although it would seem it is far more suited as an 'effect' rather than being branded as part of the 'mastering chain' process). Again, this is one of the core issues uncovered from such an examination; the distortion of a plugin's functionality by targeting marketing strategies towards the prosumer. Essentially, it is the naming of the plugin that questions current marketing strategies of audio plugin manufacturers. Aside from providing tone control and character, it seems at odds that the plugin has been marketed to prosumers as a mastering tool as it is seemingly more suited as a tone control plugin, perhaps used on mix bus by studio and mix engineers. Further to this, it was highlighted that the prosumers current demand for digitally emulated hardware devices of the past, led to the development of the Waves Abbey Road TG Mastering Chain.

While companies like Waves, Universal Audio, SoundToys and others attempt to 'engineer authenticity' and in order to define the implications of this process, technological determinism and human agency were examined to provide a framework for discussion and this, in many ways, helped bring back into focus the mastering engineer's ongoing, evolving and future relationship with technology.

This relationship, like all technological advancements, is in flux, yet as Hepworth-Sayer and Hodgson [22] observe, one thing remains the same in that 'every mastering engineer works differently, and often using different tools, even if they pursue the exact same aesthetic goal, namely, producing the best record possible from the mixes they are given'. The casting of the mastering engineer as a craftsman or craftsperson is well documented throughout popular music production histories, so much so that Katz [21] pointed out that 'old fashioned craftsmanship and attention to detail will always be in demand'. 2007 was a long time ago, particularly in technological years as associated mastering technologies and workflow processes (meta data for physical mediums like CDs) have either evolved or disappeared. From its birth as a strand within audio engineering, the position of the mastering engineer has always been shrouded in mystery. O'Grady [23,24] suggests that the 'representation of mastering as "mysterious" works to reinforce the importance of this practice and also to safeguard it from new technologies that might challenge its dominance'.

One fundamental separation between both physical and virtual mastering processes still exists; the role of human objectivity. As much as modelled mastering plugins can offer, a plugin preset cannot make an experienced and informed decision on the fly or offer years of tried and tested methodologies made by humans. Any mastering engineer knows that a new plugin or piece of hardware can bring about new possibilities but can, at the same time, bring about limitations or as Wiener (1948) suggests this 'progress imposes not only new possibilities for the future but new restrictions'. The plugin, along with online mastering, pulls into the limelight of what an objective and skilled mastering engineer can bring to finalising a project. Perhaps these levels of human objectivity are at an end or as Birtchnell (2018) suggests 'in some instances, technologies are the root cause of human obsolescence and drive redundancies in occupations, skills, and livelihoods'.

As the use of AI (Artificial Intelligence) and Machine Learning becomes utilised within the mastering process, algorithmic control of what has primarily been a human driven enterprise, will see the mastering domain drastically shift over the next decade. LANDR, an AI and Machine Learning based mastering service is one platform competing with traditional mastering services. With a slightly naive tagline of 'professional audio mastering with instant results' and 'sound like a pro without paying studio rates or learning complex plug-ins' [25], LANDR uses, based on an algorithmic analysis, a set of processing tools for the pre mastering process. For the mastering process itself, via both AI and Machine Learning, the algorithms adjust these parameters for the final output and mix. Statements such as 'create, we'll do the rest' (*Ibid*), proposes to challenge, distort and perhaps undermine the value of mastering itself as or Stern and Razlogova (2019) suggest AI and Machine Learning 'devalues the people's aesthetic labor as it establishes higher standards for recordings online'. Nonetheless, it is important to recognise that perhaps AI and Machine Learning are just variants of traditional mastering services in that 'far from the spectacular rhetoric around AI as an emergent form of nonhuman agency, in learning from LANDR we find a very familiar set of agencies–financial, corporate, technical, musical, and human–hard at work in a new setting' (*Ibid*, p. 16).

Emergent technologies have led to the development of extended mastering techniques over the last number of years and these technologies have played a role in shaping the sound of mastering today. While both Waves and Abbey Road sell the prosumer sound of Abbey Road and the EMI TG 12140 Transfer Console, both LANDR and eMastering sell similarly priced options, but perhaps it will be the latter that will become more dominant as no plugin purchase is required and further to this, an algorithm carries out the mastering process itself.

AI and Machine Learning has undoubtedly changed the way in which the world communicates and works. Analysing data, identifying patterns and making decisions with minimal human intervention is by no means a path towards 'engineering authenticity' within the mastering process. Ultimately, going forward, mastering as a human orientated artistic craft, will have to fight and strive towards remaining a human-centred process and if required, incorporate algorithmic approaches that include collaboration with humans rather than attempting to replace or outsell them.

REFERENCES

1. Toffler, A. *The First Wave: The Classic Study of Tomorrow*. Bantam Books, New York, (1980), p. 101.
2. Waves Audio Introduces Abbey Road TG Mastering Chain. Accessed October 2019 from https://www.prosoundweb.com/waves-audio-introduces-abbey-road-tg-mastering-chain/.
3. Waves Abbey Road TG Mastering Chain. Accessed November 2019 from https://www.waves.com/plugins/abbey-road-tg-mastering-chain#presenting-abbey-road-mastering-chain.
4. Waves Abbey Road TG Mastering Chain User Manual, Accessed September 2019 from https://www.waves.com/1lib/pdf/plugins/abbey-road-tg-mastering-chain.pdf.
5. Wiener, Norbert, *Cybernetics or Control and Communication in the Animal and the Machine*. MIT Press, Cambridge, MA, (1965), p. 121.
6. Cole, S. J. The Prosumer and the Project Studio: The Battle for Distinction in the Field of Music Recording, *Journal of Sociology*, Vol. *45*, No. 3, (2011), pp. 447–463.
7. Wishart, T. *Audible Design*. Orpheus-The Pantomime Press, London, (1994), p. 321.
8. Frith, S. *Taking Popular Music Seriously*. Routledge, London, (2017), p. 73.
9. Bourdieu, P. *Distinction: A Social Critique of the Judgement of Taste*. Harvard University Press, Cambridge, (1984), p. 1.
10. Birtchnell, T. *'Listening without Ears: Artificial Intelligence in Audio Mastering'*. *Big Data & Society*, Vol. *4*, No. 2 Accessed July 2010 from https://doi.org/10.1177/2053951718808553.
11. Kaiser, C. Analog Distinction – Music Production Processes and Social Inequality. *Journal on the Art of Record Production*, No. 12, Accessed September2019 from https://www.arpjournal.com/asarpwp/analog-distinction-music-production-processes-and-social-inequality/.
12. Klett, J. and Gerber, A. The Meaning of Indeterminacy: Noise Music as Performance. *Cultural Sociology*, Vol. *8*, (2014), pp. 275–290.
13. Chikofsky, E. J. and Cross, J. H. Reverse Engineering and Design Recovery: A Taxonomy. *IEEE Software*, Vol. *7*, No. 1, (1990), pp. 13–17.
14. Reynolds, S. *Retromania*. Faber and Faber, London, (2012), pp. 22–56.
15. Sterne, J. and Razlogova, E. *'Machine Learning in Context, or Learning from LANDR: Artificial Intelligence and the Platformization of Music Mastering'*. *The Journal of Social Media and Society*, Vol. *5*, No. 2. https://doi.org/10.1177/2056305119847525.
16. Dafoe, A. On Technological Determinism: A Typology, Scope Conditions, and a Mechanism, *Science, Technology, & Human Values*, Vol. *40*, No. 6, (2015), pp. 1047–1076.
17. Marcuse, H. *One-Dimensional Man*. Beacon Press, Boston, (1964), p. 11.
18. Mayr, E. *Understanding Human Agency*. Oxford University Press, Oxford, (2011), pp. 1047–1053.

19. McGregor, I. The Relationship Between Emulation and Simulation, Accessed September 2019 from http://citeseerx.ist.psu.edu/viewdoc/download?doi=10.1.1.58.2915&rep=rep1&type=pdf.
20. Moore, A. All Buttons In: An Investigation into the Use of the 1176 FET Compressor in Popular Music Production. *Journal on the Art of Record Production*, No. 6, Accessed July 2019 from https://www.arpjournal.com/asarpwp/all-buttons-in-an-investigation-into-the-use-of-the-1176-fet-compressor-in-popular-music-production/.
21. Katz, B. *Mastering Audio*. 2nd edn. Focal Press, Oxford, (2007), p. 5.
22. Hepworth-Sawyer, R. and Hodgson, J. *Audio Mastering - The Artists (Perspectives on Music Production)*. Routledge, London, (2018), p. 2.
23. O'Grady, P. 'The Master of Mystery - Technology, Legitimacy and Status in Audio Mastering'. *Journal of Popular Music Studies*, Vol. *31*, No. 2, (2019), p. 3.
24. Pras, A., Guastavino, C. and Lavoie, M. 'The Impact of Technological Advances on Recording Studio Practices'. *Journal of the American Society for Information Science and Technology*, Vol. *64*, No. 3, pp. 612–626.
25. LANDR. Accessed December 2019 from https://www.landr.com/en/online-audio-mastering/.

5

Mastering for streaming

EXPLORING A NEW LEVELLING STANDARD

Scott Harker

INTRODUCTION

Initial research in the area of Mastering for Streaming showed that there is, at present, a limit to the academic investigation into this subject area. Contrary to this, there is an overwhelming amount of information and opinion from the pro-audio community which is largely anecdotal and developed through practice-based methods and experience. As mastering engineers, it could be argued that not only do we need to be familiar with trends in consumption but also of the tonal imprint that a playback medium might have upon the music enjoyed by the listener. Research has yet to emerge showing significant sonic differences between different streaming services which could perhaps demonstrate some disparities in the algorithms being used. The interviews and research conducted here confirm that this area is in need of further exploration, not only to explore new techniques but to give some context to unsubstantiated evidence presented from the pro-audio community. The focus of this chapter is to investigate these assertions and using those results and findings to apply them to salient areas of enquiry that results in a framework that may help guide improvements in modern mastering. This is underpinned by the following question:

> Can we arrive at a framework guiding improvements in contemporary mastering technique(s) through the analysis of popular music playback on streaming platforms and the application of multiple research methodologies?

This chapter initially focusses on streaming platform habits through an analysis of recorded findings in order to assimilate and/or uncover the behaviour of commonly used music services. Further to this, related industry mechanisms and approaches towards streaming platforms are explored through discussions with industry professionals as well as exploring the rhetoric in professional and non-professional circles. The findings from these two areas of study were used to explore mastering for streaming through a practice-based approach that concludes with a framework outlining practical recommendations for mastering engineers.

The research, practice, and discussions in this chapter have been produced and collated over about a year, with the initial analysis originally carried out in May 2019 and some interviews and further research taking place in April 2020. As a mastering engineer, the end goal was that this would aid my pursuit in finding a suitable workflow for mastering to multiple loudness levels or indeed, to find an answer as to whether this might be a necessary approach. It must be noted that since gathering data for the streaming platforms in this chapter, a lot has, and continues to, change in the landscape of loudness. Evolving discussions in other research being presented both academically and anecdotally around new loudness targets and new approaches mean that the practice-based approaches exploring possible workflows in this chapter have grown in relevance. Through these discussions it has also become clear that some engineers are using –14 LUFS as a new loudness standard. Whilst this has not yet been ratified it is an area to keep an eye on in the future and seems to have shifted away from the AES recommendation of –16 LUFS [1–4].

BACKGROUND

Mastering is a process that aims to take a mixed record and build translation, cohesion and consistency into a song or set of songs for final delivery to the listener. As digital audio developed, mastering became regularly affiliated with loudness [5–7]). For some time it has been discussed by academics and industry professionals that the 'loudness war' stemmed from new digital technology that allowed for hyper compression, and the music business industry's belief that louder songs sell better [8]. As is discussed elsewhere in this book, the loudness wars origin can be traced to earlier times. Although it has been found that louder versions of the same song tend to sound better when listening for a short amount of time [9], there are other issues that hyper compression can create. Some of these include feeling reduced excitement or emotion for a song which may be attributed to less dynamic range, as well as True Peak clipping that results in distortion artefacts when audio is encoded into a different format. This also leads to a reduction in stereo width [10]. Concerns have also been raised in relation to hearing damage as it has been said that compression and limiting can protect the listener from high peak levels but increases in sustained level may actually have more of an impact on hearing damage. It has also been argued the industry's adoption of the idea that 'louder is better' is in fact an over-simplification, and suggests that listeners actually focus more on texture, harmony and melody. Not long before the adoption of streaming services, it was assimilated that loudness normalisation would remove any further motivation for the firmly held assumptions by the industry on the loudness war [11].

Most, if not all streaming services employ loudness normalisation in order to keep a consistent playback level for consumers across their entire streaming catalogue. This ensures that consumers are not having to adjust volume when navigating the service's library which inevitably results in a better user experience. As streaming became the most common way to listen to music, and went further to become the largest revenue generator for the music industry in most of the world [12,13], many professionals that were concerned with the loudness war looked forward to the positive impact streaming would have on the reduction in hypercompression in popular music. Unfortunately, it seems that the loudness war is continuing within streaming and download services despite platforms employing normalisation. As can be observed in recent discussions between industry experts [14–17], it has become increasingly clear that among various new demands on mastering engineers in the modern era, streaming platforms are doing little to combat the effects of the loudness war. In response, industry experts and professionals authored the Audio Engineering Society's technical document that outlined recommendations for loudness for streaming playback [1]. However, whilst there is relevant information in this technical document regarding music streaming platforms, overall it is largely aimed at broadcast loudness standards. With mastering engineers being the last step in the audio production process, their role in identifying problems in the ever-developing delivery technologies has never been more critical. Because of this, it is important to explore common rhetoric and unsubstantiated ideas that are being investigated in the mastering community and factor that into the conclusions of this chapter.

ANALYSIS AND FINDINGS

Analysis of streaming platforms

The tracks analysed for this part of the chapter included twelve mastered songs from four artists. All 12 tracks are available on all major streaming platforms; including the platforms that were easily accessible for this analysis. The streaming platforms used were: Spotify, YouTube, Deezer, and Apple Music. Spotify and Deezer also have a 'no normalisation' setting which has also been analysed. All masters were approached stylistically in a more traditional way; not taking into account loudness normalisation used by streaming services and/or using limiters with True Peak functionality. Some masters will have been mastered as loud as –0.1 dBFS. This is important to note as in the analysis, not only can we reflect on what the services 'sound' like, but also how they react to masters known to have high True Peak (TP) values or a low peak to loudness ratio (PLR), measured in LUFS. Two pieces of analysis software have been used to illustrate the findings. The first being MusicTester [18,19] which is used to show statistical information in relation to loudness (Figure 5.1). The second is EAnalysis, which shows a frequency versus amplitude plot of an audio file with effective visualisation.

The open source kernel extension 'Soundflower' (Rogue [20,21]) was used to route the audio internally; directly from the application to my DAW. This ensures an accurate capture as no Digital–Analogue–Digital conversion is required to record the playback from streaming platforms which could affect the results of the analysis. Whilst there is not much information on the technical workings of Soundflower online, contact was made with the original author of the software, Matt Ingalls. Ingalls ensured that as long as application volume is full and there is no drift correction present, the resultant recorded audio should be exactly the same as the output from the application [22–24].

Figure 5.1 shows what is presented when audio is uploaded into the program MusicTester. This example shows six versions of the song

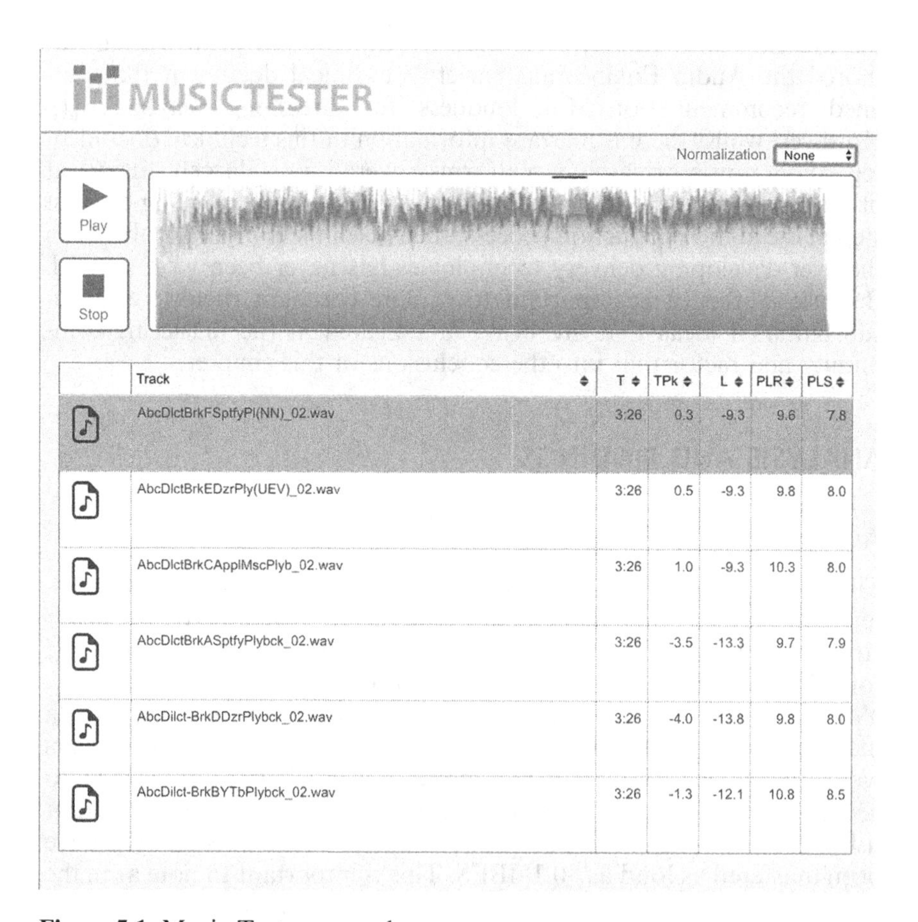

Figure 5.1 Music Tester example

'Break' all printed from the different streaming services Spotify, Deezer, YouTube, and Apple Music. The readings used from this software included LUFS and True Peak level. These figures were then analysed for individual and comparative study of the streaming platforms. The results from Music Tester have shown some significant findings in relation to loudness targets and normalisation algorithms that the streaming platforms use. It must be stated that these results only apply to this particular sample of twelve tracks across four different platforms. Larger research projects exploring similar analysis may benefit from a bigger sample size (Tables 5.1–5.3).

The results have shown that the PLR outputs of the different platforms appear not to be fixed despite what streaming service providers may quote on their respective websites. Spotify's normalised PLR levels, depending on the program material, range from –16 to –13. Apple Music's between –9 and –12. YouTube' between –12 and –14 and Deezer between –14 and –15. Spotify and Deezer's non-normalised playback settings returned figures matching Apple Music's, which were analysed without soundcheck enabled as this is Apple Music's default state. With soundcheck enabled the PLR level of Apple Music is –16 LUFS. The first thing to note is that Spotify's normalised setting returned an average PLR range of 3 LUFS, which matched that of the non-normalised versions of Spotify and Deezer, as well as Apple Music. Conversely, YouTube and Deezer's PLR range were 2 and 1 LUFS respectively. Whilst a couple of Loudness Units between songs will barely be noticeable, this information would suggest that Spotify has a more dynamic loudness algorithm than YouTube and Deezer. True Peak readings were as follows: Spotify and Deezer in normalised modes ranged from –5 to –2; YouTube ranged from –3 to –0.3; Apple Music, Spotify, and Deezer non-normalised ranged from 0.3 to 1.7. Although, Deezer read up to 0.2 dBFS less than Apple Music or Spotify on more than one occasion, these readings were as expected; with YouTube known to be a louder platform, that normalises to around –12 LUFS.

The statistics drawn from this analysis show more interesting correlations when comparisons are made between the songs and how the different platforms react to them. For example, when the original masters of Winter and Fire are compared, Winter has a lower original PLR reading and is overall a quieter song. However, Spotify's normalisation algorithm resulted in Winter being a louder song than Fire, with readings of –14.7 and –15.6 respectively. Deezer also showed a similar trend albeit with a smaller margin. YouTube, Apple Music, Deezer non-normalised and Spotify non-normalised all reflected the loudness differences of the original master. However, Deezer's non-normalised True Peak output read up to 0.2 dBFS less than Spotify non-normalised or Apple Music on more than one occasion. These examples show just how differently the platforms can react to the exact same master upload.

The figures collected from MusicTester also suggest that some of the platform's algorithms take the True Peak readings into account

Table 5.1 Music Tester analysis

Track name	Platform	Stream PLR (LUFS)	Original PLR (LUFS)	Original (True Peak dBFS)	Stream (True Peak dBFS)
Guest List A	Spotify	–13.4	–9.4	1.7	–2.2
Guest List B	YouTube	–12	–9.4	1.7	–1.2
Guest List C	Apple Music	9.7	–9.4	1.7	1.7
Guest List D	Deezer	–14.8	–9.4	1.7	–2.8
Guest List E	Deezer No Norm	9.7	–9.4	1.7	1.7
Guest List F	Spotify No Norm	9.7	–9.4	1.7	1.7
Searching A	Spotify	–14.8	–10.4	0.2	–3.7
Searching B	YouTube	–12.4	–10.4	0.2	–0.3
Searching C	Apple Music	–10.9	–10.4	0.2	0.3
Searching D	Deezer	–14.6	–10.4	0.2	–3.4
Searching E	Deezer No Norm	–10.9	–10.4	0.2	0.3
Searching F	Spotify No Norm	–10.8	–10.4	0.2	0.3
Magic A	Spotify	–13.4	–9.2	0.2	–3.6
Magic B	YouTube	–12.6	–9.2	0.2	–1.9
Magic C	Apple Music	–9.5	–9.2	0.2	0.3
Magic D	Deezer	14.2	–9.2	0.2	–4.4
Magic E	Deezer No Norm	–9.5	–9.2	0.2	0.3
Magic F	Spotify No Norm	–9.4	–9.2	0.2	0.3
Break A	Spotify	–13.3	–9.3	0.2	–3.5
Break B	YouTube	–12.1	–9.3	0.2	–1.3
Break C	Apple Music	–9.3	–9.3	0.2	1
Break D	Deezer	–13.8	–9.3	0.2	–4
Break E	Deezer No Norm	–9.3	–9.3	0.2	0.3
Break F	Spotify No Norm	–9.3	–9.3	0.2	0.5

Track name	Platform	Stream PLR (LUFS)	Original PLR (LUFS)	Original (True Peak dBFS)	Stream (True Peak dBFS)
Dreams A	Spotify	–14.2	–10.2	0.2	-3.7
Dreams B	YouTube	–12.3	–10.2	0.2	-1.1
Dreams C	Apple Music	–10.2	–10.2	0.2	0.3
Dreams D	Deezer	–13.8	–10.2	0.2	-3.3
Dreams E	Deezer No Norm	–10.2	–10.2	0.2	0.3
Dreams F	Spotify No Norm	–10.2	–10.2	0.2	0.3

when adjusting the loudness of the tracks. For example, when comparing the True Peak, Original PLR, and Streamed PLR of the tracks; the figures suggest that YouTube correlates a lower True Peak with a gain in overall loudness. Conversely, a higher True Peak value would result in a loudness penalty. This is most notable when two tracks are compared that have a similar original loudness, but different True Peak readings. For example, when comparing 'Hot Sand' and 'Not Much Left' they have a close original PLR of –11.6 and –11.7 LUFS respectively however, the higher True Peak reading of 'Not Much Left' shows a higher turn-down penalty from YouTube's algorithm. This is despite 'Not Much Left' having a lower PLR reading.

Whilst there are some variations depending on what songs are being compared, this analysis shows that the platforms react to high True Peak values by reducing loudness. The variations between the way different algorithms react to True Peak could well be down to the different codecs and limiters that are used by the streaming platforms. Depending on the codec that is used by the streaming platform, True Peak signals may cause distortion artefacts when transients are re-created. These artefacts could hit the threshold of the limiter being used by the streaming platform, which would attenuate the signal more aggressively. With this in mind, the higher the original True Peak signal, the more likely artefacts will accentuate gain reduction, which will lead to a reduction of loudness and lead to a quieter song. If thinking about the aims of this chapter and answering the research question; the findings may suggest that it would be advisable to factor in processes that diminish the chance of final masters being affected by higher True Peak signals.

To gain further insight into the way streaming services process audio and deliver it to consumers, the application EAnalysis was used to illustrate the data in a visual way. This application can show sonic

Table 5.2 Music Tester analysis

Track name	Platform	Stream PLR (LUFS)	Original PLR (LUFS)	Original (True Peak dBFS)	Stream (True Peak dBFS)
Winter (Acoustic) A	Spotify	–14.7	–11.5	0.5	–2.6
Winter (Acoustic) B	YouTube	–13.2	–11.5	0.5	–1
Winter (Acoustic) C	Apple Music	–11.6	–11.5	0.5	0.4
Winter (Acoustic) D	Deezer	–14.4	–11.5	0.5	–2.4
Winter (Acoustic) E	Deezer No Norm	–11.6	–11.5	0.5	0.4
Winter (Acoustic) F	Spotify No Norm	–11.6	–11.5	0.5	0.5
Fire A	Spotify	–15.6	–10.8	1.2	–4.6
Fire B	YouTube	–14.1	–10.8	1.2	–2.6
Fire C	Apple Music	–9.9	–10.8	1.2	1
Fire D	Deezer	–14.3	–10.8	1.2	–3.4
Fire E	Deezer No Norm	–9.9	–10.8	1.2	1
Fire F	Spotify No Norm	–9.9	–10.8	1.2	1.1
Hot Sand A	Spotify	–15.3	–11.6	0.5	–5.1
Hot Sand B	YouTube	–13.8	–11.6	0.5	–3
Hot Sand C	Apple Music	–9.7	–11.6	0.5	0.5
Hot Sand D	Deezer	–14.6	–11.6	0.5	–4.5
Hot Sand E	Deezer No Norm	–9.7	–11.6	0.5	0.4
Hot Sand F	Spotify No Norm	–9.7	–11.6	0.5	0.5
Come Back A	Spotify	–13.8	–9.2	0.9	–3.7
Come Back B	YouTube	–12.7	–9.2	0.9	–2.2
Come Back C	Apple Music	–9.2	–9.2	0.9	0.9

Track name	Platform	Stream PLR (LUFS)	Original PLR (LUFS)	Original (True Peak dBFS)	Stream (True Peak dBFS)
Come Back D	Deezer	–14.2	–9.2	0.9	–1.8
Come Back E	Deezer No Norm	–9.2	–9.2	0.9	0.9
Come Back F	Spotify No Norm	–9.2	–9.2	0.9	0.9

differences between the platforms by presenting a plot of amplitude of frequencies. When comparing the different versions of the same track, it was important to keep in mind that the normalised versions of the songs were normalised to different loudness levels, which may inaccurately portray findings. For example, as Apple Music is louder than Spotify in its normalised state, low end may appear more intense in colour, when in-fact it is simply a louder version overall. There are some differences in the visual representation that only apply to a particular track that do not repeat to other songs. With a one-off anomaly such as this, it is hard to draw a conclusion about the playback habits of a platform. Like the previous findings with MusicTester, it does suggest to a mastering engineer that there is only so much control you have over the output of the various platforms. Despite this, the EAnalysis shows that most of the platforms employ some sort of filtering. Figure 5.2 shows an EAnalysis example of the song 'Guest List' which shows Deezer (normalised) heavily low pass filtering the whole track to around 15 kHz (Figure 5.2).

This filtering is common to all songs and most platforms however, this is the most extreme example out of the 12 tracks analysed. The other tracks EAnalysis plots also confirm that Deezer filters more severely, followed by YouTube and Spotify. Interestingly, it also shows that Apple Music does not seem to employ filtering.

Industry professionals

This section hopes to gain insight from all areas and experience levels through communication with professional engineers in the industry and exploring other online discourse surrounding the topic. This, alongside the establishing analysis, will inform the practice-led research approach. As streaming has now become the primary revenue source for artists, technology that allows money to be saved in the record production process has become critical. This is mainly due to the profits from streaming being much lower than that of physical mediums which were previously the primary revenue source [25,26].

Table 5.3 Music Tester analysis

Track name	Platform	Stream PLR (LUFS)	Original PLR (LUFS)	Original (True Peak dBFS)	Stream (True Peak dBFS)
Not Much Left A	Spotify	–16.3	–11.7	0.4	–4.1
Not Much Left B	YouTube	–14.1	–11.7	0.4	–1
Not Much Left C	Apple Music	–11.7	–11.7	0.4	0.4
Not Much Left D	Deezer	–14.4	–11.7	0.4	–2.3
Not Much Left E	Deezer No Norm	–11.7	–11.7	0.4	0.4
Not Much Left F	Spotify No Norm	–11.7	–11.7	0.4	0.4
Interlude A	Spotify	–13.6	–9	0.9	–3.7
Interlude B	YouTube	–11.7	–9	0.9	–1.2
Interlude C	Apple Music	–9	–9	0.9	0.9
Interlude D	Deezer	–13.9	–9	0.9	–4.1
Interlude E	Deezer No Norm	–9	–9	0.9	0.8
Interlude F	Spotify No Norm	–9	–9	0.9	0.9
Wanna Know A	Spotify	–15.6	–10.8	0.6	–4.4
Wanna Know B	YouTube	–14.1	–10.8	0.6	–1.6
Wanna Know C	Apple Music	–10.7	–10.8	0.6	0.6
Wanna Know D	Deezer	–14.5	–10.8	0.6	–3.3
Wanna Know E	Deezer No Norm	–10.6	–10.8	0.6	0.6
Wanna Know F	Spotify No Norm	–10.6	–10.8	0.6	0.6

With streaming now being the main source of income, it makes sense that some mastering engineers might use 'mastering for streaming' as a way to market their mastering services. Whilst the large mastering houses are yet to do this, independent professional mastering engineers

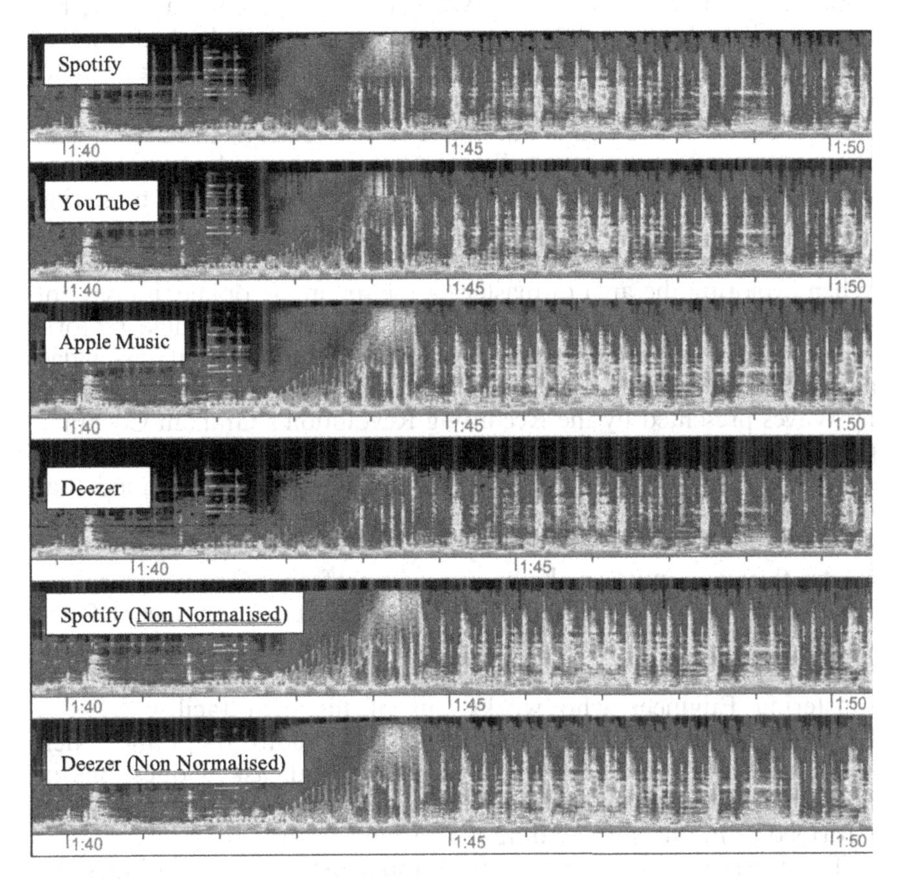

Figure 5.2 ABC Dialect, Guest List EAnalysis example

such as Mike Marra are taking advantage of this opportunity [27–29]. The way engineers such as Marra advertise separate services for different platforms is similar to how many engineers use and used the MFiT 'Mastered for iTunes' standard [30]. This was introduced to optimise masters for iTunes specifically although it was relevant for other lossy formats. It could be argued that much of what needs to be adjusted when mastering lossy formats under what is now called 'Apple Digital Masters' actually applies to all streaming services. This is because most of these platforms employ lossy codecs to convert the audio that is uploaded. Whilst some engineers may offer specific tailoring to each platform, the only tangible difference between any of the platforms is the difference in the loudness targets highlighted earlier.

In her talk at the Red Bull Music Academy, mastering engineer Mandy Parnell discusses how difficult it can be for a mastering engineer to offer services such as mastering for streaming. Parnell notes that there are so many different formats that need all sorts of adjustment and optimisation. Parnell states that 'We really need to communicate, because we don't have a solid format to deliver too' (Red Bull, 2018). She raises a valid point that with delivering a multitude of different formats, the likelihood of wrong submissions and mix-ups with different formats is a real concern. She points out that if services such as this are offered, communication between the client and engineer would be more important than ever. Interestingly, Parnell does not advertise her mastering for streaming services on her website so one must assume that this service would come about upon discussion of a mastering project [31]. When exploring the area of mastering for streaming online, it does not take long to come across a multitude of videos and online rhetoric surrounding mastering for specific platforms and other often mis-guided advice surrounding the subject. One example of this is a video published by Waves presented by the Recording Revolution's Graham Cochrane [32–61]. Whilst some of the content is relevant, the titles of videos with content similar to the Waves reference highlights the likelihood that these videos and techniques are probably not aimed at professional mastering engineers. Instead it may be aimed at those who are trying to master their own music at home in a non-professional environment.

Contact was made with professional mastering engineers Peter Hewitt-Dutton [62] who works at The Bakery in Los Angeles, Robin Schmidt [63] from 24–96 Mastering, Bob Katz [64], Scholar and Mastering Engineer who works out of his own facility 'Digital Domain', John Braddock of Formation Audio [65], and Nick Watson of Fluid Mastering [66]. They were all asked questions in relation to the subject area in order to gather current and well-informed opinions surrounding mastering for streaming.

The interviewees agree that traditional mastering techniques for the most part have not changed a great deal. Katz, who has for several decades championed a return to dynamic range, recognises that streaming has not affected his work, but has affected the artist financially in a big way. He does not recognise any 'new processes' in relation to the technicalities of the mastering processes and streaming, but he does aim to provide a master that will translate on all platforms and services. Katz is also well known for his work as editor on the 'Recommendations for Loudness of Audio Streaming' [1] from AES and is a purveyor in diminishing the effects of the loudness wars through his academic work. Whilst indeed this is an opinion, it is hard to ignore Katz's knowledge surrounding this subject and his suggestion that the biggest change in light of streaming, is the end of the loudness wars.

Hewitt-Dutton's opinions surrounding mastering for streaming are similar to those of Katz and he confirms that most professionals he works alongside and knows personally, all submit a single master for every format. He also suggests that 'tailored for streaming' is something mastering engineers are offering to enhance self-promotion and market their services in new ways. Hewitt-Dutton acknowledges that marketing in this way is now such an essential part of making a living. He also introduces the argument that it has never been a good idea to master right up to 0 dBFS as there are many other formats that will simply not tolerate a full range signal, although; he is aware of colleagues in the industry that still 'master to –0.1 dBFS regardless'. [62,67] These answers suggest that the recommendations set out in the AES document have actually been part of good practice in mastering before the likes of loudness normalisation or iTunes MFiT.

A common thought shared by the interviewees is that setting a specific loudness target to appease a particular streaming platform is unnecessary. Braddock argues that 'Setting a loudness target is musically unhelpful. The right loudness is, and always has been, where the track sounds good'. [65] Understanding the context, style and make-up of a track and the complex nuances and relationships between these elements is what mastering engineers have to comprehend and navigate in order to make decisions on many factors including loudness. Assigning the same 'loudness target' to every single track no matter the context could arguably have an effect on the decisions made by the engineer. This is not to dissimilar to the way the loudness wars compromised music, as before mastering begins there is an inevitable compromise already set before the engineer has started working on the music.

Much like Braddock, Watson highlighted that mastering engineers' tools and techniques haven't changed. He goes further to say that he's observed an adjustment in the industry that allows the mastering engineer to take control of the loudness narrative. He states that whilst the loudness wars are still somewhat at play with more commercial music, for the most part a lot of the industry has responded well to loudness normalisation and that from his experience, attitudes are going in the right direction.

> The mechanisms that we use for mastering have not changed, but the objectives are more musical. That's what we've seen developing in the last 5 or so years with loudness normalisation. [66]

Watson has also found that working in this way makes his mastering speak for itself when listening back from streaming platforms and has ultimately led to a growing, loyal client base. This also re-iterates the points of Braddock that having a number or target dictating your instincts as a mastering engineer will inevitably take you away from what the track should actually sound like. The overwhelming view from the professionals interviewed surrounding 'loudness targets' is that it is important to remember the 'master' will out–live this listening medium as it has out-lived all others before it. Treating it as such feeds into the idea that creating a great sounding master will translate to any platform at any point.

Discussions with Schmidt highlighted and reiterated the importance professionals put on the permanence of the work they do as 'levelling standards are short lived, but sound is forever. At least, until the next re-master!' [63]. Schmidt also argues that re-mastering for different platforms to take advantage of the loudness algorithm is exactly what LUFS and normalisation were introduced to avoid. Despite sharing many of the opinions already laid out by other professionals in the industry, Schmidt finds some of the new tools in relation to mastering for streaming helpful. However, these tools are only useful when a client specifies a problem with the PLR of a certain track or exact compliance with a specific loudness target.

Being that as Mastering Engineers we are often the final step in conformity, much of this chapter and its research has focused on the effect loudness normalisation has on the Mastering Engineer, whilst exploring the ways in which we can work with loudness to best serve the client and the music. Though it is argued by some that loudness normalisation has had an impact on mitigating hyper compression, and that attitudes surrounding loudness have relaxed, in the commercial pop music world; there will always be clients that want their track to be the loudest. However, it has been suggested by Watson that the responsibility of a loud sounding master that has impact in context of loudness normalisation, no longer sits solely with the mastering engineer. It is argued that in order to achieve the best sounding and most competitive music in streaming playlists, loudness needs to be thought about much earlier than the mastering stage.

> Prior to Loudness Normalisation in Streaming, the Mastering Engineer was often expected to "Make It Loud", because the loudness of the master was absolute, and the Mastering Engineer had ultimate responsibility for that. The reality, however, was that some mixes could handle being made loud better than others. So, even if a mix or arrangement was inferior and consequently would suffer more, you could still get the edge over another track (in a competitive sense – assuming that's your mindset) by making your master louder than the other track. Now, because of streaming, the absolute measurable loudness of the master is disconnected from the playback. This means that for a track to "compete" in the market (i.e. in a playlist) it is no longer necessary for it to be loud, but it is necessary for it to sound loud. This shifts the onus back onto the mixer, the producer, the arranger (along with the Mastering Engineer). Perhaps more than anything it is the arrangement now that really has to be right, in order for a commercial track to have impact in the context of a playlist. In other words, whatever the A&R or artist says, you can't 'cheat' by just pushing things a little harder. Instead, everyone has to raise their game. [66].

Whilst there are examples of some commercial pop music adopting more dynamic outlooks, it is arguable that the majority is still as loud as it has ever been. In a recent article by the New York Times,

Bob Ludwig comments that whilst mastering engineers may adopt new practices and see the obvious advantages of loudness normalisation in streaming platforms, there is nothing to stop mixing engineers 'ladling on the loudness'. [68,69]. This solidifies Watson's argument that this is a problem not fixed by everyone in the production process. Unfortunately, this also suggests that the loudness wars are not yet over.

Relevance of the traditional and the new

As the conversations surrounding this topic have grown, music technology business keeps innovating. There are many new software packages, websites, and plug-ins available to interface with mastering for streaming. Most of these are designed to help mastering engineers adhere to the algorithms (mainly loudness PLR targets), of the major streaming platforms. Whilst it could be said that these new tools are changing attitudes of mastering surrounding the loudness wars, the opinions and data gathered in this chapter show that this may not necessarily influence the seasoned professional. Nevertheless, there are software tools available to all layers of professional and non-professionals alike. One such tool is MeterPlugs' Dyameter [70] which guides the user through the loudness levels for streaming visually. This simple-to-use interface aims to show how to achieve a dynamic and loud track most suited to streaming services. In fact, it is a good tool to permit the best-fit across most streaming services from the one master.

As well as metering plug-ins there are also online browser-based tools such as Loudness Penalty Analyzer Figure 5.3 [70,71] that illustrate the way streaming platform algorithms will 'turn down' a master once uploaded. Whilst this is a good tool for engineers to point clients to when showing the benefits of not brick walling a master, it could be said that this also feeds into the narrative that we should have separate masters for each platform.

Another commonly used tool is 'Master Check' by Nugen Audio [72,73]. Master Check has the ability to specify PLR targets for each streaming platform and allow the user to accurately master/meter in accordance with that target. Despite the lack of consensus surrounding practices like this in the mastering community, this plug-in and many others like it have endorsements from professional mastering engineers. It could be argued that having this feature built into popular metering plug-ins could convince self-taught or new mastering engineers that they need to provide clients with multiple masters with different PLR's. Like Parnell, Katz argues that providing multiple masters could also prove confusing for clients and could lead to files being mixed up when providing masters to streaming platform aggregates. He also describes the benefits of using the MFiT standard as the one size fits all solution.

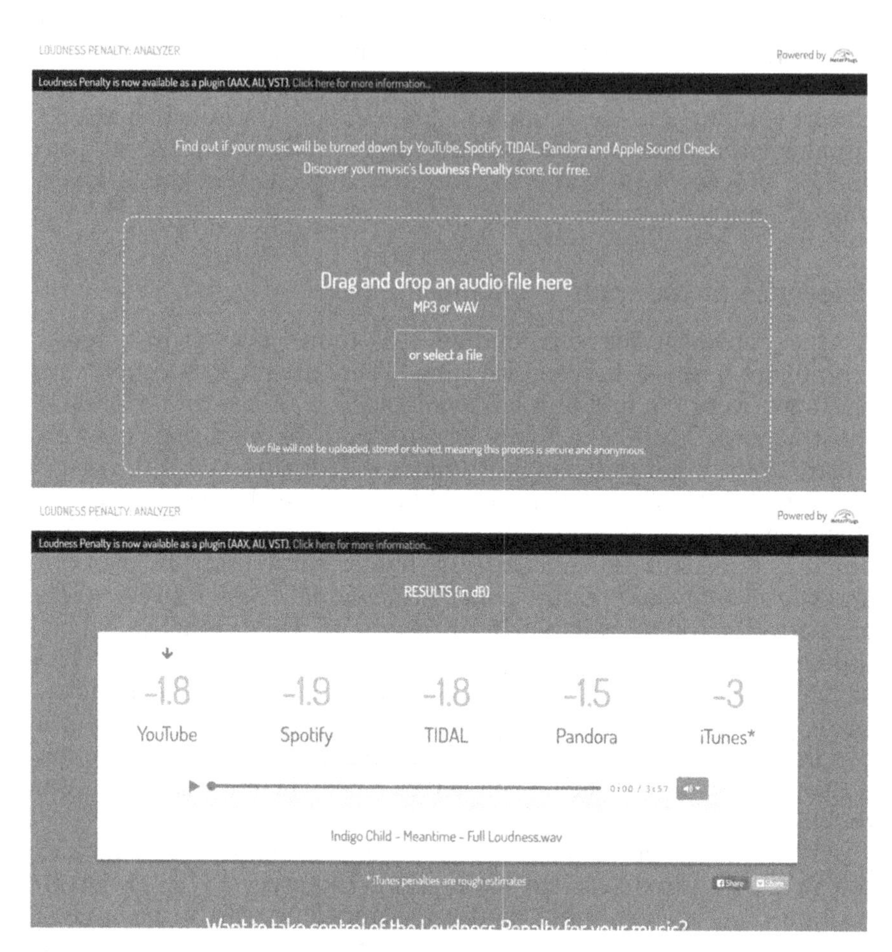

Figure 5.3 Loudness Penalty Analyzer

As for which master to present to the client, I have my client audition the CD master. However, with the streaming services and distributors currently asking for 1644 masters [16 bit, 44.1 kHz PCM standard] – except for Mastered for iTunes – it's impractical for the mastering engineer to produce a 1644 master that's optimized to prevent True Peak overshoot after coding and a separate 1644 master whose max peak level is perhaps 0 dBTP. So, the master going to streaming at least in my case is usually the same as that going to CD. And that's a mistake. I've been advocating (to no avail) for a long time that the 2444 mastered for iTunes (MFIT) master be used to send to ALL streaming services, not just Apple. Thus, all streaming services will benefit from the precautions that Apple themselves recommend to prevent 0 dBFS+ overshoots after coding. Until such time as the distributors will listen to me, and accept the MFIT 2444 masters for all streamers, then there's immediately a built-in compromise for what we send to all streamers. I resist the idea of sending two different 1644 masters at different levels as they can be easily confused as soon as they leave my place to be sent to distribution. [64,74,75]

Creative practice

The creative practice led element of this chapter was designed to explore a number of different approaches that could illustrate the effectiveness of certain workflows. The approach to the practice led element of the project does not aim to prove or disprove the theories of pro engineers and their practices as traditional mastering practices are well established. To a mastering engineer, developing an understanding of delivery formats, workflow and the relationship between the two is essential to producing the final product [9,76]. With this in mind, five varied workflow approaches have been considered and practiced with a total of 20 songs being mastered in various forms and workflows. The music used for this practice–based research includes a mixture of songs from the author's clients where consent has been given. Other songs included in this portfolio include tracks taken from Mike Senior's Cambridge Music Tech website who insists these tracks can be used under these circumstances [77].

Approaches are as follows:

1. Traditional mastering techniques (prior to loudness normalisation approaches)
2. YouTube target of –12LUFS (at the time of investigation, this was the expected level)
3. Blended approach using 1 and 2 above concurrently
4. Spotify Loudness Target
5. Nugen's 'Ideal Loudness for Streaming'

Approach 1 – Traditional mastering technique

The first approach focusses on traditional mastering techniques akin to that of a typical specification for a file optimised for download on the iTunes Store or similar. The output limit on the last limiter in the chain on all tracks was set to –0.1 dBFS. Although typically, the output limit would usually be set to –0.5 dBFS or lower if the master was destined for MFiT in order to avoid True Peak clipping. –0.1 dBFS was decided as the benchmark due to many pro mastering engineers varying between –0.1 and –0.5 dBFS [78]. Approach 1 represents the standard followed by pro engineers that have not adopted the idea that there are new practices involved in optimising masters for streaming purposes. It is also a starting point from which we can gauge the successes of approaches 2–5 and the likelihood of them fulfilling the queries of the research question.

In order to keep this approach relevant among the other four approaches, 'Loudness Meter' plug-in by Waves was employed [32]. This was used to take note of where the tracks ended up on the loudness scale. It was found that despite it not being a part of the regular workflow, WLM meter was used as another meter reference for the full loudness masters. It was also found that the Loudness Range read out was a handy secondary check on whether the track was hitting the limiters threshold too much (Figure 5.4). Readings less than about 4LU indicates that the track may have a narrow dynamic range (0LU being no dynamic range whatsoever) however, it goes without saying that rules for one track do not necessarily apply to another.

Whilst using the WLM meter throughout Approach 1 as a reference tool for LU Range, the 'Long Term' PLR readings from the plug-in also became useful. This was because of the increasing familiarity with typical LUFS numbers associated with a normal master. Added to the 'Range' read out, this plug-in makes WLM meter and other Loudness Meter plug-ins a useful tool to use alongside other metering options even when not using it for its intended purpose of mastering for streaming.

Approach 2 – YouTube target of –12 LUFS

The second approach involved mastering a selection of songs to YouTube's PLR average of −12 LUFS (the standard at the time of investigation), printing these and then proceeding to bring these masters up to a normal mastering level. This approach was created in order to examine the workflow of providing a 'streaming master' as well as a 'normal master'. YouTube was chosen due to its PLR target being the highest out of all streaming platforms. This was the case at the time the practice led research was being carried out. YouTube's algorithm will only turn down tracks that have a PLR higher than –12 LUFS however, it will not turn up tracks that are have a PLR lower than –12 LUFS. YouTube's PLR is therefore more suitable as a one size fits all streaming master. This is because a master optimised to Spotify's PLR average of –14 LUFS would sound noticeably quieter than other music on YouTube as it would not be turned up by +2 LUFS to –12 LUFS. Conversely, a YouTube optimised master with a PLR of –12 LUFS would be ideal for all the other platforms too as the normalisation from the other platforms would ensure that the master was as competitive as anything else on the platform. When starting this mastering approach, there was a lot of negotiation with limiter settings which were changed throughout each master in order to achieve the desired loudness target of –12 LUFS. It was also found that the only way to accurately achieve the loudness target was by referencing the WLM Loudness meter for visual feedback and tweaking the limiter as appropriate. It was also found that in some

Figure 5.4 Waves WLM meter

instances, creative decisions may have been affected in Approach 2 due to the reliance on visual feedback from the loudness meters. It would be hard to say exactly what creative decisions were affected as these are largely instinctive. However, it must be said that the concentration on loudness and hitting the PLR target, came close in importance to that of the tonal decisions made when mastering. As worries about loudness do not usually affect the decision-making process it is likely that this had an influence over the finished masters.

Once the YouTube master was set and complete and the full loudness master was adjusted, it was found that the impact and groove of transients seemed accentuated when brought up to a 'normal level' which was around –0.1 dBFS. This led to driving the tracks into the limiter more on the normal master versions to achieve greater gain reduction on the tracks. The purpose of this was to smooth the transients out and achieve the desired envelope/groove. It could be said that the unsatisfactory results when bringing the tracks up to full loudness were due to the differences in monitoring level in the control room as this is always set the same for consistency in playback. This meant that the YouTube loudness masters were considerably less loud in the monitoring environment than the normal master version. It was observed as a possibility that adjusting the volume and calibrating the monitor control for different degrees of loudness would lower the chances of the monitor control volume having such an effect over the final result of a master. As well as this, it may have the effect of relying less on read outs from loudness meters if we can perceive how loud a track is meant to sound, without referring to visual read outs from loudness meters. It has been encouraged, by Mastering engineers, such as Bob Katz to generally playback at the same volume in their monitoring environment for consistency. For this reason, adjusting monitoring to compensate for loudness differences, may come with its own problems and may lead to inaccurate perception of low end due to the ear's non-linear response (see Fletcher Munson curves in [79,80]).

Approach 3 – Blended approach using 1 and 2

Approach 3 felt instantly adoptable as a process that may be suitable for the traditional mastering engineer. In effect, this process allows the engineer to focus merely on the song and work to the same parameters as a normal mastering standard or similar. It was found that if using a limiter that had a separate output trim, the track could be attenuated as appropriate to the LUFS target after the master had been complete to a normal loudness. Setting the output trim as appropriate for the streaming service took around 1 or 2 passes per song in order to achieve the correct LUFS target. It was found that –1 LUFS corresponded to –1.0 dBFS on the output. Meaning, if a track at full loudness had a PLR reading of –7.5 LUFS, to achieve Spotify's recommended PLR target of –14 LUFS; we would have to reduce the output trim of the limiter by –6.5 dBFS. It was also found that whilst adjusting these tracks for different LUFS targets, a PLR target of –14 could in fact be as loud as –13.5 as it seems the figures are rounded up or down as appropriate. The added advantage of using the output trim in this way is that the likelihood of True Peak overs is much less likely due to output of the track not exceeding –2.0 dBFS once adjusted for PLR targets. The process of optimising full loudness tracks for streaming would have to take place once masters had been accepted by the client. This would mean that more time would have to be put aside after the full loudness master was finished. If a method such as this is streamlined and made part of the workflow, it could be lucrative for an engineer as this is probably something an assistant could deal with, whilst charging an extra amount for optimising the tracks for specific LUFS targets. It could also be said that this approach could only be adopted if the tracks being worked on were 24-bit as using the output trim as such is effectively reducing the bit-depth. 24-bit files did not appear to suffer whilst researching this approach, but it would be hard to recommend doing the same with 16-bit files without further testing.

Approach 4 – Spotify loudness target

Similarly to approach 2, approach 4 focused on Spotify's PLR target that could only be achieved by referencing a loudness meter regularly throughout the mastering process. What had been learnt from previous approaches led to an altered workflow approach by setting the limiter with an output trim of –3.0 dBFS to see if mastering could be approached as normal and achieve a LUFS reading closer to the LUFS target of –14 of Spotify. Whilst the effects of visual feedback with the loudness meter had been reduced through this change of workflow, more time would need to be taken to hone this new workflow for worries about levels to be diminished. Nugen Audio's Master Check was used alongside Waves WLM Meter to monitor the effects of the codecs and compare the interfaces of each plug-in. It was found that the visual feedback from Master Check made anticipating the final PLR read out easier which arguably reduced the effects of workflow disruption getting in the way of the final master. Being able to A-B codecs and different streaming platforms as well as monitoring distortion artefacts through MasterCheck is a helpful tool that can point out potential issues from inter sample peaks. Although, if mastering to a Spotify PLR target, it is unlikely that this will be an issue. However, it may well be a useful tool for checking distortion artefacts from different platforms and codecs with full loudness masters.

Approach 5 – Nugen's 'ideal loudness for streaming'

By this point in the practical aspect of the research, adapting to the workflow change of visually referencing loudness meters regularly while mastering each track became easier. It is unclear what influence the workflow change had in the creative process of mastering, but after engaging and becoming more familiar with using loudness meters; the more efficient the process became. Interestingly, MasterCheck, at the time of investigation, used –12 LUFS as the 'ideal standard for loudness'. It is uncertain if they came to the same conclusion as has been discussed in the research regarding mastering to the loudest streaming platforms PLR target. It could be said that even traditional mastering engineers could adapt to using these tools in the way approach 2, 4, and 5 have, but it may take a certain amount of re-calibration in the monitoring environment. There has been some rhetoric surrounding mastering for streaming that suggests that metering with loudness plugins enables you to be more dynamic with your masters. Although this may be a possibility, it was apparent that this workflow change altered the usual practices of the usual mastering techniques.

CONCLUSIONS OF PRACTICE AND RESEARCH

If or when clients are regularly asking for streaming optimised masters for individual platforms; it could be argued that Approach 3 will work best for the traditional mastering engineer. By using an output trim or similar at the end of the chain once the normal master is complete, it is relatively straightforward to work out the difference in dBFS to achieve the streaming target set by the platform. It is also yet to be seen if engineers mastering to streaming targets are pushing their tracks as loud as possible by having it round down to the nearest PLR target. However, this may be irrelevant as the difference in loudness would never be more than 0.9 of a loudness unit, equivalent to 1 dBFS. In discussions with Watson [66], it was also noted that mastering for streaming is becoming competitive in other ways, with some engineers using EQ to fool loudness algorithms and make things perceptively louder. The idea being that the perceived loudness gained from EQ will outweigh the loudness turn down from the platform. Obviously, this practice would not be in line with making the song sound better, only louder. Mastering to specific streaming targets for the moment seems in-practical and unnecessary as regardless of what is uploaded to a streaming platform, loudness normalisation is in effect. The research also suggests that Approach 1 may be most sensible from a logistical point of view, as dealing with multiple versions of the same master may prove dangerous for distribution as the mastering engineer relinquishes control as soon as they hand them over. This could lead to the wrong master going to the wrong streaming platform for upload, which could result in limiting on Spotify or a quieter track on YouTube. Overall, this practice-based research has helped to shed light on possible workflows that may guide mastering engineers in the event that mastering for specific streaming platforms becomes a common service that is offered. However, it must be said that using True Peak monitoring to avoid distortion artefacts in uploaded masters has shown to be even more important due to the loudness penalties shown in the establishing analysis. For this reason, Katz's suggestion that Apple's MFiT standard being implemented for all streaming platforms could arguably be the most practical solution.

Whilst some professional engineers are offering mastering for streaming services, it is unclear whether they are offering services that are specifically for a single platform. Because of this it is curious why major companies and plug-in developers are releasing software that allows functionality of mastering for specific streaming platforms. It could be argued that this software is not meant for the hardened pro, but for a new generation of mastering engineers that are honing their skills in the modern landscape of all digital consumption. It could be said that without the tutelage of a seasoned mastering engineer and the experience of working with multiple formats for a good number of years; home grown and self-taught engineers may be swayed by a tool that can give an opportunity to offer a service. Although there are many valuable lessons that can be learned from the new media of YouTube and others, it has to be said that some of the discourse can be misconstrued and backed by unsubstantiated evidence often driven by opinion. However, as there is little academic or professional discourse actively being circulated it is not hard to see why budding engineers may turn to new media from information about mastering for streaming. With the existing online discussion around the subject in relative infancy, it is probably sensible for professional engineers to stick to the workflow they are used to until further compelling discourse presents itself. Nevertheless, if clients are specifically asking for you to provide these services, the framework following this section will be useful.

It is clear to see that a new generation of mastering engineer, might see 'loudness' as a confusion when exploring the ideas around new levelling standards and the best practices to follow. The discussions with professional mastering engineers in this chapter helps to show the necessity of using your ears to judge loudness more than worrying about the numbers. However, it could also be argued that not only the new, but existing engineers who have been in the industry for a long time may need to consider alternative approaches. It could be said that because of the loudness wars, some engineers may be calibrated to master in a way that may be detrimental for audio that is destined purely for streaming platforms. With this in mind, the effects of loudness normalisation may also have an effect on what some engineers consider their loudness norm to be. For some engineers that master a lot of commercial music, especially that which gets heavy radio play, it may be necessary for them to adjust their natural loudness calibration in order to take advantage of loudness normalisation and reap the sonic advantages as discussed in the chapter. Using the new loudness tools in the creative practice element of this chapter has also showed the effectiveness of PLR based meter plug-ins if used in a similar way to that of Peak/RMS meters. Using short term PLR, PLR and LU Range gave effective results and could be an alternative metering proposition for new mastering engineers. Whilst Peak/RMS and VU are of course, very valid metering types; PLR based meters do have the added advantage of being able to keep an eye on a standard that is being widely used by all the streaming platforms.

Framework

The research conclusions from this chapter have been summed up into key points that can be looked at as a framework that helps guide modern mastering techniques:

- True Peak signals may affect the overall loudness when processed by a streaming platform's algorithm. Using a true–peak limiter and having the threshold no higher than –1.0 dBFS should ensure that a master does not receive a loudness penalty. True Peak signals may also create distortion or artefacts when streaming platforms convert the masters into their codec of choice.
- Avoid sending to many versions of the same master to the client. If this is a necessity make sure communication is very clear between engineer and client to avoid errors in distribution.

- Master as normal (Full Scale) then adjust the master for other streaming platforms. This is best achieved using an output trim at the end of your mastering chain. This will ensure no change of monitoring volume in the control room environment, with Fletcher Munson curves and calibration having less of an impact on the final master.
- If a client has asked for a single master for all streaming platforms, –14 LUFS may be most suitable as this seems to be the emerging standard for most services.
- When negotiating loudness, use your ears! And encourage your clients to use theirs. The sonic results from streaming platform playback and your instincts should speak for themselves.

REFERENCES

1. Byers, R. et al. *Recommendation for Loudness of Audio Streaming and Network File Playback*. In: Katz, B. editor. *Audio Engineering Society*, (2015), p. 6.
2. Cieslik, P. and Szybinska, K. 'Analysis of Polish Web Streaming Loudness'. In: *Audio Engineering Society Convention 146*, Audio Engineering Society, (2019). Available at: http://www.aes.org/e–lib/browse.cfm?elib=20302.
3. Coefficient Audio. *Mastering, Coefficient Audio Mastering*, (2019). Available at: https://coefficient–mastering.com/mastering
4. Cousins, M. and Hepworth–Sawyer, R. *Practical Mastering: A Guide to Mastering in the Modern Studio*. Focal Press, New York, (2013). Available at: http://capitadiscovery.co.uk/uwl/items/673887.
5. Pedersen, K. and Grimshaw-Aagaard, M. *The Recording, Mixing, and Mastering Reference Handbook*. Oxford University Press, Incorporated, Oxford, USA, (2019). Available at: http://ebookcentral.proquest.com/lib/uwestlon/detail.action?docID=5602451.
6. Perkins, J. 'The Mastering Guide to Audio Formats and Delivery Mediums (UPDATED: 2019)', *Pro Audio Files*, 20 April, (2019). Available at: https://theproaudiofiles.com/audio–mastering–format–and–delivery–guide–2014/.
7. Research, A. M. *Music Streaming Application Market to hit $6.50 Billion by 2025 – Global Analysis by Trends, Size, Share, Strategy, Service Type and Growth Opportunities: Adroit Market Research, GlobeNewswire News Room*,(2019). Available at: http://www.globenewswire.com/news–release/2019/05/14/1823459/0/en/Music–Streaming–Application–Market–to–hit–17–50–Billion–by–2025–-Global–Analysis–by–Trends–Size–Share–Strategy–Service–Type–and–-Growth–Opportunities–Adroit–Market–Research.html.
8. Talyor, R. W. 'Hyper-Compression in Music Production; Agency, Structure and the Myth That "Louder Is Better"'. *Art of Record Production Journal*, (2017). Available at: http://arpjournal.com/hyper-compression-in-music-production-agency-structure-and-the-myth-that-louder-is-better/.

9. ———. *Mastering Audio: The Art and the Science*. 2nd edn. Focal Press, New York, (2007).
10. Katz, B. 'Sound Board: Can We Stop the Loudness War in Streaming?', *Journal of the Audio Engineering Society*, Vol. *63*, No. 11, (2015), pp. 939–940.
11. Vickers, E. 'The Loudness War: Background, Speculation, and Recommendations'. In: *Audio Engineering Society Convention 129*, Audio Engineering Society, (2010). Available at: http://www.aes.org/e-lib/browse.cfm?elib=15598.
12. RIAA. *RIAA Releases 2018 Mid–Year Music Industry Revenue Report, RIAA*, (2018). Available at: https://www.riaa.com/riaa–releases–2018–mid–year–music–industry–revenue–report/.
13. Ronan, M., Ward, N. and Sazdov, R. Considerations When Calibrating Program Material Stimuli Using LUFS. In: *Audio Engineering Society Convention 140*, Audio Engineering Society, (2016). Available at: http://www.aes.org/e–lib/browse.cfm?elib=18149.
14. ———. 'Mastering in an Ever–Expanding Universe'. *Journal of the Audio Engineering Society*, Vol. *58*, No.1/2, (2010), pp. 65–71.
15. ———. 'Mastering for Today's Media'. *Journal of the Audio Engineering Society*, Vol. *61*, No. 1/2, (2013), pp. 79–83.
16. ———. 'Loudness Revisited', *Journal of the Audio Engineering Society*, Vol. *62*, No. 12, (2015), pp. 906–910.
17. Rumsey, F. 'Recording in the Light of New Technology', *Journal of the Audio Engineering Society*, Vol. *63*, No. 12, (2016), pp. 1053–1057.
18. MusicTester. *MusicTester – Analyzer Demo, MusicTester Analyzer*, (2019). Available at: http://musictester.net/demo/.
19. Nugen. 'How loud is too loud? | NUGEN Audio', (2019). Available at: https://nugenaudio.com/how–loud–is–too–loud/.
20. Amoeba, R. *Rogue Amoeba | Soundflower*, (2019). Available at: https://rogueamoeba.com/freebies/soundflower/.
21. AMP Tracks. *Mastering for Streaming, AMP Tracks,*(2019). Available at: https://amp–tracks.com/music–production/mastering–for–streaming/
22. Ingalls, M. *Matt Ingalls: Composer, Clarinetist, Computer Musician, Matt Ingalls*, (2019). Available at: http://sfsound.org/matt.html/.
23. Mastering, J. *Jerboa Mastering,* (2019). Available at: https://www.jerboamastering.com.
24. Johansson, A. et al. *Spotify Teardown: Inside the Black Box of Streaming Music*. The MIT Press, Cambridge, MA, (2019). Available at: http://capitadiscovery.co.uk/uwl/items/704712.
25. Digital Music News. 'What Streaming Music Services Pay (Updated for 2019)', *Digital Music News, 25 December*, (2018). Available at: https://www.digitalmusicnews.com/2018/12/25/streaming–music–services–pay–2019/.
26. Hepworth–Sawyer, R. and Hodgson, J. *Audio Mastering: The Artists*. 1st edn. Routledge, New York, NY, (2018).
27. Marra, M. *'Mastering for Online Streaming'. Mike Marra Mastering*, (2019). Available at: http://mikemarramastering.com/mastering/mastering–online–streaming/.
28. Meterplugs. *'What's the Ideal Loudness Penalty Score?'* (2019).

29. MeterPlugs. *Dynameter: Dynamics Metering for AAX, AU and VST*, (2019). Available at: https://www.meterplugs.com/dynameter.
30. Apple. *iTunes – Mastered for iTunes – Apple (UK)*,(2019). Available at: https://www.apple.com/uk/itunes/mastered–for–itunes/.
31. Parnell, M. *Black Saloon Studios, Black Saloon Studios*, (2019). Available at: http://www.blacksaloonstudios.com/.
32. Waves. *Loudness Meter Plugin – WLM Plus | Waves, waves.com*, (2019). Available at: https://www.waves.com/plugins/wlm–loudness–meter.
33. Wyner, J. *Audio Mastering: Essential Practices. Pap/Com edition.* Berklee Press Publications, Milwaukee, (2013).
34. Young, J. D. *Home Studio Mastering*. 1st edn. Routledge, London, (2018). Available at: http://capitadiscovery.co.uk/uwl/items/705001.

DISCOGRAPHY

35. ———. [digital release], *Self Titled, Self Release*, (2019).
36. ABC Dialect. [digital release], *Real Life EP, Casablanca Sunset*, (2019).
37. Arise. *Run Run Run, Unknown*, (2010).
38. Blac, B. *If You Want Success, Unknown*, (2010).
39. Bran, J. *To Save Noel, Unknown*, (2010).
40. COTTON. [digital release], *AM4 EP. Self Release*, (2019).
41. Dido. *Still on My Mind*, BMG, (2019).
42. Giddens, A. *Get Blown, Unknown*, (2010).
43. Indigo Child. *Self Titled EP, Self Released*, (2019).
44. Gerrard, D. *Never Stop, Unknown*, (2010).
45. ———. [digital release], *Best In Town (Single)*. Heist or Hit, London, (2016).
46. Honeymoon. [digital release], *Magic (Single)*. Heist or Hit, London, (2019).
47. Jeffries, M. *Sikth Remix, Unknown*, (2015).
48. Kruger, D. *En Dance, Self Release*, (2010).
49. Lie, E. *Space Highway, Library Music*, (2019).
50. Lightcliffe. *Drapes, Failure by Design Records*, (2018).
51. ———. [digital release], *Fire Single, Beatnik Creative*, (2018).
52. MAVICA. [digital release], *Hot Sand (Single), Hidden Track Records*, (2018).
53. Maxwell, S. *Mild Cheddar, Un-released*, (2019).
54. Obscure, S. *Infernal Machine, Unknown*, (2010).
55. RNA. *Dream, Self Released*, (2006).
56. Rubin, J. *Fly Away, Self Released*, (2015).
57. Slaves. *Hypnotised, Virgin EMI Records*, (2016).
58. Vane, C. *So Raw, Unknown*, (2010).
59. We Fell From The Sky. *Not You, Unknown*, (2010).
60. Whitehead, P. [digital release], *Winter (Acoustic)*. Self Release, London, (2019).
61. Whittingham, J. *Abbey Road. Un-released*, (2018).
62. Hewitt-Dutton, P. *'Discussions on Mastering for Streaming'*. (2019).

63. Schmidt, R. *'Discussions on Mastering for Streaming',* (2019).
64. Katz, B. *'Discussions on Mastering for Streaming'.* (2020).
65. Braddock, J. *'Discussions on Mastering for Streaming',* (2020).
66. Watson, N. *'Discussions on Mastering for Streaming',* (2020).
67. IFPI. *IFPI — Representing the Recording Industry Worldwide, Global Music Report 2019,* (2019). Available at: https://www.ifpi.org/downloads/GMR2019.pdf.
68. Milner, G. 'Opinion | They Really Don't Make Music Like They Used To'. *The New York Times*, 7 February 2019, (2019). Available at: https://www.nytimes.com/2019/02/07/opinion/what-these-grammy-songs-tell-us-about-the-loudness-wars.html.
69. MusicTech.net 'Mandy Parnell Interview – A Master of the Industry'. *MusicTech*, 19 January, (2018). Available at: https://www.musictech.net/features/mandy–parnell–interview/.
70. MeterPlugs. *Loudness Penalty*, (2019). Available at: http://www.loudnesspenalty.com.
71. MIDIA. 'Mid–Year 2018 Streaming Market Shares'. *MIDiA Research*, 13 September, (2018). Available at: https://www.midiaresearch.com/blog/mid–year–2018–streaming–market–shares/.
72. Nugen. 'MasterCheck | NUGEN Audio', (2019). Available at: https://nugenaudio.com/mastercheck/.
73. Owsinski, B. *The Mastering Engineer's Handbook: The Audio Mastering Handbook*. 2nd Rev edn. Thomson Course Technology PTR: Cengage Learning, Boston, MA, (2007).
74. Magrini, E. 'Current Trends in Mastering', *Warp Academy, 19 June,* (2017). Available at: https://www.warpacademy.com/current–trends–in–mastering/.
75. Malecki, P., Czopek, D. and Sochaczewska, K. 'Music Streaming Platforms—Quality and Technical Comparison'. In: *Audio Engineering Society Convention 145*, Audio Engineering Society, (2018). Available at: http://www.aes.org/e–lib/browse.cfm?elib=19752.
76. Katz, R. A. *iTunes Music: Mastering High Resolution Audio Delivery: Produce Great Sounding Music With Mastered for iTunes*. Focal Press, New York, (2013). Available at: http://capitadiscovery.co.uk/uwl/items/621667.
77. Senior, M. *Studio Services & Training (Cambridge Music Technology)*, Cambridge Music Tech, (2019). Available at: http://www.cambridge–mt.com/AboutCMT.htm (Accessed: 20 September 2019).
78. Walker, M. *Mastering CDs On Your PC, Sound on Sound*, (2003). Available at: https://www.soundonsound.com/techniques/mastering–cds–your–pc.
79. Smithers, B. *'Square One: Can You Hear Me Now?'*, *Electronic Musician; New York, N.Y.*, June, (2009), p. 48.
80. Spotify *Mastering & loudness – FAQ – Spotify for Artists,* (2019). Available at: https://artists.spotify.com/faq/mastering–and–loudness#will–spotify–play–my–track–at–the–level–it's–mastered.

6

Mastering audio analysis

TEACHING THE ART OF LISTENING

John Paul Braddock

ABOUT THE AUTHOR

John Paul 'JP' Braddock is best known as a mastering engineer focused in both the analogue and digital domain. His audio journey started over 30 years ago establishing 'Rubber Biscuit Studio' (RBS) where he first discovered his passion for restoration mastering. This led to the creation of the 'British Music Archive' (BMA) heritage restoration project. Delivering hundreds of albums over the last three decades from a wide range of musical perspectives helped in his development of 'Full Dynamic Range High Definition' (FDRHD) audiophile delivery format. Alongside commercial work at his latest facility 'Formation Audio' (FA), he has established himself as a skilled educator delivering lectures and seminars for industry events and universities across the United Kingdom. This includes the design of mastering modules at Nottingham Trent University (NTU) and De Montfort University (DMU). He is also an active member of the Audio Engineering Society (AES) as co-chair of AES Mastering Group [AES:MG].

FOREWORD

The author sets out to evidence observations and knowledge accumulated in teaching students to listen over the last decade focused purely in the field of mastering audio. This chapter therefore constitutes an expert opinion piece, with statements based on personal observation and professional experience.

INTRODUCTION

Considered music creation references previous creative outcomes, whether consciously or unconsciously, creatives absorb a great deal from the music around them and the soundscapes of everyday life. From drawing our first breath, we are listening, building a bank of sonic imprints. Some music professionals develop by self-directed learning and reach a familiarity with a wide range of musical sources through first-hand exposure, immersed in their music scene; constantly referencing what they feel 'quality' music is; often inspirational audio from their perspective. In contrast, some music professionals developed through a commercial avenue, attaining exposure by the variety of work they're presented, referencing other acclaimed commercial examples as a guide to quality. Both are equally sonically busy in mind and practically active, listening and creating. That ability to listen and critically assess audio to a very high level is essential to any audio engineer's career development [1].

IN EDUCATION

In 2006, I was approached to design a final year module for an audio engineering degree focused around mastering. I was excited, not only about sharing knowledge with a new generation but also what analysis and evaluation of my own approach would bring to my practice. The opportunity to develop and broaden a student's comprehension of differing genres and their own music's historical referencing is one I relished, and still do.

In defining 'audio mastering' in concept, there's a general disparity between new learners' perception of mastering and the actual role. Phrases like 'The Dark Arts' do not help in addressing the technical outcomes required [2]; neither does the media's focus around high-end 'tools'; these can seem unattainable, reinforcing the mysticism of the field. The reality is, it's all about listening and not the tools; the equipment is helpful, but not required [3]. The link between analysis and application of tool and type correctly has the largest affect on outcomes.

In the many discussions I've had directly with mastering engineers or observed in interviews, the majority have said it takes at least 10 years of listening to qualify with appropriate monitoring before being able to clearly identify what quality is [4]. In my own journey into mastering it took at least that period before I was professionally confident enough to use the term 'Mastering Engineer' in commercial work. For me, it was establishing a sense of average, comprehending a clear overview of a given piece of music's sonics and its required translation in the wider context of all music and its delivery.

This collective self-analysis from mastering engineers for at least a decade of listening before being able to fully engage in the task of mastering leads to an obvious question. 'How can a mastering engineer teach 'mastering' in a module over an academic year or even over a whole degree of three or four years?' The simple answer is 'you can't', there isn't the available listening time, but you can teach best practice in how to listen and methods to assist in developing learners critical listening. This is in-part why previous generations of mastering engineers to my own, often cite the apprenticeship as the best route to becoming an effective mastering engineer (Hepworth-Sawyer 2019, personal communication).

These opinions were clearly in focus when initially devising a programme of learning: not trying in any way to create the impossible but focus on the development of students' ability to assess and analyse audio. That learning outcome: 'Develop critical listening skills to identify and analyse components of audio work such as dynamics, spatial and frequency content, digital error, distortion' has been a mainstay of students learning year on year. Most importantly making clear they're not suddenly going to become a mastering engineer at the end of the course, though they will be able to hear significantly more detail in the music they love. For some, they'll find they're no longer able to listen to music without dissecting it, maybe even in some contexts they stop enjoying music as much as they did previously.

The latter point is an interesting one, in the many years of delivering content there has always been a broad set of abilities with a new cohort of learners. One consistent aspect I've observed is those students who feel advanced in their music production skills are often the ones who find it the most difficult to start in the process of deconstruction in audio analysis. Anecdotally, the emotional attachment with the music creation they love makes breaking it into sonics of tone and dynamic noticeably more difficult than with others. Equally, prejudice towards genre being another common barrier; not listening to something because it's outside of their musical grouping, socially.

I feel it's important to take note of other approaches to teaching listening which can be productive. The work of Dave Moulton or Jason Cory are clear examples [5,6]. These programmes and approaches have many positives to 'train' the ear but they do not focus on complex programme material. They do not approach 'ear training' by referencing finished commercial music as sonic exemplars in the development of a learner's auditory analysis skills. Without this approach, we are listening to the parts of the musical soundscape and not the overview required to master. This has to be the key focus in ear training for the fielding of mastering, in my opinion.

The barriers towards developing our ability to engage critically are often derived from assumption, learned behaviour or musical prejudice. Often the easiest way to break them down is by listening to those very aspects. This can be easily evidenced in session by asking students to assess a reference of their own choice (one they have musical affinity with; they feel it has kudos) and repeating the task with a given musical reference from a genre they dislike. The majority of the time the students themselves will recognise they're not listening to both pieces of music in the same way. But rather, being emotionally swayed to one in analysis. This is especially evident in the focus of the descriptive language used to explain critical observations, especially when assessing flaws of the given material.

I've found using the work of European Broadcast Union (EBU) listening guides a useful way to start students on a path towards analysis in breaking a given piece of music into its component parts [7]. Just discussing 'depth of field' or 'stereo image' is not enough to focus students to unlearn how they listen. Given the clear statements for each sonic parameter they have to respond both descriptively and numerically. The latter enabling students to start analysing comparatively between tracks in gauging quality. Using the EBU guides, for example, if they give a piece of music a high score for 'stereo impression' and their next track has a clearer stereo image; the previous can't be as high as they originally thought. This numerical addition to analysis facilitates comparative listening, moving away from genre and musical likes or dislikes to the sonics of the music. Something mastering engineers, do all of their lives, having spent tens of thousands of hours comparing music across genres, whether professionally in session or socially exposed to music environmentally. This all contributes to their knowledge of quality and a sense of where the average sits within this. Therefore, that average is necessary to master effectively, but to gain this knowledge is a time-consuming learning process.

ESTABLISHING LISTENING PRINCIPLES

It is challenging to teach critical listening, one of the hardest aspects is helping learners develop an ability to hear transient shape clearly. In-part because much of their formative listening lacks transient detail. Low quality playback devices, lossy audio and hyper-compressed music can all be contributing factors.

It is well known that equal loudness is a fundamental key in a mastering engineers' listening practise, but to start this engagement without practical evidence is something I have realised students find difficult to grasp. Practical reinforcement in all listening comparatives is a must, to enable learners to evidence their own sensitivity to volume difference in analysis.

I use the following simple test to evidence issues of loudness and human perception in seminars or lectures with new students. It's quite effective on a general listening system but more so on a high quality outcome:

Take a piece of commercial audio, one generally known to the learners (current commercial track) and label 'A'. Level for correct listening loudness in the environment and make a copy 'B' turning it down 1 dB. Avoid any visual guide e.g. that there's a difference in amplitude on fader or graphic. State there is a difference and ask learners to analyse the two using a simple comparative AB whilst playing. Make clear there is a difference but a subtle one. Ask the learners not to over think it, it's not a trick, just listen to how they sound different. After a considered listen, the theme of 'A' sounds better, it's brighter, more transient, clearer, fuller bass, deeper; while B lacks punch, is duller and so on. A correct analysis, but as we know one based around loudness difference and not a true representation of the difference in the audio.

Invert the polarity on one of the audio tracks while playing to make a sum difference and removing the gain difference to evidence nulling clearly proves to learners that the pieces of audio are identical. Any sonic difference we heard were purely our perception, all our perceptions, and the fact this is reaffirmed by all learners in the room, no matter their level of analysis is a powerful learning outcome. It's critical to instil a core comprehension of how we hear, in order to dispel myths and promote investigation.

LEARNING TO LEVEL

Practical engagement with equal loudness concepts requires a fixed listening level. The application of calibrated metering in various commercial standards is a technically helpful starting point [8,9]). Though from implementation in a simple practical perspective, Bob Katz K-system [10,11] integration is a positive way to explain. I've found students find the familiar format of peak/Root-Mean-Square (RMS) metering easy to comprehend, hence learning the principles of maintaining calibrated listening level while also being able to adjust the usable dynamic range. Often helping to make the context of RMS being related to loudness more tangible, their focus previously being peaking based from use of digital recording systems. With the change in focus it becomes clear that the peak is really a distraction with regard to loudness.

Given that the K-system is integrated into most modern Digital Audio Workstations (DAW) iterations, learners are more willing to start to use this in everyday metering practise especially when mixing – becoming more aware of average level/loudness rather than digital peak. Something I've observed is ingrained for some developing engineers from constant exposure to digital peak based metering. General learning has been focused towards the avoidance of clipping the digital system. As opposed to an awareness of the power of a signal to achieve the best signal consistency. As was common practise when engineering was only in the analogue realm, increasing the signal-to-noise ratio becomes a priority. Refocusing awareness of the visual measurement of the signal becomes that crucial key to learners' comprehension of loudness.

Once the principle of peak/RMS measurement for dynamic range has been established, the introduction of current modern approaches to loudness, utilising commercial standards from International Communications Union (ITU) and (EBU) are contemporary to assist learners appraise the wider uses of commercial metering systems, thus giving context to loudness over time and its use in broadcast and post production. Despite this, in many ways, a simple analogue Volume Unit (VU) meter is still a great way to visualise the power of a track from a mastering perspective. The levelling over time in loudness should be about listening, not responding strictly to the numbers given by a metering system. These numbers are not music.

Bringing the learning back around to establishing the simple principles of listening helps learners develop at an exponential rate. Without this basic framework all the other practical outcomes in learning critical evaluation can be flawed or bias, taking students considerably longer to develop their critical listening and thinking. I cannot express enough the value in using a fixed listening level and equal loudness analysis in the progress of learners. Anecdotally, over years of watching students engage in their development, there's a clear progress gap that opens up between those who fully embrace equal loudness practise and those that do not.

AUDIO REFERENCING

As previously considered, listening to superlative music to understand what makes 'quality' has been a fundamental part of mastering engineers' development for decades. Enabling this focus, alongside; listening level; levelling those references at equal loudness when working comparatively; and digesting in a considered listening environment, has to be a key focus to assist learners progress.

To hear quality, learners need to seek out audio reference material [3]. Students tend to be swayed by their emotional response to the song and personal attachment to the music rather than the sonic qualities. These two aspects are not the same. A great song is still great whether it sounds sonically good or not. Making the separation between the music and sonic quality is key.

Differentiating this aspect can generally be overcome by group analysis of the properties of a given track. Directed peer observation and comment can be a powerful tool in analysis and help a group realise their attachments to the material, as learners will often have disparate musical tastes across genres. This allows learners to assess and point out discrepancies with sonics or equally praise positives where they would not normally assess them. Once this process has been started, I've observed groups of learners often proactively start to build a body of reference material across their musical tastes. Exploring the sonics and appreciating the aspects they feel contribute. Positive listening outcomes for all learners.

An issue for students born in the mid 1990s onwards is that many, if not all, have grown up listening mainly to new media delivery; with easy access to lossy based audio delivery on growingly poor, albeit more convenient, listening devices. Hearing higher quality outcomes was a normal consumer activity in previous generations as there wasn't another choice. Even though there were still poor audio playback systems they tended to have more bass delivery than the comparatively smaller modern playback systems.

The comprehension of quality from exposure has been lacking in current learners' everyday development. I have observed students nowadays when initially challenged with listening in a high quality environment, some learners preferring the sound of a lossy audio version to its 'lossless' counterpart, as the latter sounds too aggressive or forward. On exploring these initial observations, learners quickly reverse their opinion after more focused listening to lossy negatives. Listening to encoded side information on the monitor path and switching between lossy and lossless to clearly hearing the 'aliasing' present and lack of ambient detail on the lossy. Thus, when learners are listening back in stereo this lack of detail plus aliasing can be heard in the stereo. This method of focusing on an aspect we want the learner to perceive, once heard, most can now easily hear it in the complex programme material playback as their ear has attuned.

Simple aural lossy and lossless comparatives are telling when listening in a quality environment. This audible evaluation often helps learners understand the complexity of their audio journey so far. Making the directives laid out towards listening more tangible. Technical learning through exposure to quality.

Once these principles of quality are established, exploration of differing genres can be a powerful tool in further developing listening skills. Much of the issues in initial analysis comes from the focus on listening to the music rather than the sonic qualities. Obviously both these aspects are valid but those learners who are advanced in their music production tend to be biased towards the musical aspects from exposure in their creative production, recording and/or songwriting. When listening to a song they veer towards these, often expressing emotional responses to the song. Not a bad thing: but their focus on individual elements; the like or dislike; such as the sound of the snare or the reverb on the vocal; rather than stepping back and hearing the overview of the music together, perceiving its overall timbre/spatial impression.

A positive way to assist in exploring genre is peer-to-peer analysis of reference material. This promotes listening to music outside of personal preference and exploring prejudice towards genres. Listening to music they do not enjoy, enables focus on overview with more ease. This deeper view into their ability to analyse can be transferred to more familiar tracks to shed new light on their makeup. This sharing of musical genre often benefits in widening learners' appreciation of genres they may have previously overlooked. Starting to see more than just a liking for the music but gain an appreciation of the sonic quality in music they do not necessarily enjoy. New sonic knowledge that could be transferred to their own productions.

Another critical aspect in this learning process is the environment. I think the basic principle; 'if you can't hear it, you can't change it' has to be at the forefront of our mind in the development of a student's critical listening. Establishing an awareness of true transient detail present in a recording is only possible through appropriate monitoring and a neutral listening room. Something taken as normal in previous decades; the principle of listening to completed masters. Quality Control (QC) was a mainstay of many mastering engineer's development. Hence if one is indeed to teach and develop a high level of critical analysis in their learners, the listening environment must be of the highest order possible.

LISTENING FOR OVERVIEW

A proportion of learners struggle with the switching in listening style, changing 'hats'; from inspecting a mix in its individual aspects, to taking in the whole picture in tone and dynamic; the overview required to master effectively. The difficulty generally centres around initial interpretation of a mix and the instant emotional response, and a like or dislike in the sound of a given mix, e.g., the kick not being a sound they would use; or guitars too distorted; harmony parts not having enough reverb. All of which are aspects for the mix engineer and producer to address; the composite parts of the mix or production. It's not the mastering engineer's job to mix it, but enhance all the good work and hours spent in creating the production so far. We don't get to choose the snare sound, but collaborate with others intention in their production.

This can be very challenging for some, to learn how to make this separation in listening. A positive listening exercise is to expose those people to production styles they're not familiar with. Hence, feeling challenged especially in focus on sound selection and balance as they don't have a reference point. This approach helps the learner to focus on the whole production and not making ascetic choices about the individual tracks production. The learning that comes from this experience over time becomes transferable. Often this can take a couple of months of exposure in the course of listening sessions. The constant reaffirmation of the direction for analysis eventually allows learners to find that switch in perspective.

THE PROBLEM WITH VISUALS

The prevalence of digital tools to give 'visual' feedback is a barrier to students' critical listening development and not a help as would be suggested by the large array of tools available. Their use of colourful responsive graphics and visual analysis aids are a distraction from actually engaging at the task at hand – listening [12]. A general exception to this is basic peak/RMS metering.

A simple justification for this statement is: even we as audio engineers have not yet designed a loudness meter that can level music to the same subjective appropriateness as a professional mastering engineer can by ear. Yes, the power/loudness can be measured accurately and these measurements used comparatively, but this will only get most mastered audio within a dB or so of critical evaluative levelling by ear. Most mastering engineers spend quite a lot of time fine tuning finites of 0.1 dB. A full 1 dB is a chasm in comparison! These differences in measuring loudness to subjective levelling become much larger as the dynamic range of the program material increases. Meaning a more complex meter of frequency-based level does not again evidence subjective levelling outcomes. Technically correct in measured decibels in a given band but that doesn't mean we've now a visualisation of music from a musical perspective.

Digital tools do not listen to complex musical material but simply measure an aspect of sound. There isn't a meter type available that can interpret music comparatively. This is something we as humans are intrinsically able to do; our exposure to many differing forms of sound over our lifetimes whether direct or indirectly, all adds up to our sense of average; comprehension of music and form.

The further we try to focus on the detail of the sound, the more distracting the visual element can be. Levelling audio loudness to a meter to achieve a positive calibrated room listening level is helpful, using it then to justify the final levelling between different pieces of audio isn't good practice or helpful. Students need to be encouraged to use their ears – hide the meter once the audio is in the ballpark and listen in to the musical aspect, to hear the relative translation in power firstly. Focus on the loudest mix elements, such as vocal or lead line. This is the aspect the consumer tends to focus on when listening, the melody and or lyrics. The power of the low end in genre can often be a distraction, a wide dynamic acoustic and vocal song will sound very loud, relative to a dense band track, if levelled just on a RMS meter. In many instances, the vocal needs to balance in level across tracks more than the instruments do. This simple task is not assisted by a visual aid, it just distracts and delivers an inaccurate outcome. Lack of listening leads, eventually, to incorrect interpretation in analysis of the audio as they're fooled by loudness.

Students, especially in the last couple of decades, have been brought up with visual aids delivered via computer generated simulations. When initially finding their place within the world of audio many are fully exposed to the DAW/plugin/visualiser aspects and without realisation or intent, they are biased by them. I've seen many students who find it difficult to equalise without a visual aid – this is very telling indeed. Looking and not listening.

Once a learner has discovered that the simplest of meters don't evidence completely correct outcomes from a musical perspective, it's an easy transition to question others visualisation, especially in a reliance on frequency-based analysis tools. A good way to learn the link between frequency and equalisation is by repeatedly assessing what frequency area is to be addressed, set the equaliser, making note of the frequency/range as it has been analysed; applying without sweeping, a straight switch on/off at equal loudness; if this is not in the correct range, move towards the required frequency direction. Once located, learners will make note of the actual frequency as they've been working with the numbers, rather than simply listening to the sound being swept around whilst looking at it on an analyser. They are working comparatively, switching between frequencies and thinking about the numeric values linking it to their critical audio analysis.

The more frequently this method is used, the faster the learner becomes able to find the correct frequency balance. Importantly in the context of musicality of programme material, the continued reliance on visual analysis will not support their development. Frequency balance isn't music, all music has different musical frequency balance. If there was a correct amount of frequency balance for every track a simple pink noise match equalisation would function correctly, a more advanced analysis of this principle against genre examples can facilitate some musical context to the balance, but again it's not listening to the musical context. This is far more complex, something we as humans, the creators and creative elements are able to intuitively assess.

HEARING TRANSIENTS

The final aspect of developing a learner's critical analysis and the aspect which is the most challenging to master is starting to hear transient shape. Without the ability to correctly evaluate the envelope shape in the programme material audio engineers cannot effectively apply dynamic control. Being attuned to peak and average levels to perceive shape and hence comprehend the requirement musically to be manipulated: the engineer can make the link from evaluation of analysis to the application of tools and importantly tool type.

In the application of tools, an ability to clearly hear the envelope in the programme material facilitates the musical use of dynamics to enhance and blend the audio to our creative directive. Not by luck, or random application. One of the best ways I found to help learners develop this skill is not by discussing the use of the dynamic control itself, but by starting to observe how the envelope shape changes in theory, linking to previous learning around synthesis and use of attack, sustain, decay, release (ADSR) envelope.

Trainee engineers comprehend simple ADSR applications and outcomes. A very fast attack, short release creates a click, strong transient with no sustain. Open out the release and the sound becomes more sustained, denser. That latter link to tone is critical. The short sound has little bass. Sustained has significantly more bass, meaning a transferable thought process would be; increasing the sustain of complex programme material (by upwards compression) will increase the amount of perceived bass in an equal loudness compared to the original. Whereas, fast attack downwards compression will reduce the transient, hence the amount of treble, dulling the tone. Upwards expansion will lift the transient height and consequently increase the treble, downwards expansion reduces the sustain, thinning the audio and hence removes perceived bass sustain in comparative. The latter process would be rarely used in a modern mastering scenario, but is useful for a full set of outcome in theory.

The most successful way I found to practically expose the learner to these outcomes is not initially with full programme material ready to master, but with a stem drum multitrack. Having strong transient material and a wide dynamic range, it's easier to evidence dulling via downward compression using a limiter and increase bass from upward compression, using a simple parallel compression outcome. Once the learners can hear these tonal changes it helps their ear to focus on the transient's shape change. They can then start to transfer this critical listening to make observations in complex programme material. This can take some time for some learners, but generally they will appreciate the principles first time around.

Understanding of envelope helps learners start to comprehend the macro and micro dynamic of the audio manipulation with compression and those processes secondary outcome with tonal shift. A mainstay in most mastering transfer paths. This also clearly links the use of manual compression for the macro-dynamic manipulation of the audio [11].

BACK TO RECORDING

A key element to focus on in any discussion around mastering, is the part the whole recording and mixing process has to play. We know as a mastering engineer that our goal is to enhance all the previous positive work. We are also fully aware of meaningless statements, 'we'll fix it in the mix' or 'the mastering engineer will sort it'. Yes, mastering engineers can enhance the source programme material but if that source is poor, the outcome is never going to be amazing! Better yes, but not something to aspire to!

I feel it's crucial to instil a sense of striving for excellence in all learners. If they're to engage in any part of the creative process it must be with purpose and outcome. Just recording or mixing something for the sake of it is a poor rational. Every piece of work completed will be judged, whether they want it to or not, by their peers or other musicians in the scenes they're involved with.

It's critical that aspiring engineers involve themselves with projects where the aspirations are high and where the directives are clear in outcome. To make the best music they can. A band or artist that's not interested in taking their performance to the next level isn't worth the efforts involved. Everything about the recording chain comes down to source and performance [13]. With a great performance and consideration of the source for recording makes for a great recording, hence also a great mix and master. Without awareness of how all the parts of the production process interrelate, an engineer cannot truly develop. Without awareness of how each part of the processes has a direct relationship to overall quality outcomes, the engineer's listening is never going to truly comprehend overview. Gain appreciation of the final outcome and where their creativity fits in to it. Knowing how we want it to sound in the end clearly assists in achieving the best outcomes from the recording and mixing process.

At all stages in learning to listen, this awareness of source and the quality of, is crucial in the development of a comprehension of what is acceptable and what just needs to be addressed again. Seasoned mastering engineers know immediately, on hearing a mix, what the issues are in that regard. So, in approaching a piece of material, they're clearly aware of what could have been improved upon in its development. This has to then be factored in, you cannot make it into something it isn't, but you can enhance what's there.

I have found a good way to build an appreciation of this link back to the recording and mix process is by the learners attempting to comparatively, with reference material, engage in the process of mastering their own project mixes. There are several positive learning outcomes; an awareness of what is wrong in terms of their mix approach and the source material; an appreciation of the source recording and performance; it's quality or possible lack of; the processing they've now had to apply that could have been avoided, by better attention in the previous aspect, of the process. Valuable self-assed feedback from the mastering process to the beginning of the chain.

This works most effectively with material they've completed at least three months before. Hence, they've an ability to focus back in overview without the attachment to the recording process or mix, generating a more reflective process.

I've also found in general, as students mixes are not to a commercial expectation in level, they can correctly summarise the best path to 'fix' with less indecision, as the aspects they need to address are more obvious than with commercial mixes. This gives learners confidence in their analysis skills and also their ability to link to correct tools to enhance. Hence when engaged with more challenging material, recorded and mixed to a high standard, they make better judgements in analysis and evaluation around how to improve.

LOUDNESS AND LIMITING

The theme of loudness, loudness normalisation and 'the loudness wars' are constantly present tropes in the music press and in educational discourse. This is an important aspect to allow students to explore with a fully informed outlook. Much of the discussion around these tropes talks about the lack of dynamic range in programme material as can be seen in the many articles and themes from across the years referenced on the 'Loudness War' Wikipedia page [14]. How material is hyper-compressed, which could be true from one point of view. Another perspective is the engineers and producers involved in the creation of that music made a reasoned choice in overall dynamic range based on their artistic judgement, which is a view I hold in contrast to a lot of those tropes promoted.

Professional engineers will listen to music and make it sound appropriate for their intended outcome. Whether something is right or wrong isn't a collective judgement. Music, if not all art, is judged as much by the individual experiencing as judgement is passed by those with positions of influence. The latter influence may swing this debate for some, but in the cold light of day with no measurement of visual analysis to distract away from listening to the audio. Just listening to music at an appropriate listening level, what judgement would be made then? Is music judged by the value of the dynamic range or by the quality of the emotional response experienced by the listener?

With best intentions many write about the problems with loudness, we as engineers should conform to a standard even if it isn't a standard but a company's suggested ingest [15]. This detracts from the artist's key directive in delivering their outcome for the music. You could argue people are already restricted by delivery formats themselves, true, but a limit on resolution doesn't dictate where the audio should sit within it. Equally the post application of loudness normalisation in most streaming services means multiple issues can arise. The playback app doesn't have normalisation; the user doesn't turn it on; if they did, they can select different levels of 'loudness'. All of this means if the engineer decided to set their dynamic range relative to the ingest requirement, the audio playback may be diminished by it. All these themes are important to the debate to balance against the tropes laid out in the media. The apparent current dilemma of loudness normalisation is one that some may agonise over and some will just keep doing what they always have, listening to the audio. The learner is the winner in all of this, discussing loudness can only promote these learners to explore through listening, make their own conclusion and outcomes with audio.

The simple and important final lesson in comparative once a track has been mastered. Listening at appropriate listening level and equal loudness: asking the question, 'have we improved the audio in all aspects from the original?' A simple honest test to know whether 'loudness' has negatively affected the audio or the dynamic reduction has enhanced the music's outcome.

CONCLUSION

Learning to listen is a lifelong journey of which an audio engineering degree can be a positive introduction. You're not going to teach someone to be a mastering engineer but you can teach students how to listen, become aware critically how they're affected when listening and evaluate their observations. Hence start to make the link in comprehending how audio tools and their types affect the sonics. All of this can be taught in an educational setting by a skilled mastering engineer with the appropriate listening environment. The quality in audio outcomes for the learners will come with time and effort, on an engineer's individual journeys post education. A degree is a spring board we can prime students for their next development phase, but it's their time that needs to be invested in their own audio future.

As I say to a new cohort of students; I can teach you the technical skills needed and the principles of engagement in processing, guide you in developing your critical ear, but I can't give you the three decades of critical listening I have and the sense of average that comes from that. You'll have to invest in that on your own. Audio, is a journey that can bring much joy to one's musical heart, but you have to put the time in to gain the experience. There are no shortcuts or visual aids to assist in this aspect.

Mastering is a very fulfilling practise. Being given the responsibility of finalising someone else's art is a privilege. The skill that can be gained in learning to engage with it as a process can be used in all aspects of engineering. After all, what happens at the end can, in hindsight, always teach you something about what you could change at the beginning. This, I feel, is one of the most valuable lessons that can be taken from a mastering course designed effectively alongside the learners developing a more critical ear. These are the outcomes I observe in my students year on year. Something I feel very fortunate to be able to behold.

We, as practitioners, can teach the skills required for analysis and the techniques used to manipulate audio but the critical insight required is in the hands of the learner, to develop in their own audio journey over many years through critical listening.

As a final thought, teaching audio engineering boils down to a simple fact for me. If you can't hear it, you can't effectively change it.

REFERENCES

1. Walzer, D. A. (2015). 'Critical listening assessment in undergraduate music technology programmes' *Journal of Music, Technology & Education*, 8: 1, pp. 41–53, doi: 10.1386/jmte.8.1.41_1.
2. Corrao, S. Mastering Isn't a Dark Art, It's Just Misunderstood - TuneCore. [online] United States, (2020). Available at: https://www.tunecore.com/blog/2018/10/mastering-isnt-a-dark-art-its-just-misunderstood.html [Accessed 8 March 2020].
3. Parnell, M. *Interviewed by John Braddock*, Confetti ICT, 5 March, (2015).
4. Proper, D. *Interviewed by Robert Toulson for AES*, 22 September, (2018).
5. Corey, J. *Audio Production and Critical Listening*. 2nd edn. Focal Press, (2013).

6. Moulton, D. Moulton Laboratories:: About. [online] Moultonlabs.com., (2020). Available at: http://www.moultonlabs.com/page/cat/About/ [Accessed 8 March 2020].
7. ———. *'Tech 3286, Subjective Evaluation of Quality - Music Programme', European Broadcasting Union*, (1997).
8. EBU. *'EBU Technical Recommendation R128 – Loudness Normalisation and Permitted Maximum Level of Audio Signals', European Broadcasting Union*, (2014).
9. ITU-R Rec. ITU-R BS.1770-4, *'Algorithms to measure audio programme loudness and truepeak audio level'. International Telecommunications Union*, (2015).
10. ———. 'Integrated Approach to Metering, Monitoring, and Leveling Practices, Part 1: Two-Channel Metering'. *JAES*, Vol. 48, No. 9, (2015), pp. 800–809; September 2000.
11. Katz, B. *Mastering Audio*. 3rd edn. Focal Press, Burlington, MA, (2015).
12. Mycroft, J., Reiss, J. and Stockman, T. *The Influence of Graphical User Interface Design on Critical Listening Skills*, (2013).
13. Platt, T. *Lecture*, De Montfort University, 3 March, (2017).
14. En.wikipedia.org. Loudness War, (2020). [online] Available at: https://en.wikipedia.org/wiki/Loudness_war [Accessed 8 March 2020].
15. Artists.spotify.com. Mastering & Loudness – FAQ – Spotify For Artists, (2020) [online] Available at: https://artists.spotify.com/faqmusic [Accessed 10 March 2020].

Part Two

Mastering: music and people

7

Interview with Darcy Proper

Rob Toulson

INTRODUCTION

Darcy Proper is an award-winning mastering engineer and owner of Proper Prent Sound, who is currently operating out of Valhalla Studios in Auburn, NY. Prior to her recent return to the U.S., Darcy was based at Wisseloord Studios in The Netherlands and Galaxy Studios in Belgium. Darcy began her career at Sony Music Studios in New York City in the classical music department and has since gone on to work in all music genres, with a particular specialism in surround sound and immersive mastering. Darcy is known for her work on historical reissue projects for prestigious artists including Billie Holiday, Louis Armstrong, and Frank Sinatra, as well as for high-resolution surround/immersive mastering for the likes of Donald Fagen, Patricia Barber, and Engine Earz Experiment. She has won four Grammy Awards, most recently in 2018 for Best Surround Sound Album for her work on saxophonist and composer Jane Ira Bloom's *Early Americans*. This article is an edited transcript based on Darcy Proper's keynote interview at the Audio Engineering Society Mastering Conference in 2018 with Conference Host, Professor Rob Toulson (Figures 7.1).

BACKGROUND AND CAREER

How did you become a mastering engineer?

I ask myself that question a lot, but it all started in High School. I loved playing music; I was a saxophone and clarinet player, but I didn't have that passion or that drive to really be a performer. I loved playing in ensembles, I could read music really well, but I was not the person to be playing an improvised solo and that kind of thing. Being in the spotlight was difficult for me. Nowadays I'm used to working very often alone in a room and that's where I feel most comfortable,

Figure 7.1 Mastering engineer Darcy Proper.

so that personality trait is what prompted me. In addition to loving music, I was a good student in math and science, and when the high school band was giving a rock and roll show to earn some money for a field trip or band uniforms, I suddenly became aware of sound engineering. A sound reinforcement engineer who was a friend of the band leader showed up with a little mixing console with 16 channels; it was certainly nothing elaborate, but I saw that thing with all the knobs, the lights, the VU meters moving, and I thought 'that looks cool, that looks really interesting, how does that work?' That prompted me to look at the field of audio engineering. As much as I'd listened to music for all my life, up to that point I'd never thought about the technical process behind music. I looked up some college programs and tried to find out some more information about the field. There were not many schools in 1986 with programs that specialised in any form of audio engineering, a lot of times it was paired with communications rather than being music focused, but New York University (NYU) had a good programme that was relatively new at that time, and I ended up studying there.

When I graduated from NYU I still didn't really know what mastering was. It's not something you step into initially, it's a path that you find as you progress, I think. So I left school thinking I was going to move into research and development, but ended up being an assistant technician at a dance remix studio. While I was there, I started doing part-time quality control (QC) work for Sony Classical, which meant that I was one of a few people who was called to help out in the evenings to make masters. In those days, the masters were on three-quarter inch video and they needed six copies of every master, because they went to various record manufacturing plants around the world. My job as QC was to make the copies, analyse them and then listen to each of the six copies to make sure that there were no audible faults, either from the tape itself or from something that might have been missed by the engineer. So this involved hours and hours of listening to all sorts of different music, created by all sorts of different engineers and I would say that lowly position is probably the experience that is most responsible for me ending up in mastering – all those hours of having the opportunity to listen and having to quantify what I heard to explain to an engineer if I thought there was a problem.

I moved a bit further up the food chain at Sony, becoming an assistant recording engineer and an editor for classical productions, which lead to also doing remastering of reissue material, first for Sony Classical and then for Sony Legacy which holds their pop, rock and jazz back-catalogue. Eventually, I left the classical side of things to work on basically everything else, because the client base was moving that way. There were fewer and fewer true classical productions being made and there was a team big enough to handle those. In the meantime, in about 1998, I was getting fascinated with surround sound and there was more developing at that moment in the pop and non-classical genres for working in surround. And it just went from there.

With those years of having to listen over and over again on a daily basis to random things coming through at Sony, did that give you an essential grounding to work as a mastering engineer?

Oh, definitely. I think mastering engineers all develop an internal dictionary of sound, of how music should sound, and how to bring that forward through mastering. Having the opportunity to listen to music productions from all eras of recorded music in a wide variety of musical styles was priceless in terms of expanding my 'sonic and musical dictionary'.

You obviously have quite a diverse portfolio of formats and genres. Is there anything you would say you gravitate towards, perhaps a particular genre or a particular audio format?

There's no one particular genre that I would say that I specialise in. I would say that an awful lot of what I do is commercial pop but that's just because there are more of those kind of productions out there, if you're working in the non-classical world. I tend not to do very much in electronic dance music (EDM), not because I don't like the music, but because I'm probably not the best person in the world at pushing masters to a dynamic range of just two decibels. Some people do that with pleasure and enjoy it, but it's not really my thing. There's not a lot of hip hop and urban music in the Netherlands (where Darcy was based at the time of the interview) and I don't master a lot of really strictly classical music. I master a fair amount of classical crossover music, but strictly classical productions don't typically have a separate mastering process, it's generally done in the process of editing and mixing.

Your successes and experiences with the Grammy Awards are interesting because they span over a number of years. Could you give an overview of your Grammy Awards please?

Oddly enough I've never won a Grammy Award for a stereo project. I have one in mono that was for Best Historical Album in 2001 for a Billie Holiday compilation. This was a ten CD set of catalogue material from the original lacquer masters or metal discs, depending on what was available to make transfers from, from the Sony/Columbia archives. This was done with a team of people including a couple of reissue producers and a team of at least five mastering engineers, so I did not do this all on my own – it was a team effort and we all had the good fortune to win the Grammy in that category in that year. Then in 2006, I won for Best Surround Sound Album for Donald Fagen's *Morph the Cat*. That album was mixed by Elliott Scheiner and, if you know anything about Elliott's work, you can imagine it sounded pretty good when it came in, so it was a pleasure to master. In 2013, Patricia Barber, who is a jazz singer from Chicago, released a surround sound album called *Modern Cool*, which I mastered. That record won the Grammy Award for Best Surround Sound Album, thanks to the mixing talent of Jim Anderson.

Then I mastered two surround albums for Jane Ira Bloom, an amazing composer and soprano saxophonist, the second of which won also for Best Surround Sound Album in 2017. It's a stunning album of a jazz trio with just acoustic bass, drums and soprano saxophone. Having been a clarinet and tenor saxophone player, I truly believed before meeting Jane that the soprano saxophone represented everything that's wrong with a clarinet combined with everything that's wrong with a saxophone. The first project I did for her was an album of ballads; engineer/producer Jim Anderson called me and asked 'Are you available in April?', to which I replied 'Yeah, sure. I'd be happy to be of service'. Then he explained that it was a 78-minute album of ballads on soprano saxophone, and I'm thinking 'Oh, damn, did I just tell him I'm available?' But, then I heard Jim's great mixes and I was blown away by Jane's talent; she really specialises on that instrument and she knows exactly what to do with it, so I was delighted to be a part of it. The second album called *Early Americans* has a bit more up-tempo content, and for that one we had the good fortune to win the Grammy.

MASTERING APPROACH

Could you describe your preferred studio set-up?

I have six channels of gear that I can use, so I can utilise the signal path for mastering surround (5.1) projects. I typically play mixes back from Pro Tools with Prism digital-to-analogue converters and then I have an SPL MMC1 mastering console and various outboard processing gear. I have some Dangerous Music equalisers, some SPL PQ equalisers, a Millennia compressor, some nice Maselec limiters and a Neve surround compressor, which is a bit like the Neve 33609, but a specific version for mastering. Then I will usually go back to digital through Lavry convertors into Pyramix software at whatever the target sample rate is for the project. The nice thing about working with the two separate workstations (Pro Tools and Pyramix) is that the sample rate of the mixes I receive can be independent of the target sample rate for the release format. If the primary release format is just going to be standard (44.1 or 48 kHz) resolution, I'd rather master directly via analogue to the standard resolution rather than mastering at a higher resolution and relying on a sample rate conversion to get back down. I know standard sample rate converters are quite good, but I still feel that I get a better sound if 44.1K is going to be the primary release format. In terms of acoustics, I use Eggleston Works monitors and Krell power amplifiers for surround, and, when working in 9.1, I use a matched set of small PMC AML2 monitors.

Do you have any different approaches to working on a remaster project verses a new release project?

With remastering, it's important to keep in mind what the original was. People will have been in love with the original and presumably that's the reason why the record company or the artist may want to reissue it. They want to see if it's possible to achieve a better sound with the newer technology, but they don't necessarily want to dismiss what was done the first time around. The most important thing is to make sure that you've got the right content to work with – to be sure that one of the tracks didn't come from a different reel of tape or a different lacquer disk. If any speed adjustments were done on the originals, for example, you need to incorporate that, or if you choose not to do it exactly the way that they did originally, you should at least be doing it consciously, being aware of the fact that you're changing things. There is a significant listening public for these types of reissues, and the audience know every single artefact on every one of these old recordings. If you dare to change something you better be prepared to justify that decision.

It's interesting to know how you approach more modern projects, working with artists or producers. Do you have any kind of personal interaction, or a process of getting to know the artist or the producer, or to discuss a shared goal or objective for the record?

Absolutely, I think like most mastering engineers, I really love to have some interaction with the people that I'm working for.

Sometimes, in a conversation that maybe doesn't have much to do with the topic of the mastering per se, some casual remark suddenly gives some insight into their character and maybe into their expectations of what they're looking for from the mastering. When I first moved to Belgium and spoke not a word of the language, I found it really felt like my radar wasn't working anymore. If people were in the room, they would address me in English but then they would talk among themselves in their own language, and bccause I couldn't even subconsciously process that information I really felt like I was missing out on something essential. I really believe that communication is a very important part of the mastering process. It's not my music, it's my job to bring the artist's intentions forward as much as possible and connect the listener to what the artist wants to convey. Therefore, I have to understand what the artist is looking for and not try to fit their music into a particular 'box' or style. That's one of the real disadvantages I think with these automated and online mastering systems where the client doesn't have the opportunity to have any personal contact with the mastering engineer. I think we feed on the information that we get from each other when dealing with something creative like music, and I think we can all do a better job and have a better chance of getting it right the first time, if we take a few minutes beforehand to try to figure out what the goals are.

Do you find that mastering is a particularly sensitive part of the process for an artist, given that this is this the final sign off to the record being finished?

There is a reason why my favourite mastering room I've worked in is red and deep wood-tones – to give it the feeling that this is not a dentist's office, and this is not a laboratory; that this is a warm, embracing environment where you can finish off your project. For those few clients that attend sessions these days, I want them to feel that they are not just sitting there waiting for the 'bad news'. Sometimes clients think your job as mastering engineer is to be critical of everything that they've done up to that point – and it may have taken many years of their lives to record and mix their album. They know this is the last moment that they can adjust things before 'the baby is born' and goes out into the world, and often they're just scared to death that there will be disappointment – that you'll say 'Sorry, I can't get you to where you want to go with the material that you've delivered to me'. So, I like to do anything I can do to put them at ease and make them feel like 'Hey, you should be celebrating! You've been working your butt off on this thing for years and we're here to put the finishing touches on your masterpiece'. I think artists are very courageous to even bring their music out to the public, which is generally very happy to criticise and often not so inclined to praise. It takes a lot of strength to put yourself out there like that. The fact that some artists bother to come in and be a part of the mastering process is great, and if they do attend, it should be joy they are feeling as we're working on their material, not worry.

A lot of the music that you have mastered has foreign language lyrics. Does that make a difference for mastering? Do lyrics become something that you sometimes understand in your own language and hence they have any emotive role in the mastering, and how do you overcome that when the material is in a language you understand?

Well, yes and no. It certainly helps to know a little bit what the song is about, but even when I'm working on a track in English, I'm not really listening word to word for the lyrics while I'm in the initial process of figuring things out. I need to know enough of what the message is to know how best to convey it, but a lot of that comes across regardless of the language. For example, I don't speak German very well but I've worked on a lot of German material and studied some German in school, so I can generally follow the meaning enough to know what is essential to the song. I worked on something recently in Slovenian that was a little bit more challenging because I really couldn't make any sense of it at all, so I made contact with the production team and asked what the song is about. But usually the music and the tone of the voice and the way it's delivered tells me enough about what the artist wants me to connect with, and I can tune into that and deliver it. Maybe towards the end of the process, when I'm doing a final listen through, I might really start to listen and appreciate the lyrics if they're in a language that I understand.

When you're mastering for stereo, do you tend to provide a streaming version of a master and if so what target levels do you aim for?

I have to say that most of the time my clients, for budget reasons, are looking for a 'one size fits all' approach to mastering, so the audio for the streaming version is likely to be the same as for the CD or digital download. As the normalisation levels of streaming platforms have become a hot topic in the industry recently, it's beneficial that the sonic issues surrounding over-limited music are becoming more known to the public. Also, the renewed interest in vinyl allows me to give some justification for talking clients 'off the ledge', inspiring them not to subscribe to the 'loudness wars' for their CD/digital release, since it won't work well for the vinyl. In deciding on level, typically there are two factors: first, I want the music to sound how it sounds best, so if it's something that needs to be packed in a bit more and tighter compressed, I'll try to achieve the sound that I'm looking for first. I'll just master with my ears, without looking at numbers and meters too much. Then I'll have a look and see if I got a little bit 'limiter happy' and might have pushed it too far, so I can back it off. After so many years working in the same room on a console with my VU meters and my monitor pot in a certain position, I do have a pretty good idea how much sound should be coming back at me, so usually if I'm crossing the line, I know it already by the time I start looking critically at the numbers. I would say my target is around −14 dB integrated as an average loudness measure, which is a typical standard across most platforms right now. I actually aim to be a little bit above that just because many of the streaming services don't raise the level of quieter material, but they do drop the level if it's too loud, so if you can give something a good healthy sound and you're just a little above that target, it will get dropped down a bit to end up right at the normalisation target. Then you haven't sacrificed a lot of dynamic range to get there and your material will be at a good apparent loudness compared to other material when played back by the listener. Every once in a while, if you have very broadly dynamic stuff, the numbers might come out quite low, and then you have to decide what to do. Personally, I wouldn't push something too hard just to reach that magic number if it's going to destroy the music. But, in general, I think right now it's safer to try to sit just a little bit above the target levels and let the streaming services bring the song down a little to their particular normalisation level. Otherwise, if you are below the target you can't be sure what the streaming service will do to normalise. Do they bring the song up to the natural peak that's in the material? Or are they going to bring it up by adding limiting? If they do use limiting, then that's going to change what it sounds like, so it's something to consider.

SURROUND MASTERING

There are a lot of people who are quite comfortable with the concepts of stereo mastering, but wouldn't quite know what to do if they were given a surround sound mastering project, not least because they may need to set up a surround studio for the first time. So, what's different for you between a stereo project and a surround project, from either a technical or a creative perspective?

Technically speaking it's not so radically different. A well-balanced mix in stereo and a well-balanced mix in surround still have a lot in common and will feel 'right' to a trained listener. The fun thing with surround is that it can be even more exciting on the creative front - the artist has a lot more space to play around with, the recording engineer has more options, the mixing engineer has a bigger palette to work with. As far as the technical side of things is concerned for mastering, my room is set up so that I can work with my traditional analogue chain if I want to, which I do most of the time. They're a pretty basic (but high quality) set of processing tools, and I can get the sound that I'm looking for most of the time by using them differently for different types of projects. So there is the initial investment in adding multiple channels of gear when working in surround in the analogue domain, but in the digital domain, many workstations and plug-ins are quite capable of working in multi-channel modes. You just have to learn to listen in a few different directions at the same time while following your usual good mastering instincts and practice. While working, I might sometimes pull things apart to get a sense what's happening where – maybe I'll work across the front initially with the left-centre-right speakers and then listen to the rear speakers and see what's going on there. You can potentially make each channel or each pair of channels sound as good as possible, but it's important to make your final decisions by evaluating the results as a whole. Mix engineers will recognise this from their work too. You can make the perfect guitar sound and the perfect drum sound, and the perfect bass sound, and the perfect vocal sound, and then when you unmute all of the channels and everything plays together, it feels like pure chaos because the elements don't have their own places; they don't claim their own territory and they don't leave room for any of the other components. So instead, you might slim a big fat acoustic guitar sound down by filtering off everything under 200 Hz and everything over 2 kHz for example. You do these sometimes radical things to individual instruments and, when you put them all back together, everybody's there and their instrument sounds as it should, but now everybody's got their own space in the mix and there is space for everyone else there too. In surround mastering, you can have a similar experience. You can get those channels all sounding very nice when you have them in pairs or soloed, but you have to be prepared to adjust afterwards for how it sounds when the whole thing is put together.

Surround sound music releases seem to come in and out of fashion, and it seems you have convinced some artists to embrace surround in their projects. Do you feel that artists are excited by surround sound when they realise what they can do creatively with it?

The artists who are introduced to surround and also immersive audio, seem really to take to it like fish to water. It truly expands the possibilities for how they can present their music to their listeners. For example, I mastered an album in stereo for Belgian artist Ozark Henry, who creates what I would describe as alternative pop music incorporating interesting soundscapes and effects. While working with him I said 'You should consider doing your stuff in surround, because I think your music really lends itself to that kind of listening experience'. He responded by saying he'd never even heard any music in surround, so when we finished mastering I sat with him for about an hour and just played through all sorts of material; stuff that I had mastered, stuff that other people had mastered, including Beck's 'Sea Change album (mixed by Elliott Scheiner and mastered by Bob Ludwig), which is a fantastic album that is awesome in surround and it struck a chord with Ozark Henry. So, when I saw him again a year later, he was planning a new project of his music with an orchestra. This time, not only had he decided to do it in surround, he'd leapt ahead and composed the whole album with immersive in mind, because he said, 'as an artist, I just can't say no to this huge canvas that's now become available to me'. So I think for artists it means a lot to not be trapped in the stereo format.

So it's valuable for them to make that decision early in the production process?

The earlier in the process, the better, I believe. You can certainly repurpose an existing album for surround and, if you have good multi-tracks, a creative mix engineer will be able to make something quite compelling out of it. But if the material is artistically composed and arranged with surround in mind, then an artist can really take full advantage of the format and give the listener a truly special listening experience.

Do you find that you can do more as a mastering engineer with surround? For example, if the bass is a bit challenging, maybe you've got more control over that, or if maybe you're looking to make a vocal more prominent.

I would say so, yes, because things are dispersed over more channels allowing for a bit more separation of territory. One of the challenges in mastering is that we generally get material delivered to us which is already mixed, and everything that we adjust can do something in a positive direction for one instrument perhaps, but it can do something counterproductive to some other aspect of the mix. In surround, because things are split up a bit more, adjustments in one or two channels may be able to create the desired result for a particular aspect of the mix, while creating fewer unwanted side-effects on the rest. For example, a lot of productions use the centre channel for lead vocals and that can give you the opportunity to make some significant changes in the vocal itself without messing up everything else too much. Given that vocals are very often the most sensitive aspect of a production, that's a valuable thing to be able to adjust.

Do you feel that actually having a speaker in the middle rather than a phantom image of a vocal kind of gives it some greater presence or realism?

Yes, when I work on the stereo version of a track and the surround version of the same track, having the stability that the centre channel provides really makes a difference, especially as the listener moves around the room. The centre channel serves as a kind of anchor to keep the vocal in its place.

Do you receive mixes that make extreme use of the surround field? For example, Jane Ira Bloom's Early Americans, has more surround channels than instruments, so the mixer might decide to put the double bass on one side and the saxophone somewhere else, but I assume that it's usually more subtle than that?

For that album in particular, it was quite subtle. Engineer/producer Jim Anderson and Jane herself made the decision to keep it simple and real. Jane wanted her listeners to be able to experience her music the way she herself hears it when she's with the musicians in the studio, and Jim is a master at knowing exactly what to do and what not to do. This was a recording of an acoustic jazz trio and it wouldn't make a lot of musical sense to take a very natural and pure recording like that and start moving musicians dynamically around the room and that kind of thing – it would completely detract from the listening experience. Rather than falling into the music and being part of that scene, and enjoying the rich reality of it, it would become a bizarre fantasy which wouldn't have worked to convey Jane's intention for that particular album. On the other hand, I've worked on some electronic dance music (EDM) releases that are highly creative in their use of the surround or immersive sound field, because there is no natural environment to try to recreate. There is no authentic relationship to adhere to, there's nothing from microphones that's picking up the room reflections of a real space or anything like that – it's all artificial. In that case the mix engineer can have a blast and just go for it – things can jump up and spin around, hit the roof and come back down again. A lot of electronic music is intentionally repetitive – it's loops, it's beats, it's things that continue in a subtly-shifting pattern, and then these spatial highlights, these fireworks that go off can really keep your attention and make it an interesting listen. So when it serves the music, you should be as wild as you want to be and if it gets in the way of music then tone it down and make it real. The nice thing with surround and immersive audio is that it can be very real and very simple or it can be a wild rollercoaster ride.

What advice would you give to a mastering engineer who's doing their first surround or immersive project?

Don't be afraid of it, just go for it – you're a mastering engineer, you have good ears, you have good instincts, so just follow those. The same rules that apply in good stereo audio apply in surround as well. You have to watch out for phase and timing relationships between channels and that kind of thing, but most of those things – if they've gone wrong – they've gone wrong in the mix process and there's not a whole lot you can do to fix them in mastering. Naturally, as a mastering engineer, you have to be careful that you don't introduce any of those problems. You have to make sure your system is properly clocked, that your gear is properly calibrated, that your gain structure is healthy, but these are things that you do on a daily basis in stereo anyway – so there's nothing new there, just more channels to watch out for. I would also say give yourself some time before you start making decisions to listen to the material that you're working on and listen to some other examples in surround. We all have this sort of instinct that makes us perceive things behind us as being louder than things in front of us. We're naturally programmed to listen for things sneaking up behind us and you can react too heavily to that instinct and potentially suck the life and excitement out of a surround production by being too conservative with the rear channels. This is especially the case if it's a mix where things in the back are not just space and audience, but actual integral parts of the composition and the balance of the music. You don't want to push those down too much and keep all the focus up front just because that's what you're used to. Give yourself time to get used to listening in surround, it doesn't take very long. For example, when I first started working in immersive audio (surround with height), it helped me initially to put my chair a bit higher so that the lower layer was a bit further below my ears than normal, and I was a bit closer to the top layer. That approach helped me to be able to understand and quantify which sounds were coming from above and which were coming from below, and which elements were shared between the two. After a little bit of listening and getting used the fact that there was now a vertical component to listening, I could go back to my usual listening position. Our brains are remarkably adaptive. We listen to real life in immersive every minute of every day and yet we've gotten used to making the translation for stereo when listening to music and still feeling the space and believing it. Likewise, if you sit in your car and you sit next to the left speaker in stereo, you don't really think about the fact that you're not hearing the right channel as much as you should anymore – your brain kind of automatically makes that adjustment based on your prior listening experience. If you start working in surround, you find as you gain experience that you can move around more and still accurately judge when things are well balanced and well EQ-ed without having to always be in exactly the sweet spot. And it's actually very interesting to walk around the surround field and experience the music differently from different listening positions.

Do you often get sent the stereo mastered files before you start mastering a surround project? If so, will you work with that as a reference and try to match what you're doing to the stereo version?

It depends on the project. Sometimes I'm hired to be work on only the surround version of an album and will receive already-mastered stereo files. When that's the case, I try to align the stereo and surround versions as much as possible. In one case, I mastered an immersive reissue of Tiesto's *Elements of Life* album, which was a very successful electronic dance album from 10 or 12 years ago. There was an existing stereo master which was 'EDM-loud', and would need to be included on the release for the new edition. In terms of the layout of the album (pauses, relative levels, etc.), I used the stereo as a guideline so they would match exactly, but I could not master the immersive version nearly as loud as the original stereo release. If I had pushed the immersive stream to that level of loudness/limiting, it would have been unbearable with nine speakers 'barking' at the listener from all directions. So the decision was made, in conjunction with the artist and the label, to keep the stereo master as it was, i.e., not do a new stereo mastering, but to drop its level to match the immersive stream so nobody would get hurt if they switched between the two versions on the disc. After all, listeners should be able to switch between the streams on a disc and not get their heads blown off doing it.

If I'm mastering both the stereo and the surround versions for a project, I like to start with the stereo first, because, if I begin with the surround version, it very often becomes a kind of disappointment to go back to stereo, and 'disappointed' isn't really the best frame of mind for working on someone's music. Before I'm fully committed to finalising the stereo, though, I might check the a track or two of the surround just to be sure I can get the two versions to sit well with each other. Another advantage to doing the stereo version first is that it's very easy for the artist, who might be unable to attend the mastering session in person, to listen to references and give feedback in terms of the pauses between the tracks, relative levels, and how the album feels to him or her in general. With many important decisions already made, and having experienced the album already in a simpler form, so to speak, you can then go on to master the surround version, free to focus on the unique elements of the surround.

REMASTERING

Concerning your remastering or reissue work, have you ever had to turn down or reject a project because you couldn't get good enough sound sources?

I would have sometimes liked to have said 'no, I'm not going to remaster if unmastered mixes or the original sources are not available', but the fact of the matter is, if the client wants to remaster from existing CDs because they can't find better sources for the songs, it's my job to do the best I can with whatever I have to work with. If we do end up having to work from previously mastered material, it makes a difference to me if they explain to the public why that decision was made so the end listener knows what they are getting and why. The frustration is that, very often when working from mastered material, there's not a whole lot you can do to improve it. It might already be 44.1 kHz/16-bit created with old converters, while the original analogue master tapes may have sounded much better. Or if the source material comes from mastered digital files in the prime days of the 'loudness war', they may already be so hyper-limited that you can't help them much. It's always best to go back to the original sources, of course, if they are locatable and in playable condition, but when nothing else can be found, I just do the best I can with what is given to me.

Archiving seems to be a very important thing which is ever-changing because formats are always changing. Do you see any issues with the approaches taken for archiving at this moment in time?

To be honest, I think reissue projects are going to be a nightmare in the coming decades. There's an anniversary coming up for a production from 2001 that I was involved in, and just going back to try to restore the original stereo and surround mixes required pulling old computers with old operating systems out of storage and finding old AIT (Advanced Intelligent Tape) drives, and hoping that they didn't eat the tapes while restoring the session data. Then old versions of the workstation software had to be found in order to open up the mix sessions. It was a process for one album that lasted many weeks.

Also, documentation these days is not as thorough as it used to be – it's easy to say that the details are all contained in the DAW session files, but if that session doesn't open, you're left with folders containing dozens or even hundreds of loose un-named sound files. At least when working on old tapes and old lacquers, the tapes are generally very well labelled. You could pull out an old half-inch three-track master from an archive and, more than likely, you will find all the information you need about the contents of that tape on or in the tape box. If not, you could play that half inch tape on a stereo half-inch playback machine, and even though it's a three track tape, you'll at least know what music is on it. You wouldn't be getting the advantage of the centre channel, which you'd certainly miss, but that would probably lead you to investigate further, and eventually you'd be able to play it back correctly. But a DAW session, which might have two hundred tracks called *audio 1*, *audio 2*, *audio 3*, *audio 4*, is a big, hard-to-solve mystery to someone looking at it for the first time. It will be interesting to see how future reissues will be handled in these kinds of situations.

EDUCATING ON AUDIO MASTERING

How would you approach educating on audio mastering and enhancing the process of learning?

I think as mastering engineers we learn to play our gear and equipment as musical instruments. At some point you stop thinking about the numbers on the dials and stop thinking about the technology. I want to hear 'this' come out of the speakers and so I instinctively reach for a particular adjustment, in the same way pianists don't have to think about putting their fingers on the piano keys; they want to play a certain chord and their fingers go there. The other half of the equation is that it took me many years to figure out that it's not my job as a mastering engineer to fix things. It's my job to enhance things and to bring out the best of what it is. If beginners come at mastering with the approach of wanting to fix things, then it assumes you have an image in mind or you have a kind of 'box' that the music has to fit into. There's a lot of music out there now, particularly with artists putting out their productions independently – I think the diversity in music is broader than ever and if, as a mastering engineer, you start trying to make music fit into pigeon holes, then you're not doing justice to what you have in front of you. If you're fixing things, you're focused on the negative rather than the positive so it's a much tougher job. I'd personally much rather be focusing on what I love about a project and bring that forward, which minimises the flaws itself.

Are you doing some teaching yourself nowadays?

That would be a glorified term for it I think. There are a couple of schools including the Pop Academy in Mannheim where I teach at least once a year, giving a masterclass for the senior students and a lecture on what mastering is for all of the new incoming students. They have a three-hour workshop where they want everybody who comes into that school to know what the various phases of a production are, even if they're going to be managers or composers and maybe not necessarily working on the technical side. I've also taught at Abbey Road Institute in Amsterdam, where I deliver mastering lectures. Some groups and schools also come to the studio for a day of actual hands-on mastering if we can arrange it.

Are there any things that students are surprised by when you talk about mastering in the real-world, compared to how it had been introduced to them as part of a course or degree?

I think a lot of them are surprised at how simple mastering is, in its ideal form. I think there is a common misconception among students that no matter how good the mix is that they're working on, it's their job in mastering to really change it and make it something else – to put their own thumbprint on it. If, in the course of their classwork, they need to master a track 'in the box', there's often a tendency to put, say, twelve plugins on the track. Then when we evaluate the mastering process together, I'll suggest, 'now listening back, tell me what you were trying to accomplish with this plug-in'. They might say 'the track felt a little bit dull to me'. Then we check the original mix and notice that it actually wasn't all that dull to start with. 'So how did it get dull?' With a bit of further investigation, we realise it's one of other plugins in the chain that maybe widened the mix a bit but had the side effect of making the track feel a bit muffled overall. The solution lies in discovering which tools are having the necessary positive effect and eliminating those whose unwanted side-effects outweigh their benefits for that particular track. I think one of the most important and perhaps difficult concepts to learn as a new mastering engineer is 'if it isn't broken, don't fix it'. Experienced mastering engineers know and understand that every piece of gear we put in the signal path creates potential distance between the artist and the listener, so if it doesn't have to be there and it doesn't serve a purpose, it shouldn't be there.

8

'Past' masters, present beats

EXPONENTIAL SOUND STAGING AS SAMPLE-BASED (RE)MASTERING IN CONTEMPORARY HIP HOP PRACTICE

Michail Exarchos (a.k.a Stereo Mike)

INTRODUCTION

At the end of his article, 'Considering Space in Music', William Moylan poses the following questions [1]:

> How do we define the activities and states of spatial qualities as musical materials (concepts) or as ornamental embellishments within the musical texture? How do we calculate their impact on the music, their functions and significance?

His call for further 'inquiry … of how space functions in recorded music' [1] follows the proposition of a methodology and theoretical framework that consider the spatial qualities, perceived distance locations, and lateral imaging of both individual elements and the overall sound of records. In response, this chapter examines the implications of the spatial architectures that are constructed within records, in terms of their function as source material in sample-based hip hop practice. Furthermore, the study takes advantage of creative reactions to the legal and financial landscape surrounding the diminishing use of copyrighted samples (for a considerable proportion of the beat-making community), and explores contemporary approaches to the creation of sample material. The investigation focuses explicitly on the contemporary practice that involves hip hop producers creating and imbuing original source content with convincing – often vintage – phonographic qualities, to facilitate subsequent sampling processes in pursuit of a sample-based production aesthetic. The underlying hypothesis is that, unlike Moylan's pop/rock phonographic examples (e.g. The Beatles and Pink Floyd) that are founded on a track-based approach towards the creation of mix architectures, sample-based Hip Hop depends on the juxtaposition, interaction, and mixing of full *masters*. The approach leads to a form of *exponential sound staging* that sees beat-makers carefully negotiating and reshaping often multiple instances of layered master segments and, it will be argued that, this phenomenon is a defining aspect of the sample-based sonic aesthetic. As an issue that has not yet received sufficient attention, it complicates existing discourse relating to the notion of *staging,* necessitating further inquiry. The questions this study pursues, thus, are:

- How do sample-*creating*-based practitioners construct and merge spatial illusions contained within 'masters' used as source material in hip hop production?
- What are the dynamics of this interaction? In other words, how do beat-makers negotiate the dimensions of *depth*, *height,* and *width* imbued into masters as part of the creative sample-based process?
- And what is the meaning of these exponential staging strategies for the sonic narratives communicated by the end artefacts?

In order to answer these questions, the chapter deploys a bricolage methodology that combines literary and aural analysis with auto-ethnographic interpretations of creative practice. The aim is to allow for the study, respectively, of literature dealing with the notion of staging, previous hip hop discography containing relevant case studies, and creative practice functioning as an applied context.

STAGING LITERATURE AND HIP HOP SONICS

Staging is a notion that has emerged from theorisations by a number of scholars over the placement of musical elements within the perceived or virtual space of a (popular) music mix. In essence, it suggests conceptualising a music mix as a 'stage' where the placement, but also the dynamic movement and manipulation of musical elements (mediation), assumes thematic and narrative implications (meaning) for both listeners and producers. The concept was first introduced by William Moylan with a focus on the spatial implications of mediation possible within a mix [2]. Serge Lacasse explored it further, investigating the effect of textural and dynamic manipulation specifically on the voice in rock production [3]. Simon Zagorski-Thomas extended the definition to include functional and media-based staging, respectively taking into account 'the function to which the recorded output will be put' [4] and the effect of how 'particular forms of mediation associated with audio reproduction media have been used to generate meaning within the production process' [5]. Michael Holland expanded the concept to include the use of acoustic spaces captured in tracking as a form of staging mediation [6]; and Aaron Liu-Rosenbaum has been tracing musical and narrative meaning in recording studio aesthetics offering 'an ... expanded notion of staging which applies not only to the voice, but also to instruments' [7].

As staging heavily references a visual metaphor for the representation of sonic phenomena, a number of authors have developed intuitive graphical strategies to illustrate the placement, movement, and manipulation of sonic objects within contemporary music mixes. Popular examples include David Gibson's conceptualisation of mix layers as sonic objects represented in three dimensions (using a vertical/height axis for pitch/frequency, a horizontal/width axis for lateral position, and a depth axis for distance location) [8]; and Moore and Dockwray's 'sound-box' illustrations, which add 'temporal continuity' to their conceptualisation of a four-dimensional virtual performance space [9]. Moylan, however, clarifies that 'aligning pitch/frequency with elevation … is not an element of the actual spatial locations and relationships of sounds, but rather a conceptualization of vertical placement of pitch' [1]. Nicholas Cook goes beyond metaphor and considers the merits of data-driven visual representation for audio analysis, whilst warning against solely empirical or statistical readings of recordings. His position balances the promise of 'a visualization based on objective measurement [that] can act as a prompt to further critical study' with a question of whether 'empirical … approaches [can] really help us understand music as a cultural practice' [10]. Visual analogy is, therefore, widely deployed to enrich literary theorising on the spatial aspects of recordings and the meaning of staging strategies, but the pursuit of thematic, narrative or cultural implications favours metaphor over objective data representation (as a bridge between textual reification and sonic manifestations of mixing practice). As will be shown next, conceptual visualisation forms a key means of extending staging theory to cover sample-based phenomena, and the strategy will focus on illustrating how the (multi)dimensional space of full masters is (re)staged within hip hop constructs – a notion that is referred to as 'sample-staging' in the remainder of this chapter.

The central motivation behind pursuing an extension of staging theory to cover sample-based phenomena is that existing discourse uses, as the basis for the development of analytical frameworks, a binary lens focusing predominantly on two levels: that of the overall sound of a record, and that of individual sources. Moylan asserts that 'these two levels of perspective or detail are what separate the mastering engineer … and the mix engineer' [1]. But when full phonographic master segments are utilised as building blocks in sample-based composition/production, this has profound ramifications for the meaning(s) of the practice. The sample-based producer additionally assumes a *mastering* perspective, working with the overall sound stages of full masters (record segments), yet *mixing* them as individual elements within the sample-based 'collage'. Beat-making practice, therefore, does not only blur the lines between production and mixing – see, for example, Matt Shelvock [11, p. 170] – but mastering as well, necessitating a rethinking of sample-based source elements as multidimensional sonic objects.

When it comes to understanding the sample-based aesthetic from a sonic perspective, nevertheless, the heavy focus on 'musical borrowing'

[12] in hip hop literature does shift the analytical lens to surface phenomena (such as the layering and re-arrangement of musical motifs). Conversely, the chapter will attempt to demonstrate the interrelationship between staging mechanics and the essence of the aesthetic. Notable attempts that offer a useful basis to start from include Sewell's sampling typology, which categorises hip hop samples in terms of their layering/structural function [13], and Krims's notion of the 'hip-hop sublime' which acknowledges *timbre* as the essential organising factor at the heart of the music [14, pp. 41–54]. Neither of these crucial theoretical perspectives, however, cover the mixing mechanics underlying the timbral processes and – by extension – the staging phenomena responsible for the 'architectural' organisation of the respective sampled layers. Although Schloss does highlight beat-makers' 'ability to juxtapose the qualities of different recording environments', he defaults back to a *motivic* understanding when he defines 'chopping' as the 'practice of dividing a long sample into smaller pieces and then rearranging those pieces in a different order to create a new *melody*' (emphasis added) [15, p. 151]. Could this be then why Goldberg, instead, attributes the hip hop aesthetic to the 'spatial modification' inherent in the 'exploding kicks', 'echoing snares, and the sometimes terrifying sonic manipulations of DJ scratches' [16, p. 130]? The following case studies drawn from discography illuminate such mixing/staging phenomena identified in masters used as samples in hip hop production.

(ILLUSTRATING) SAMPLE-STAGING IN DISCOGRAPHY

Width, height, and media-based staging

Starting from a sample-staging strategy dealing with a practical conundrum first, the following excerpt from a recent article on low-end stereo placement, illustrates how Melba Moore's 'The Flesh Failures (Let the Sunshine In)' [17] has been (re)staged in Mos Def's 'Sunshine' [18] by Kanye West [19]:

> Hip Hop producers ... often face the problem of adding a more powerful bass element to a historic loop containing a bass part ... Kanye West solves this by applying mid/side processing to the sample, thus creating ultra wide stereo with a significant dip in low end frequencies in the middle of the image. Into this he places low bass, often only occupying the sub-bass spectrum ... the careful application of the mid/side processing allows for acceptable mono reproduction. (p. 89)

Following the textual analysis with a visual representation, Figures 8.1a and 8.1b respectively portray the sampled record's perceived stage, and the way it has been reshaped within the space of West's beat (and Superstar Dave Dar's mix):

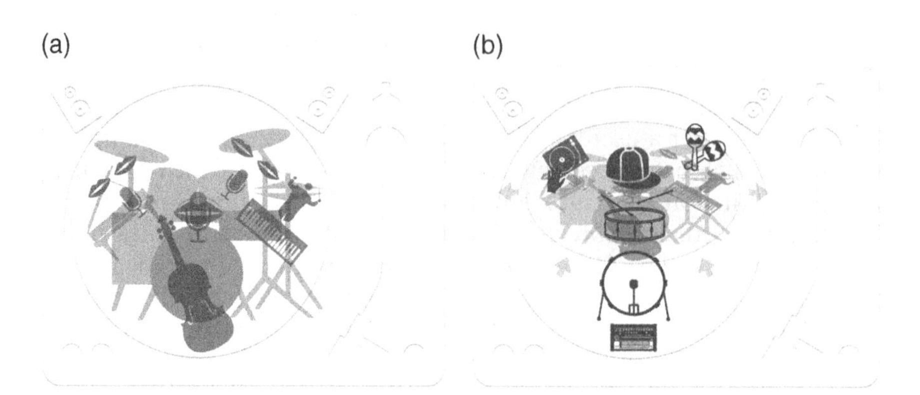

Figure 8.1 (a) A schematic representation of the perceived staging of the chorus in Melba Moore's 'The Flesh Failures (Let the Sunshine In)' and (b) its reshaping in Kanye West's production of Mos Defoo 's 'Sunshine' (the new beat elements enter at the end of the chorus, while a different segment from the original is used for the verses.

Although already notably wide – featuring a 'diagonal' [9] image with piano on the left; organ and orchestral elements on the right; lead vocals, drums and bass in the middle; and different registers of backing vocals spread both left (for low parts) and right (for high parts) – the 1970s master has been further widened on the lateral axis, but also pitched/sped up. The pitch adjustment results in a frequency shift, pushing the spectrum higher, whilst additional equalisation may have been deployed as part of the mid-side processing. Whether the processing has taken place in the beat-making stage by West, the mixing stage by Dar, or as a combination of both, the resulting sonification is equivalent to a series of (re)mastering artefacts: the weakened middle image and shifted frequency spectra may not have made sense as mastering decisions for an actual or standalone release, but in the context of the new sonic environment they function both in terms of mix architecture and, as is discussed next, in a narrative sense.

A discernible amount of vinyl crackle can be heard on the resulting introductory section of the hip hop production, which may be the result of a particular combination of record player, stylus, and vinyl record deployed, enhanced by the pitch/equalisation adjustments, or even added in post-production so as to accentuate the vintage qualities of the source. Sample-based Hip Hop has been founded upon the use of past phonographic sources, and therefore featuring the sonic past – in an audible, exaggerated, or even artificial sense – within its contemporary artefacts has become part and parcel of its aesthetic [20]. Going beyond the functional rationale, thus, it can be argued that the combination of lateral, vertical, and media-based staging for the sampled record has thematic and narrative implications, too. The layers of old elements (sample) and new additions (Mos Def's rap, Kanye West's drum hits and sub bass) are communicated as *distinct streams* via their vintage-contemporary sonic signature binaries. The striking spatial staging enhances the effect and although it may have been initially conceived of as a pragmatic strategy, creating mix space for the new elements (in the lateral and vertical sense), it remains congruent to the sonic interplay of 'past' and 'present' – a dynamic which overwhelmingly characterises the sample-based aesthetic. Moylan asks in relation to image width: 'Does the size of the source establish a context or reference for other sources?' [1]. This example illustrates that, in a sample-based context, the phonographic object *does* indeed, and it does so in a stylistically-defining sense: its *poly*-dimensional (not just spatial, but also media-based) staging utterances establish both a functional (mix-architectural) and narrative (communicative of the sonic past) referential canvas, against which the new elements may be positioned.

The idea of 'sonic narrative' is used here in Liu-Rosenbaum's sense of the word 'where changes in spatial or timbral qualities of an excerpt could conceivably convey a sense of goal-oriented movement' [7]. It is also worth noting that this particular version of the song sampled from the 1970 release is difficult to source beyond second-hand vinyl, and not readily accessible from streaming or download services. Therefore, it is safe to assume that what we are hearing on 'Sunshine' is a unique sampling occurrence of particular variables (equipment and vinyl record) that have taken place in West's process. As a result, the type of vinyl noise that is audible and the specific rhythm of its manifestation, become unique signifiers of the sampling ephemeron on hand – a processual 'footprint' of sorts. Mark Fisher explains in his article, 'The Metaphysics of Crackle' [21]:

> Crackle unsettles the very distinction between surface and depth, between background and foreground… The surface noise of the sample unsettles the illusion of presence in at least two ways: first, temporally, by alerting us to the fact that what we are listening to is a phonographic revenant; and second, ontologically, by introducing the technical frame, the material pre-condition of the recording, on the level of content … we are witnessing a captured slice of the past irrupting into the present. (pp. 48–49)

Depth/proximity

The case study above demonstrates how the lateral and vertical dimensions of width and height are restaged when a full phonographic master is manipulated in the context of a sample-based composition, and how media-based staging can further 'stamp' and accentuate the narrative ramifications. Of course, relative volume reduction, which often occurs as a result of the source's recontextualisation, presents implications also for the depth or distance perception of the sample's position in the new mix architecture. Alongside the control of ambience/

reverberation and high frequency content, volume manipulation is one of the three essential mediation strategies that engineers deploy to communicate the proximity of a source (Moore and Dockwray often refer to this dimension as 'prominence' [9, p. 219]. To explore a dedicated case study, Figures 8.2a and 8.2b (taken from a recent article entitled 'Sample Magic', which deals with Hip Hop's unique recipe for phonographic juxtaposition) illustrate the effect of perceived depth on the interaction of sampled and new beat-making elements [22].

The sample in question is 'A Theme for L.A.'s Team' by the Thomas Bell Orchestra featuring Doc Severinsen [23], used as a source within Marley Marl's production 'Musika', featuring KRS-One [24]. Although the source track's pitch/frequency spectrum is shifted – again – slightly higher, perhaps the most striking effect here is how the rich construction of a multi-layered depth illusion on the original, becomes a discursive feature in Marl' sample-based juxtaposition. Toby Seay provides a fascinating analysis of the textural and spatial characteristics of Philadelphia's Sigma Sound productions, and the sampled track in question subscribes to these [25]. Using Seay's technical findings, the following excerpt from 'Sample Magic' highlights the respective staging interactions [22].

> … the original recording … carries with it a number of sonic illusions: … superimposed acoustic spaces (echo chamber) upon the actual spaces captured owing to reflections during recording; [and] re-amplified instrumental sections (and their reflections) captured owing to bleed during overdubbing … The strings are very rich in texture as a result of the overdubbing approach, occupying a wide stereo image and implied depth (illusion), which is typical of the Philly sound …

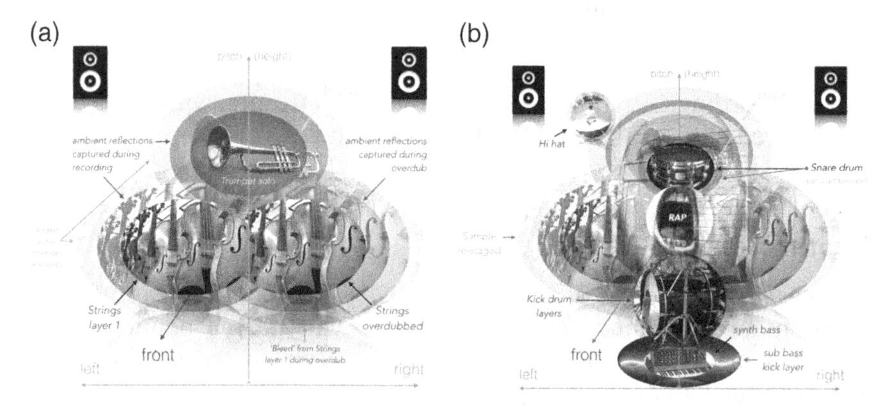

Figure 8.2 (a) A schematic representation of the sonic space occupied by the sample from 'A Theme for L. A.'s Team'. (b) A schematic representation of the exponential staging illusions on track 'Musika'.

It is difficult to discern whether Marl has added any further reverberation to the sample, therefore superimposing yet another space upon the 1979 spatial illusions, but this – again – is common sample-based hip hop practice aiming to 'glue' all of the borrowed elements within a new implied 'stage'. The low-frequency sounds (kick drum, sub-bass, and bass synthesiser) come across as completely 'dry' ... which places them rather 'forward' in the staging illusion'. (p. 44)

For a hip hop song with mystical thematic allusions (KRS-One raps: 'Marley Marl on the musika, KRS on dem lyrics da/On the side I teach meta-ta-ta-physica' [24]), the dynamic discourse taking place on the depth axis of 'Musica' indeed suits Reynolds's characterisation of the 'sample collage' as 'the musical art of ghost co-ordination and ghost arrangement' [26, pp. 313–314]. Marl's discreet negotiation of the 'Philly' sound's textural and spatial vintage signatures allow him to juxtapose his contemporary sounds (and KRS-One's rap) against a sonic object that feels like an 'echo' of a past perspective – painting, so to speak, his present (pun intended) boom bap sonics against a three-dimensional canvas that communicates the past. Of course, the perception of proximity, distance, or depth is a negotiation of sonic perspective on multiple levels – for example, between listener and source, and between source and other sources. 'The listener can be ... drawn into becoming part of the "story" (music) or observing the "story" (music) from some distance' [1]. A sample-based composition, however, can additionally carry a story within a story, providing a meta-vantage point so to speak, as it presents the possibility of featuring a record within a record. But how does one go about constructing such staging interactions within newly created source material?

(CONSTRUCTING) SAMPLE-STAGING INCREATIVE PRACTICE

The autonomous sonic object

Two practice-based scenarios are reviewed next, where a sample-based composition has been created out of originally produced source/sampling content. Excerpts from the accompanying reflective journal of the process are analysed as a means to reflexively build upon developing interpretations of the practice. The first practice-based case study concerns the manipulation of a single instrumental element, demonstrating how phonographic processes related to mix staging and mastering help transcend its perceived quality from a mere 'recording', to a 'record', in the context of a sample-based creative process [27]:

> I came across a grand piano recording I had self-captured about a year ago. I had used two Neumann U 87 [microphones] over the sound holes of the piano and a stereo ribbon AEA R88 Mk2

> facing the piano lid from some distance, giving me both a solid, clear stereo image of the instrument, as well as a warmer, mellow room tone that I could blend in to change its staging. Reacting to the source, I quickly reached out for a vintage (spring) reverb emulation and applied it only to the close mics. I was aiming for a more distant tone and I also wanted to make the piano more three-dimensional on the Z [depth] axis … I guess I was making it feel *further away,* both in terms of physical illusion but also conceptually. I was chasing that phonographic 'otherness', quite consciously attempting to make it feel more mysterious. (emphasis in original)

In terms of creative intent for the piano source, the rationale and process relayed in the excerpt mirror Reynolds's characterisation of the sample-collage as 'ghostly', and the sample in 'Musika' as a distinct, three-dimensional sonic object. It is clear that both the recording techniques and the spatial mixing decisions were aimed at creating a sonic object of notable depth and width. The following reflection demonstrates how the sample was 'distanced' even further through a series of mastering processes and conscious media-based staging choices [27]:

> Synchronising the sampling drum machine to the DAW multitrack playing back the piano tracks, I loaded it up with banks of drum samples and sampled vinyl crackle (that I often capture from the end of vinyl LPs). I wanted to distance the piano even further. So, I programmed a combination of vinyl noise samples that made the four-bar piano patterns running in parallel feel like they had been lifted off vinyl. I scanned the 35-minute recording of the piano improvisation for inspiring moments and decided to give the piano mix itself some 'colour' reminiscent of past recording eras … I applied multitrack tape machine emulation to the individual looping piano subgroups and then ran the full piano mix–including the reverb returns–through a mastering equalizer, a mix-bus compressor, and both master tape recorder and vinyl cutting lathe emulations.

Two essential strategies can be extracted from this process, which aim at infusing the source 'master' with a phonographic footprint and distancing it enough against new elements within the sample-based context: first, the selection and layering of convincing vinyl-crackle patterns and textures placed over the instrumental source – 'there is … no myth without a recording surface which both refers to a (lost) presence and blocks us from attaining it', writes Fisher; second, the colouring of the 'master' via the simulation of a vintage-informed mixing and mastering signal flow, reminiscent of 'a time when recording technology had developed sufficiently to achieve a kind of sepia effect …' [21, p. 49]. Inevitably, the distancing effect pursued is also related to ideas of perceived authenticity and authority tied to the sample-based aesthetic. Zagorski-Thomas elaborates: 'Playing, sampling and pressing a performance to vinyl as part of the creative process were important statements of authenticity within the Bristol sound of artists'; while for British indie rock in the early to mid-1990s 'the notion of authority stems from ... the sound of analogue tape and valve or tube amplifiers ... used to *distance* the sound of Oasis ... from the sound of the 1980s' (emphasis added) [5].

The first of the two tracks showcased in Video 1 [28] corresponds to the end sample-based artefact built from the piano source production and sonifies the interaction between the 'constructed' sample and the new beat elements. A noteworthy utterance created by the 'chopping' process performed upon the source master highlights yet another important characteristic: at the fourth bar of every A-section four-bar loop repeat, a reverberant, 'ghostly' texture can be heard, rhythmically interrupting the main piano part on the off beats. This is the result of a motif performed on the pads of the sampling drum machine, some of which have inadvertently been assigned with soundbites of just reverb decay, as opposed to actual piano notes or chords. The monophonic, legato-style mode enabled on the sampler (a staple of the boom bap approach – for more on this, see [29, pp. 36–43]) means that moments of fully staged 'architectures' from the piano 'master' are played as if they were notes on a monophonic synthesiser, each new segment muting the previous one still playing. This performing mode – in combination with other programming and swing quantisation characteristics unique to particular sampling drum machines (see [29] for more on this as well) – results in striking *staging rhythms*. These could be described as rhythmical shifts between momentary, or at least short, staging architectures 'frozen in time' on the micro-structural level. Holland cites Lacasse to describe the effect in a macro-structural sense [6]:

> In Lacasse's terms, the use of multiple reverberant signatures as the track's narrative develops ... are directly related to the piece's structure ... the changes in reverberant character function as an example of diachronic contrast, as the various levels of reverberation are experienced relative to others unfolding within the frame of the recording.

Moylan applies the idea to shifts in lateral imaging, elaborating that 'patterns of locations ... and the repetitions and alterations of these patterns can create musical interest just as the patterns of changing pitches, timbres or harmonies' [1]. In that sense, *staging rhythms* become a unique musical utterance in sample-based styles, with a narrative-structural function; but the *diachronic contrasts* unfold on a micro scale and within the time domain of the 'loop'. Of course, the effect can take an exponential character when the juxtaposition of momentary 'stages' involves multiple sources, rather than multiple sections from the same source.

The multitrack sonic object

The second practice-based case study illustrates the construction of an original multitrack source for subsequent sampling, highlighting a layered, developmental approach to the creation of a number of staging manifestations. The source production in this case has been built by overdubbing acoustic drums, electric bass and guitar, Nord organ, and Fender Rhodes electric piano, followed by the juxtaposition of vocals taken from another source production. A guide beat was also programmed on a sampling drum-machine in sync with the developing multitrack, to enable an ongoing evaluation of the evolving 'samples' within a sense of the end context. The following journal excerpt describes how the instrumental performances were recorded with a range of spatial enhancements and timbral shaping gradually committed. As an archiving strategy, the track/file names used during recording disclose the range of effects – serially – applied [27]:

> A [track] name such as 'Tele Wah Stone 63 55 Neve Tape' indicates, for example, a Telecaster guitar, played through a Cry Baby Wah Wah pedal, into an Electro Harmonix Small Stone phaser, and finally a Boss Fender '63 spring reverb pedal. The remainder of the name relates to software emulations [also committed during tracking, such as]: a Fender '55 Tweed Deluxe amplifier, a Neve Preamp, and a Studer A800 multichannel tape recorder … [Performing through] both the pedal reverb being tracked and an AKG BX 20 spring reverb emulation [used only as foldback] inspired the performance, but I could also envision the *staging* of the guitar in the final mix architecture, whilst making complimentary timbral and musical decisions … I then reached for my Lakland Jazz bass with the LaBella flats [strings] and played very close to the neck (emulating Aston 'Family Man' Barrett's reggae tone). To compliment the resulting tone, I run the signal 'hot' through a tube preamp, boosted the low frequencies slightly, and hit an optical tube compressor followed by a VCA [compressor] shaving off the peaks … [The end result was] tracked through a Studer tape emulation, effectively mimicking a complete classic signal flow for the referenced era.

The tracking of the guitar and bass highlight the conscious timbre-shaping decisions committed, on the one hand, ensuring a complimentary tone to sonics gradually being recorded (functional aesthetic) and, on the other, communicating stylistic/era signatures of a non-specific, yet vintage quality (narrative aesthetic). A similar approach was deployed when tracking the keyboards, while the vocal parts were recorded with a Shure 520DX 'Green Bullet' microphone slightly saturated through guitar-amplifier and tape-recorder emulations (a typical blues harp recording signal flow). As Zagorski-Thomas explains: 'The other common reason for using media based staging in record production is to evoke the sound of a particular (or more commonly just a vague) historical period' [5]. These creative strategies are consistent with an aspect of Justin William's intertextual argument on musical borrowing in Hip Hop. He points out that we may be moving towards a focus on sampling *stylistic topics*, where 'generic signifiers ... become more important than the actual identity of the sample' [12, p. 201]. This argument can extend beyond the musical and the abstract, however, to the materially sonic and concrete, as the following excerpt also illustrates [27]:

> Once I found a one-bar [drums] phrase that was sitting well ... I looped it around with all mics active and started mixing it. Auditioning it with and without the beat running in sync, I tried to decide which overheads I should use (I tracked multiple options, so that I could push the drum aesthetic toward different 'eras') ... The drums had been recorded through my choice of hardware preamps with some compression and EQ already committed ... [I] run a parallel send of the whole drum mix into a pumping VCA compressor, followed by a passive vintage EQ [both emulations]. The highlighted recorded ambience, enhanced air, and tonal glue achieved by the New-York-style parallel layer gave the drums a 'phonographic' quality that was complimentary to the programmed drum hits, providing a sense of 'glue' and achieving that live/non-live fusion that felt stylistically relevant. Taking the beat out, I was surprised by how few of the drum mics I actually needed for the drum-layering effect to work. I ended up with only the stereo overheads and a little kick support ... The overall mix-bus was going through mastering equalization, mix-bus compression, and master tape emulations, [so] I had been reviewing and working on the drum mix with the 'hindsight' of auditioning it in this more finalised (end-format) fashion.

The journal excerpt indicates that the drums had been recorded prior to the multitrack subscribing to a 'stylistic topic', using a strategy that deployed multiple microphone choices/techniques, which in turn allowed a degree of sonic-signature shaping flexibility later in the process. The drum production approach demonstrates particular attention paid to expanding the captured ambient characteristics and testing the interaction between the acoustic sonic and the programmed beat. It can be argued that any convincing phonographic sample qualities achieved were the result of the source operating as a blended, yet distinct sonic 'world' or mix architecture contained underneath the beat – courtesy of complimentary staging decisions; shared colorations pertaining to deliberate signal flow choices; a conceptual 'inhabiting' of an aesthetic/era that drove both musical and tonal decisions; and the 'glue' achieved by both tracking and mix-bus processing choices. The mix-buss equalisation, compression, and tape emulation gave the underlying master of the recorded performances a tracked-to-a-particular-recording-medium coherence, which both unify it as a mix of performed elements *and* separate it as a phonographic entity from the – new – beat (elements). Figure 8.3 features a collage of photographs depicting the recording sessions responsible for the production of the constructed 'sample'. The following section discusses its use for, and incorporation into, the second sample-based composition under examination.

Exponential staging in sample-creating-based hip hop practice

In a similar vain to the sampling and chopping processes described for the piano-based production, the beat built out of the multitrack has been constructed by isolating multiple 'staged' moments from the lengthy (approximately 25-minute) blues-funk 'jam' deconstructed above; pitching/slowing down the samples by -1.63 semitones (about 8.25 beats per minute); rhythmically performing various combinations of the resulting momentary 'masters' using the sampling drum-machine's pads; and further manipulating the segments using the sampler's internal mix functionality. It will be useful to provide some summarised definitions arising from Sewell's sampling typology before proceeding to deconstruct the sample layers used and their sonic interaction within the sample-based production. Sewell categorises samples according to their structural function in hip hop productions and splits them into:

- 'Percussion-only' structural types containing 'sampled drums [that] are looped throughout the new track';
- 'intact' structural types that include 'every element from the source material, usually drums and at least one other instrumental line';
- 'non-percussion' structural types which are 'very similar to an intact structural sample, except that [they do] not contain sampled drums'; and
- 'aggregate' structural types which consist of component 'layers … sampled from different sources' or '*different parts of the same source*' (emphasis added).

To these, she adds 'surface' and 'lyric' sample types (the former, with further subcategorisations), which have a more intermittent/ornamental layering function and which can be delineated from each other by their intended lyrical intelligibility [13, pp. 36–54].

The final piece's main A, B, and C sections are, thus, created predominantly out of *percussion-only* and *non-percussion* layers, while the breakdown section uses an *aggregate* structure made out of layering multiple component elements sampled from the source multitrack – see Video 1 [28]; the vocal samples could be described as functioning either as a *surface* or *lyric* type. Of course, access to the

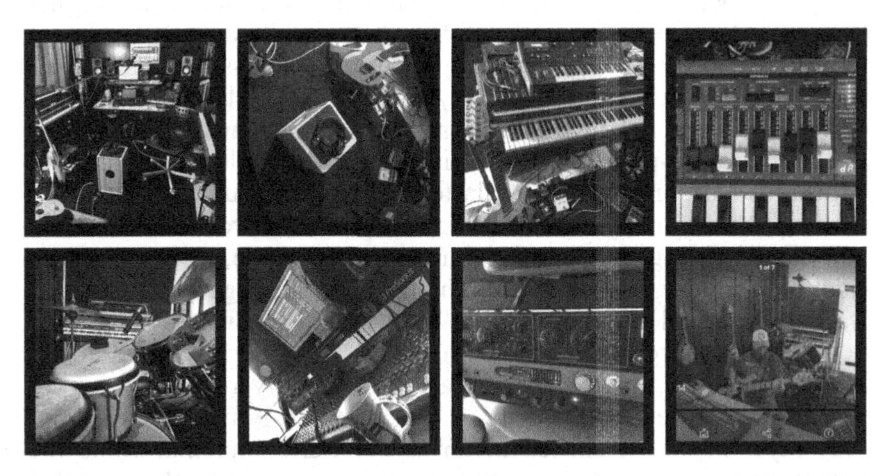

Figure 8.3 A collage of photographs from the recording sessions responsible for the production of the constructed multitrack 'sample'.

component layers is ensured by having created the multitrack production oneself, which differs from the possibilities presented by sampling copyrighted, previously-released phonographic material. The rationale behind working with a range of structural types here, is fuelled by wanting to test the limitations of access to near-*intact* scenarios, the compositional freedom presented by access to *aggregate* components, but also – importantly – the sonic implications of either approach. As DJ Bobcat explains in Sewell: 'A lot of times when somebody samples a bass and a guitar riff or a horn from the same song, it's because sonically they're the same. They're taking it because they already sound the same' [13, p. 44]. But could this sonic 'sameness' be further unpacked and is it the result of an underlying 'staging harmony' (i.e. a spatial architecture to which all the component layers adhere, even when isolated)? To illustrate the underlying architecture, Figure 8.4a schematically represents the staging of the multitrack used as the foundation for section A of the sample-based production. Figure 8.4b represents four component layers extracted from different sections of the source multitrack (but retaining their staging placements): a non-percussion layer that includes Rhodes piano, bass, lead, and rhythm guitar (top left); and three percussion-only groupings of cajon-and-bongos (top right), shaker-and-tambourine (bottom left), and drums (bottom right). Note that under the representation of each layer there are opacities overlaid of the missing instruments' positions in the implied mix architecture (illustrating the staging 'harmony').

In order to reinforce the bass part, a matching bass-only layer has also been 'chopped', equalised, and layered beneath the resulting structure. The perceived effect is of a louder and more prominent bass placement in the main non-percussion layer (the isolated bass layer enables separate equalisation and therefore a complimentary reinforcement of the otherwise harder to access bass sonic in the almost intact, non-percussion layer). During the last (eighth) bar of every A-section, two two-beat non-percussion segments (equivalent to the main non-percussion layer in terms of included elements, but with added organ parts – the latter of which features a glissando) interrupt the main layer on beats one and three to provide a variation and climax (using the sampler's monophonic mode, as in the piano example). Figure 8.4c represents the resulting aggregate structure, as well as the vocal sample juxtaposition, plus the new beat additions (kick drum, snare drum, and hi-hat). Note the sepia colour added representing the vinyl crackle that has been layered over the aggregate structure. Additionally, the cajon-and-bongos percussion-only layer has been shifted in terms of lateral imaging, while the arrow pointing down from the new snare towards the sampled drums' snare represents side-chain compression dialed in to reduce the latter's volume on every new snare hit – the strategy aiming at both a balancing *and* rhythmic interaction between the percussive elements, thus creating complementarity between two initially unrelated samples/drum sonics. A more dynamic visual representation of the staging phenomena is showcased in Video 1 [28]. Finally, the breakdown section is based on an aggregate structure made exclusively out of component layers, which also include individual Rhodes and lead guitar samples (pitched approximately ten semitones up from the original, which results in an octave interval over the aggregate structure, and twice the tempo).

The aggregate way of working facilitates a refined (re)staging strategy for the component layers. In a sense, the already staged individual, non-percussion, or percussion-only layers are (re)mixed as elements within the sampler's mixing environment. For example, four send effects are deployed (short and long reverb, synced tape delay, and a parallel VCA compressor), which allow sharing/groupings of ambient spaces, mutual rhythmic effects, and common dynamic movement. A number of layers are also individually balanced, equalised/filtered, and compressed to negotiate the available 'space' more effectively in the resulting sample-based stage.

The intact structural approach – which is more representative of phonographic sampling as with West's and Marl's examples analysed – implies a committed stage that can only be renegotiated through a form of (re)mastering within the sample-based context. Most pragmatic sample-based creative scenarios fall somewhere between the intact and the aggregate extremes. The added layers (drum hits, etc.) have to interact in a congruent sense in terms of spectra, depth, and width against the three-dimensional frame(s) presented by intact, percussive, or near-intact samples. One of the methods for enhancing this interaction is by integrating side-chain and parallel dynamic processing between the samples, the overall mix, and additional beat elements (a strategy

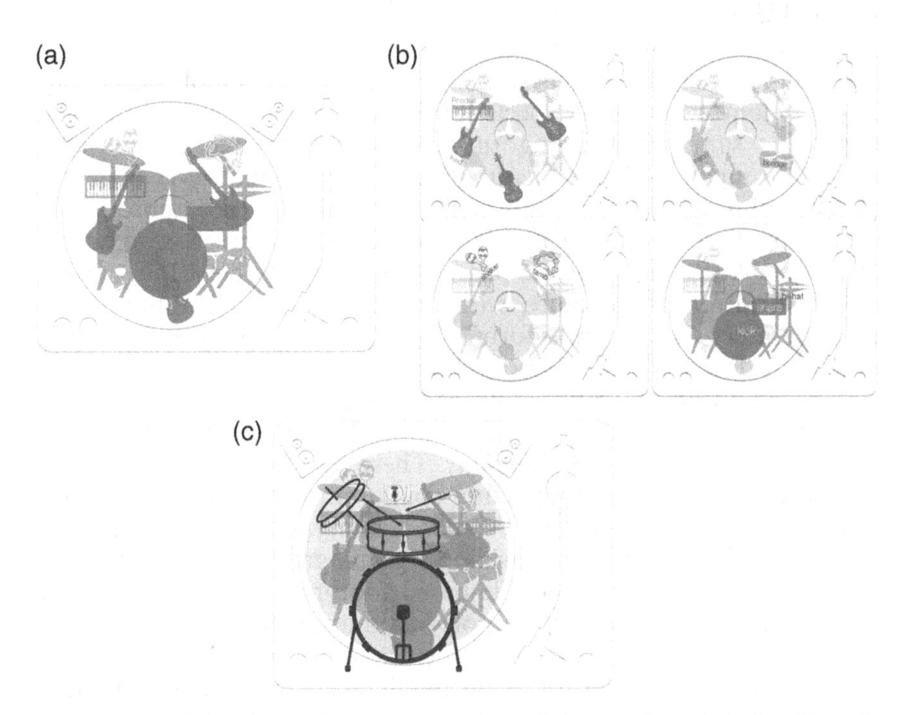

Figure 8.4 (a) A schematic representation of the staging of the multitrack used as the foundation for section A of the sample-based production. (b) A schematic representation of the four component layers extracted from different sections of the source multitrack: a non-percussion layer that includes Rhodes piano, bass, lead, and rhythm guitar (top left); and three percussion-only groupings of cajon-and-bongos (top right), shaker-and-tambourine (bottom left), and drums (bottom right). Note that under each layer's representation there are opacities overlaid of the missing instruments' positions in the implied, original mix architecture. (c) A schematic representation of the resulting aggregate structure, plus the new beat additions: the sepia colour added represents the vinyl crackle that has been layered over the aggregate structure; the cajon-and-bongos percussion-only layer has been shifted to the right in terms of lateral imaging, while the arrow pointing down from the new snare and towards the sampled drums' snare represents side-chain compression applied upon the drum layer.

indeed championed in this production: a compressor inserted on the overall sampler mix is triggered by the new kick drum sample, while multiple elements are routed to the sampler's parallel VCA-style compression bus). The artform's balance hangs in the tightrope between contrast communicated by previously-constructed and new elements, and integration achieved through spatial, timbral, rhythmical, and dynamic re-contextualisation.

CONCLUSION

The chapter has illustrated staging mechanics in sample-based hip hop phenomena across a spectrum of creative contexts: from phonographic sampling utilising full, previously-released master segments, through to sample-*creating*-based practices that – via extended access to multitrack elements – facilitate a multi-layered approach to the shaping, control, and manipulation of the source's staging dimensions. At the heart of the process, lies a sonic object that carries an extended mix architecture, with the potential to not only provide raw sonic content for this form of material composition (sample-based Hip Hop), but also a poly-dimensional referential canvas that can communicate narrative notions, such as representations of the past, diachronic contrasts, and striking genre-defining utterances such as syncopated *staging rhythms*. These perceptual effects depend on the construction of convincing spatial and media-based staging artefacts for the hip hop practitioner creating their own source material; and these, in turn, translate to phonographic signatures contributing to an authentic sample-based footprint. It can be deduced that the essential aesthetic of sample-based music forms – and the key differentiation between a sonic element and a *sample* at the heart of their processes – can be traced in this interaction with staged sonic objects carrying *markers of phonographic process*. It is a manifestation of a phonographic process interacting with previously (even if very recently) committed phonographic processes. This kind of layering can, therefore, become exponential, and sample-based music forms deal not with mixing elements, but with mixing and manipulating full 'masters'.

REFERENCES

1. ———. *Considering Space in Music. Journal on the Art of Record Production*, Vol. *4*, (2009).
2. Moylan, W. *Understanding and Crafting the Mix: The Art of Recording*. 3rd edn. Focal Press, Oxon, (2014) [1992].
3. Lacasse, S. *'Listen to My Voice:' The Evocative Power of Vocal Staging in Recorded Rock Music and Other Forms of Vocal Expression* [Ph.D. Thesis]. University of Liverpool, (2000).
4. ———. The Stadium in your Bedroom: Functional Staging, Authenticity and the Audience-led Aesthetic in Record Production, *Popular Music*, Vol. *29*, No. 2, (2010), pp. 251–266.
5. Zagorski-Thomas, S. *The Medium in the Message: Phonographic Staging Techniques that Utilize the Sonic Characteristics of Reproduction Media, Journal of the Art of Record Production*, Vol. *4*, (2009).
6. Holland, M. *'Rock Production and Staging in Non-studio Spaces: Presentations of Space in Left Or Right's Buzzy'. Journal on the Art of Record Production*, Vol. *8*, (2013).

7. Liu-Rosenbaum, A. *The Meaning in the Mix: Tracing a Sonic Narrative in 'When The Levee Breaks'. Journal on the Art of Record Production*, Vol. *7*, (2012).
8. Gibson, D. *The Art of Mixing: A Visual Guide to Recording, Engineering, and Production*, 2nd edn. Course Technology, Boston, MA, (2008).
9. Moore, A. F. and Dockwray, R. The Establishment of the Virtual Performance Space in Rock, *Twentieth-Century Music*, Vol. *5*, No. 2, (2008), pp. 219–241.
10. Cook, N. Methods for Analysing Recordings, In: Cook, N., Clarke, E., Leech-Wilkinson, D. and Rink, J., editors. *The Cambridge Companion to Recorded Music*. Cambridge University Press, Cambridge, (2009), pp. 236–242.
11. Shelvock, M. Groove and the Grid: Mixing Contemporary Hip Hop, In: Hepworth-Sawyer R. and Hodgson, J., editors. *Perspectives on Music Production: Mixing Music*. Routledge, New York, (2017), pp. 170–187.
12. Williams, J. A. Theoretical Approaches to Quotation in Hip-Hop Recordings, *Contemporary Music Review*, Vol. *33*, No. 2, (2014), pp. 188–209.
13. Sewell, A. *A Typology of Sampling in Hip-Hop* [Unpublished Ph.D. Thesis]. Indiana University, (2013).
14. Krims, A. *Rap Music and the Poetics of Identity*, Cambridge University Press, Cambridge, (2000).
15. Schloss, J. G. *Making Beats: The Art of Sampled-Based Hip-Hop*, 2nd edn. Wesleyan University Press, Middletown, (2014).
16. Goldberg, D. A. M. The Scratch is Hip-Hop: Appropriating the Phonographic Medium, In: Eglash, R., Croissant, J. L., Di Chiro, G. and Fouché, R., editors. *Appropriating Technology: Vernacular Science and Social Power*. University of Minnesota Press, Minneapolis, (2004), pp. 107–144.
17. Moore, M. *The Flesh Failures (Let The Sunshine In), Living To Give [Vinyl LP], Mercury*, (1970).
18. Mos Def. Sunshine, *The New Danger [CD], Geffen Records*, Island Records Group, (2004).
19. Exarchos, M. and Skinner, G. Bass-the Wider Frontier: Low-end Stereo Placement for Headphone Listening, In: Gullo, J-O., editor. *Proceedings of the 12th Art of Record Production Conference – Mono: Stereo: Multi – Stockholm 2017*, Us-AB, Stockholm, (2019), pp. 87–104.
20. Exarchos, M. (Re)Engineering the Cultural Object: Sonic Pasts in Hip-Hop's Future, In: Hepworth-Sawyer, R., Hodgson, J., Paterson, J. and Toulson, R., editors. *Innovation in Music: Performance, Production, Technology, and Business*. Routledge, New York, (2019), pp. 437–454.
21. Fisher, M. The Metaphysics of Crackle: Afrofuturism and Hauntology, *Dancecult: Journal of Electronic Dance Music Culture*, Vol. *5*, No. 2, (2013), pp. 42–55.
22. Exarchos, M. Sample Magic: (Conjuring) Phonographic Ghosts and Meta-illusions in Contemporary Hip-hop Production, *Popular Music*, Vol. *38*, No. 1, (2019), pp. 33–59.

23. Thomas Bell Orchestra featuring Doc Severinsen. *A Theme for L.A.'s Team, The Fish That Saved Pittsburgh [CD]*, Lorimar Records, (1979).
24. KRS-One & Marley Marl. *Musika, Hip Hop Lives [CD]*, Koch Records, (2007).
25. Seay, T. *Capturing that Philadelphia Sound: A Technical Exploration of Sigma Sound Studios*, *Journal on the Art of Record Production*, Vol. 6, (2012).
26. Reynolds, S. *Retromania: Pop Culture's Addiction to its Own Past*, Faber and Faber, London, (2012).
27. Exarchos, M. *Insights: A Research Journal*, Unpublished, (2020).
28. Stereo Mike. 'Past' Masters, Present Beats. Accessed January2020 from https://youtu.be/wZiy8FU0cko.
29. Exarchos, M. Boom Bap Ex Machina: Hip-Hop Aesthetics and the Akai MPC. In: Hepworth-Sawyer, R., Hodgson, J. and Marrington M., editors. *Perspectives on Music Production: Producing Music*, Routledge, New York, (2019), pp. 32–51.

9

Remastering Sunnyboys

Stephen Bruel

INTRODUCTION

Since the inception of the compact disc format there has been a push by record companies to reissue digitally remastered versions of iconic studio albums in various digital formats [2]. Albums originally released on vinyl from artists including The Beatles, Jimi Hendrix, Abba, The Rolling Stones, and Led Zeppelin have all been digitally remastered and rereleased through their respective record companies [3–5]. This phenomenon has led to discussion and research into the elements of remastering including increased perceived loudness (made possible through the technological aspects of the digital CD format) and diminished dynamic range, which has been well documented under the terms 'loudness wars' and 'hyper compression' [3,6–8]. Building upon the existent research regarding the systematic aspects of mastering [9,10] and the aesthetic decision-making process of remastering [11,12], this study explores the social and cultural implications surrounding remastering practice, in particular the digital replication of an existing analogue musical artefact as it applied in a case study featuring iconic Australian band Sunnyboys.

REMASTERING DEFINITION

Remastering is the practice of manipulating older recordings to make them sound optimal using modern playback systems and evolved through the introduction of the digital CD as a replacement for the analogue vinyl record format [10]. Initially record companies simply transferred or stamped the original master tape used for vinyl reproduction on to the CD with minimal intervention believing this was adequate [13]. However, as digital modern mastered recordings evolved alongside digital playback devices to produce a sonic quality consisting of a stronger emphasis on EQ and dynamics to optimise the format, early digital remasters of older recordings appeared to be lacking in these areas. As a result, remastering developed into the practice of using modern digital tools on older recordings to make them sound less lacking in these areas and sit more comfortably sonically beside the modern mastered ones. Remastering could therefore be described as the practice of creating a new digital replica or creative work from an existing analogue music artefact. When we consider the original musical artefact may be rich in musical history with an associated sense of cultural heritage, the amount and type of manipulation undertaken to create the digital replica becomes of crucial importance, particularly if the aim is to maintain the context, meaning, and significance of the original work.

CULTURAL HERITAGE

Within the fields of rock and contemporary music there appears to be an initial challenge of associating and classifying the significance and impact of cultural heritage using traditional definitions. These traditional definitions include the idea that the cultural artefact contains 'representations of custom, tradition and place that coalesce within the cultural memory of a particular national or regional context and fundamentally contribute to the shaping of the latter's collective identity' [14, p. 476–477]. The mass-produced commercial properties of rock music appear to be in contrast to these traditional definitions, however it can be argued that with regards to an ageing baby-boomer generation, rock and contemporary popular music is embedded within its cultural memory and generational identity [14]. This embedded cultural memory appears to be of a constructive nature and have a positive value as the preservation of culture (both material and intangible) has been described as a benefit to the local community [15]. An example of preserving material or tangible popular music culture was the recent 'listed building' status afforded to the pedestrian crossing at Abbey Road (immortalised on the cover of The Beatles album of the same name) by the official authorities [16]. Furthermore, it is the local communities and cities that both benefit from the preservation of cultural heritage and produce, validate, and contribute popular music heritage to national attention and status [17].

However, Hennessey's research into the 'transformation of intangible expression into digital cultural heritage', in particular the digitisation of 40-year old ethnographic archives detailing oral traditions and culture of the Indigenous Athapaskan community of the Yukon Territory in Canada, found that this process 'illuminated tensions in the transformation of intangible expression into digital cultural heritage' [18, p. 33]. Hennessey mentioned the key themes of cultural representation, ownership, and copyright as being central to these discussions and tensions. It is therefore plausible to suggest that the production of a digitally remastered cultural representation of an existing analogue popular music recording can also potentially 'illuminate tensions' within the associated community regarding its appropriateness as a cultural representation. For example, reference was made to the remastering engineers who worked on the 2009 Beatles remastered CD release as having 'been bestowed with the onerous task of sonically cleaning what is surely not just pop music but cultural heritage', thus highlighting the significance to the community of the digital cultural representation to be made [19, p. 2]. In the context of this research 'cultural heritage' is therefore defined as a positive value or sense of meaning attributed towards a musical artefact by members of a community, and this study will explore whether this cultural meaning and heritage is potentially impacted through the creation of a digital cultural representation. In particular, this research will focus on the Sydney-based 'Guitar-Pop' subculture of the early 1980s, described by Easton as emulating a rising sense of Australiana where 'clean, jangly guitars soundtrack accented vocals that can celebrate or sardonically mock the extremes of Australian culture' [20, p. 45]. Consisting of bands including The Church, Hoodoo Gurus, The Cockroaches and Ups and Downs, this research examined the perception of cultural heritage bestowed by the community upon the 'Guitar-Pop' band Sunnyboys and their 1981 eponymous vinyl release.

With reference to the creation of a potentially different sonic and cultural digital representation of an original music artefact, it is perhaps important at this stage to introduce the concept of remixing. Often confused with remastering, remixing involves using digital technology to extrapolate sections and elements of an original music artefact in order to produce a new creative work that often bears little resemblance to the original musical artefact [21]. In Bennett's study into remastering, he revealed a view from industry professionals that remixing alters the meaning of the original work and interferes with cultural heritage, identity, and significance associated with the original work [14]. Furthermore, remastering was seen as primarily a restoration service used to enhance the perceived cultural value and meaning of an original music artefact by making a digital replica that is more assessable to society through modern playback devices and with less noise and other sonic interference [14]. Artist Jeff Lynne's decision to remix as opposed to remaster existing released songs from his back catalogue appears to challenge a need or desire to maintain any sense of cultural heritage that may have been associated with the original recording [22]. Additionally, Lynne's choice to re-record and re-release various songs from his back catalogue, completely removing the original musical artefact from the production process, appears to eliminate perceived cultural heritage aligned with the original recording altogether [22]. The desire by society and artists to maintain a sense of cultural heritage and meaning associated with an original musical artefact through its digital cultural representation and replication is therefore not conclusive.

METHODS

This research employed both quantitative and qualitative methods. The quantitative methods used closely followed the audio data analysis techniques [23] used to compare The Beatles re-releases. This involved gaining an initial comparison of amplitude and shape through the graphical representation of the audio waveforms within the Pro Tools DAW environment [23]. Digital audio files were loaded into the TT Dynamic Range Offline Meter 1.1 digital audio plug in from Pleasurize Music Foundation to gain peak level readings and root mean squared (RMS) measurements. The[1] RMS values were incorporated primarily to display the 'average' level of loudness overall and therefore offered lower level values than the peak measurement. According to Owsinski, RMS metering is useful for indicating if two or more songs are approximately the same loudness level [24]. Furthermore, dynamic range (DR) values were acquired through this same method. Similar to the analysis undertaken by O'Malley [22], Voxengo SPAN Plus Fast Fourier Transform (FFT) audio spectrum analyser plug in was used within the Pro Tools environment to attain a visual representation of signal level across the frequency spectrum. The Real Time Average (RT AVG) measurement type was used as a means to produce averaged spectrum representations. It is important to note that these images only provide a brief snapshot in time on the various frequency/amplitude levels across all of the releases, although it is still useful as a frequency plot for comparison. In order to create this digital data for analysis it was critical to obtain a flat and comparable digital transfer of the 1981 vinyl LP *Sunnyboys* into the same 16 bit/44.1 kHz WAV file format as the 1991 and 2014 CD remasters for comparison. This process involved remastering engineer Rick O'Neil playing a first generation copy of the album on an AVID turntable with the signal encoded into digital using a WEISS analogue to digital (A/D) convertor. O'Neil was adamant the vinyl record used for digital transfer had to be a first-generation copy as this would have emanated from the same original master tape that was used to create the 2014 remastered CD. This digital signal was then imported into Pro Tools via an AVID HD I/O 8X8X8 analogue and digital audio interface with 10 dB of gain added to result in a directly comparable digital representation of the original vinyl record.

The qualitative methods used consisted primarily of semi-structured interviews (face-to-face, email, phone) with band members Peter Oxley (bass) and Richard Burgman (guitar) from Sunnyboys, remastering engineer Rick O'Neil and Sunnyboys' band manager Tim Pittman. These interviews were designed to gain an understanding on their understanding of remastering practice and their views on possible impacts on cultural heritage as a result of this practice. Furthermore, Oxley and Burgman participated in a comparative listening exercise to help establish perceptions of sonic differences between the original and remastered versions so that this data could be compared with the quantitative data generated through the digital audio analysis.

DIGITAL AUDIO ANALYSIS

Figure 9.1 below depicts the loudness/amplitude range over time of the respective waveforms for each release of *Sunnyboys* within the Pro Tools digital audio workstation (DAW) software environment.

The first track (blue) represents the 1981 vinyl transfer, the second track (green) depicts the 1991 CD release and the third track (purple) portrays the 2014 remastered CD release. The visual representation of the 2014 remastered CD release (purple) clearly shows a louder and more compressed signal identified by the 'block' shape of the files as compared to the other two tracks that exhibit greater shifts between softness and loudness. The waveform differences between the 1981 vinyl transfer and the 1991 CD release appear to be less severe and more closely aligned with only minor variances. However, the 1991 CD appears slightly louder overall compared to the vinyl which would lend support to the likelihood that the 1991 CD release was simply the 1981 vinyl master tape cut straight to CD with minimal manipulation.

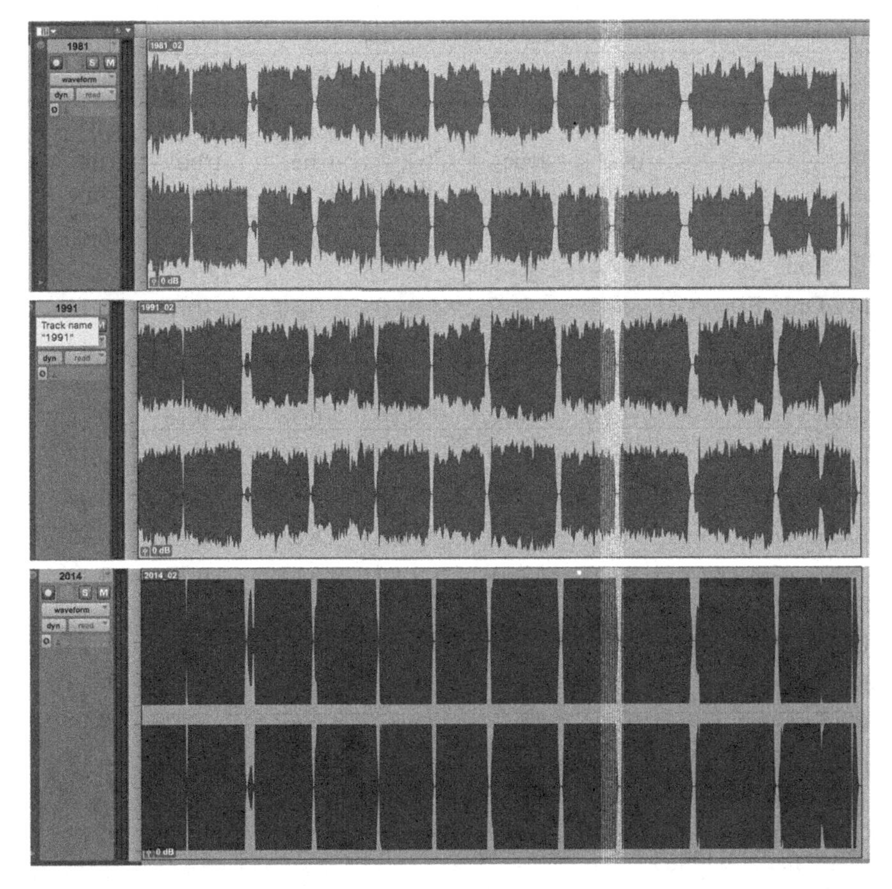

Figure 9.1 Waveform view of 1981, 1991, and 2014 *Sunnyboys* releases

Table 9.1 represents the left and right true peak level values for each individual album track as well as the album in its entirety across the three versions. The consolidated WAV file of the whole album measurement suggests that both the 1981 digital vinyl transfer and the 1991 CD are distorted as indicated by the value 'over'. If we pay particular attention to the song 'I'm Shakin'' we can see it is distorted on both the left and right channels on both the 1981 digital vinyl transfer and the 1991 CD. However, the 2014 remastered CD release is exactly the same peak level throughout the whole album –0.13dB on both channels. These results support O'Neil's claim of no distortion or clipping evident in his remastering practice due primarily to his complex use of limiters.

Table 9.2 displays the RMS levels for both left and right channels across all three releases. The difference in average loudness between the 1981 digital vinyl transfer and the 1991 CD for the consolidated whole album file is approximately 2 dB. However, the 2014 remastered CD is significantly louder on average overall than the other two releases, around 8 dB louder than the 1991 CD and approximately 10 dB louder

Table 9.1 True peak level measurements *Sunnyboys*

Song	1981 Vinyl		1991 CD		2014 CD	
	Left true peak	Right true peak	Left true peak	Right true peak	Left true peak	Right true peak
I Can't Talk To You	–0.22	–0.04	–1.78	–0.9	–0.13	–0.13
My Only Friend	–1.47	–1.17	–0.69	–0.30	–0.13	–0.13
Trouble In My Brain	–2.55	–0.85	–1.55	Over	–0.13	–0.13
Gone	–1.88	–0.96	–0.54	–0.38	–0.13	–0.13
It's Not Me	–1.56	–1.29	–0.58	–0.29	–0.13	–0.13
Happy Man	–0.56	Over	–1.18	–0.84	–0.13	–0.13
Alone With You	–1.95	Over	Over	Over	–0.13	–0.13
Tunnel Of My Love	–2.40	–2.21	–1.07	–0.61	–0.13	–0.13
Liar	–2.32	–2.78	–0.63	–0.8	–0.13	–0.13
Let You Go	–1.75	–1.58	Over	Over	–0.13	–0.13
I'm Shakin'	Over	Over	Over	Over	–0.13	–0.13
I Can't Talk To You (Reprise)	–8.65	–7.45	–4.52	–4.35	–0.13	–0.13
Consolidated WAV file of whole album	Over	Over	Over	Over	–0.13	–0.13

Table 9.2 RMS level measurements *Sunnyboys*

Song	1981 Vinyl		1991 CD		2014 CD	
	Left RMS	Right RMS	Left RMS	Right RMS	Left RMS	Right RMS
I Can't Talk To You	−15.12	−14.7	−14.72	−14.24	−6.60	−5.80
My Only Friend	−16.17	−15.52	−15.08	−14.30	−7.37	−5.88
Trouble In My Brain	−16.34	−15.78	−15.32	−15.02	−7.12	−6.32
Gone	−17.08	−16.72	−15.80	−15.75	−7.01	−6.15
It's Not Me	−15.87	−15.79	−14.37	−14.21	−6.67	−6.06
Happy Man	−18.80	−18.22	−16.35	−15.82	−6.93	−7.03
Alone With You	−17.42	−17.33	−14.05	−14.03	−6.58	−6.97
Tunnel Of My Love	−17.71	−17.88	−15.46	−15.66	−7.36	−7.88
Liar	−17.29	−17.30	−15.23	−15.24	−7.20	−7.61
Let You Go	−16.61	−16.84	−13.93	−14.15	−6.44	−6.87
I'm Shakin'	−19.01	−19.11	−15.02	−15.15	−7.03	−7.60
I Can't Talk To You (Reprise)	−24.42	−24.20	−19.94	−20.30	−11.23	−11.64
Consolidated WAV file of whole album	−17.41	−17.20	−15.27	−15.15	−7.27	−7.09

than the 1981 digital vinyl transfer. Focusing on the last two tracks 'I'm Shakin'' (1981) and 'I Can't Talk To You (Reprise)' (1981), this difference increases to between 12 dB and 13 dB. According to O'Neil, this relatively large amount of variance in RMS is actually quite normal in the pursuit of creating a modern sounding digital remaster from an old analogue recording so that it can sit comfortably alongside a new recording (Rick O'Neil, personal communication, November 9, 2015).

Table 9.3 depicts decibel measurements of dynamic range (DR) across the three releases as individual tracks as well as versions in their entirety. The consolidated WAV file of the 1981 flat digital transfer has the largest DR score of 15dB with the 1991 CD release slightly lower at 13dB. The remastered 2014 CD version has the lowest DR score of 5dB. According to O'Neil, modern sounding recordings have a dynamic range value between 3dB and 4dB, and he strives to achieve this value in his practice. O'Neil's approach appears to contradict the notion that decreased dynamic range automatically results in a poorer sonic experience for the listener as depicted in the 'loudness wars' [5,7,9].

Similar to O'Malley's work, I was keen to examine the frequency spread of the three different *Sunnyboys*' versions in an attempt to

Table 9.3 Dynamic range measured in decibels (dB) *Sunnyboys*

Song	1981 Vinyl	1991 CD	2014 CD
I Can't Talk To You	13 dB	12 dB	5 dB
My Only Friend	13 dB	12 dB	5 dB
Trouble In My Brain	12 dB	11 dB	5 dB
Gone	13 dB	13 dB	5 dB
It's Not Me	12 dB	12 dB	5 dB
Happy Man	14 dB	13 dB	5 dB
Alone With You	12 dB	13 dB	5 dB
Tunnel Of My Love	13 dB	12 dB	6 dB
Liar	13 dB	13 dB	6 dB
Let You Go	13 dB	12 dB	5 dB
I'm Shakin'	16 dB	13 dB	6 dB
I Can't Talk To You (Reprise)	12 dB	11 dB	7 dB
** Consolidated WAV file of whole album	15 dB	13 dB	5 dB

identify where these dynamic differences existed [22]. It is important to note that the images below only provide a brief snapshot in time on the various frequency and amplitude levels across all three releases. It would therefore be questionable to draw comparisons of a whole album based upon a brief snapshot so I decided to focus on a selection of single songs instead in an effort to attain a fairer and more comparable determination.

Figure 9.2 represents a snapshot image of frequency and its relationship to amplitude regarding the first song in the track listing, 'I Can't Talk To You' for the 1981 (light grey), 1991 (medium grey), and 2014 (dark grey) versions. I chose the opening track as O'Neil stated 'it was important to spend as much time on getting the opening track right and then these settings will be pretty much the same for the rest of the album' (Rick O'Neil, personal communication, 9 November 2015). The figure depicts a 'general' similar shape overall between the three releases with the 2014 CD significantly louder than the other two releases. However, there are some differences regarding EQ. For example, at around 60Hz the 1981 vinyl digital transfer is louder than the 1991 CD and has a flatter drop off at around 80Hz. Additionally, the 1991 CD appears to have a significantly louder midrange than the 1981 vinyl digital transfer at around the 800 Hz to 4 kHz range. This louder midrange is consistent with sonic issues raised by Jeff Lynne concerning CD releases in general around this time period [22]. Oxley found the 1981 version to have a 'bit edgier sound' and the proportionately less change in level between the 1.5 kHz and 4 kHz peaks in the 2014 remaster may support this view. Oxley also perceived the 2014 version to have a

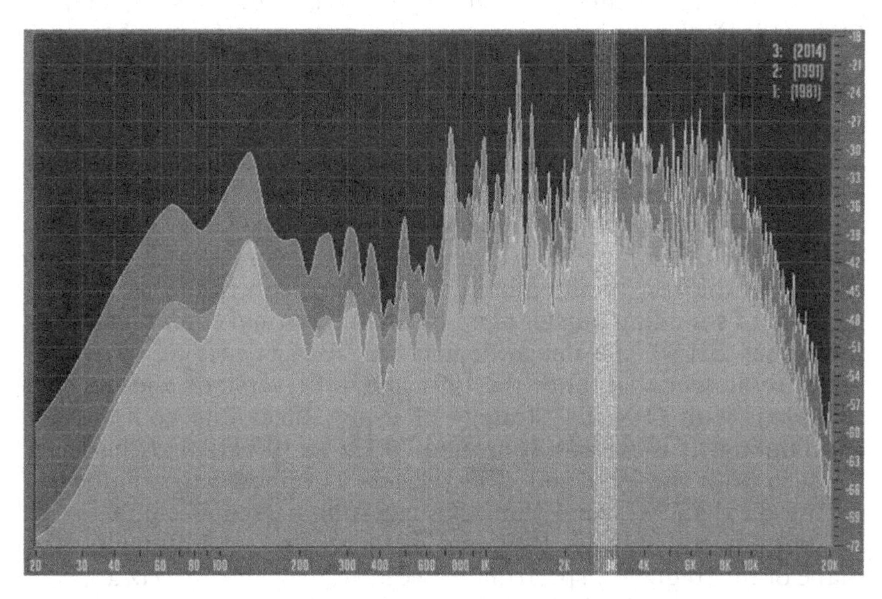

Figure 9.2 Spectrum analysis image for I Can't Talk To You – 1981 (light grey), 1991 (medium grey), and 2014 (dark grey)

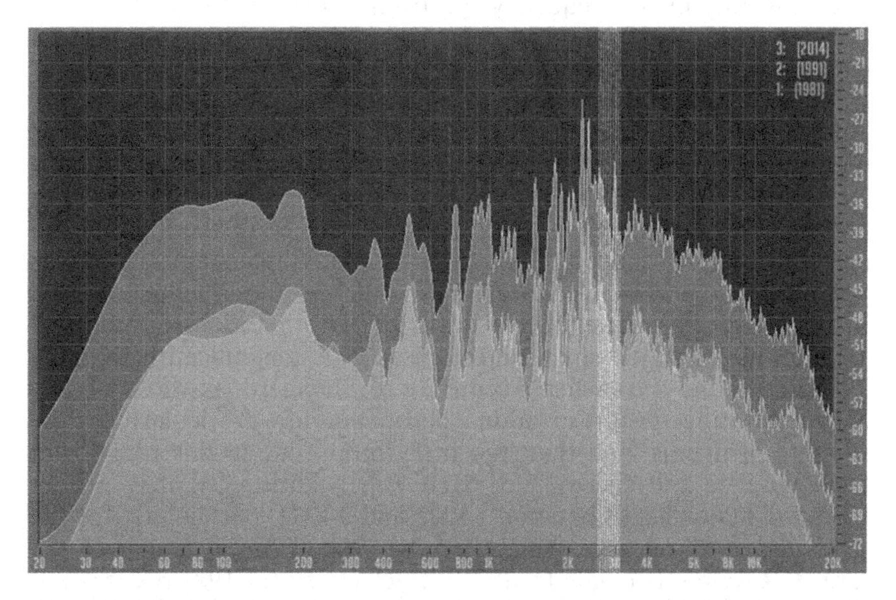

Figure 9.3 Spectrum analysis image for Happy Man – 1981 (light grey), 1991 (medium grey), and 2014 (dark grey)

'heavier and fuller guitar sustain with better separation' which may be attributed to the proportionate increase between 4 kHz and 10 kHz which adds greater clarity to the instruments, and the high spike around 4 kHz boosting the guitar sustain.

Figure 9.3 represents the frequency spectrum for the song 'Happy Man' (1981), the sixth track on the album. I chose Happy Man because as a fan and avid listener of the band I always believed this song sounded 'quieter' on the 1981 vinyl release in comparison to the previous track 'It's Not Me' and the following track 'Alone With You'. The RMS levels for the 1981 digital vinyl transfer confirm this as 'Happy Man' is approximately 3dB quieter than 'It's Not Me' and around 1dB quieter than 'Alone With You'. According to Pittman, this version of 'Happy Man' was originally recorded at a different session to the rest of the album, which may contribute to the impression of sounding quieter (Tim Pittman, personal communication, 2 October 2015). The image depicts the 2014 release as being significantly louder than both the 1991 and 1981 versions and appears consistent with O'Neil's' strategy of using substantial compression and limiting. It is evident at around 70 Hz to 100 Hz that there is a boost in both the 1991 and 2014 releases as opposed to a flatter response on the 1981 vinyl transfer suggesting a greater emphasis on the bottom end for the 1991 and 2014 releases. Additionally, the shape of the frequency spectrum between 10 kHz and 15 kHz appears to show a boost to both the 1991 and 2014 releases as opposed to the 1981 digital vinyl transfer, proposing an extra 'shimmer' to these releases. The image also reveals that the 1981 digital vinyl transfer does not fully fit the frequency spectrum (20 Hz to 20 kHz) and lies between 25 Hz and around 16 kHz. This lack of energy at the bottom end and missing 'brilliance' at the higher end of the frequency spectrum may also contribute to this song sounding 'quieter' overall. Additionally, this lack of bass energy of the 1981 version is consistent with Oxley's perception of it sounding 'happier and lighter'.

Figure 9.4 represents the frequency spectrum for the song 'I'm Shakin'' (1981), the second last track on side B off the album. The spectral analysis shapes of all three versions again appear similar overall with the 2014 releases exhibiting the most loudness. However, in the frequency range 6 kHz to 10 kHz both the 2014 and 1991 versions have a flatter shape and slope away less significantly than the 1981 digital vinyl transfer. According to Shepherd, this extra boost across this range results in 'adding clarity and life, particularly for the top end of drums' and that 'too little (amplitude in this range) will lack presence and energy' [25, para. 10]. Additionally, it is evident there are major peaks between 1 kHz and 3 kHz, particularly for the 1981 and 1991 releases, which may be a contributing factor to the large DR scores associated with this track (16dB for 1981 release) and the associated 'bite and aggression for guitars and vocals' [23, para. 9]. Oxley described the 1981 version as 'wild' and this summation may be supported by both the left and right channel true peak measurement levels being recorded as 'over' or distorted. Additionally, the boost between 4 kHz and 10 kHz on the 2014 CD frequency spectrum image appears to be consistent with Oxley's claim of 'better separation of instruments' and greater clarity.

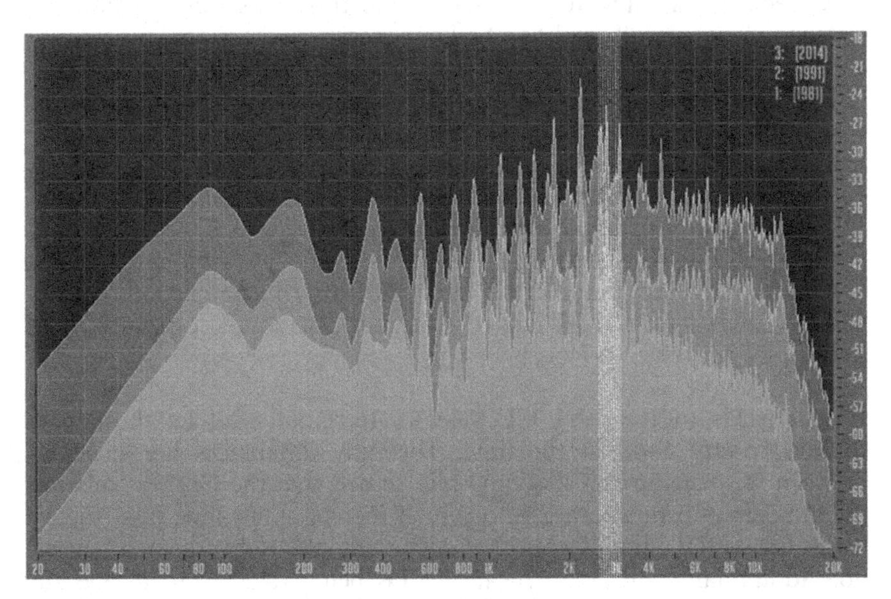

Figure 9.4 Spectrum analysis image for I'm Shakin' – 1981 (light grey), 1991 (medium grey), and 2014 (dark grey) *Sunnyboys*

LISTENING COMPARISON

Sunnyboys' band members Peter Oxley and Richard Burgman listened to a selection of songs from *Sunnyboys* and described the sonic differences they perceived between the original 1981 vinyl LP, the 1991 CD and the 2016 remastered CD. Burgman preferred to compare the sonic differences between the different versions of *Sunnyboys* in their entirety as he claimed comparisons between one era's medium and another in terms of single songs to be too subjective. However, Oxley was happy to focus on the selection of songs from the album in an effort to provide greater detail on these individual tracks. According to Burgman, he first heard the 1981 recording master tape played back in the control room at Alberts' Studio 2 through professional studio quality equipment and speakers. For Burgman, this was the best way to listen to Sunnyboys as 'audio tape has a dynamic range and frequency response that was superior to anything else at the time, vinyl records paled by comparison and don't even talk about cassettes' and he was 'blown away with the huge sound, clarity and presence of the instruments, the blending of voices, the interplay of the guitars, the work of the rhythm section, and that overall it had space, dynamics and energy' (Richard Burgman, personal communication, 29 March 2017). Concerning the 1981 vinyl release, Burgman said he generally liked what he heard but perceived it was not close to the sound he had heard in the recording studio as the vinyl record playback technology at the time was not the best, as O'Neil previously alluded to. As he writes;

> The average turntable in the 1980s was not very good in that turntable speeds varied, the cartridges and needles were poor and the amps, speakers and wiring attached to them were often of similar low quality. Most people, including myself, had lots of records and we played them all the time on stereos we cobbled together. Part of listening to music is losing yourself in it and suspending your inner critic as it performs its magic on you. This is what we did. (Richard Burgman, personal communication, 29 March, 2017)

With regards to the 1991 CD release, Burgman said he thought it sounded pretty good at the time. The only differences he perceived between this version and the vinyl release was that the CD was missing the familiar pops and crackles of vinyl, there was no tape hiss, the silences between the tracks appeared deeper and the treble seemed pushed up. Additionally, Burgman recalled CDs being referred to as the superior format at the time and similar to O'Neil's account, thought they still had their limitations in 1991. In particular, Burgmen believes that CD mastering was not what it should or could have been with regards to sound quality, and many CDs at that time were released with minimal mastering to quickly capture the market of people replacing their vinyl. Burgman was also critical of the home stereo systems at the time, particularly those that had multi CD players built in, claiming they were lightweight and cheap, and lamented the loss of the 'big old heavy good sounding amps and speakers of the 1970s systems' (Richard Burgman, personal communication, 29 March 2017). The 2014 CD was much better sonically than the 1991 CD according to Burgman, primarily due to the increased responsibility and importance of mastering and remastering in the current era as opposed to 1991. As he writes;

> The 2014 CD is louder, the dynamic range is better, the bass is clear and tight, and the highs have been mellowed to make it a more pleasant listening experience. Overall it's better balanced, cleaner and clearer, and the songs sound more alive as a result. I am very happy with it (Richard Burgman, personal communication, 29 March 2017).

However, with reference to the songs chosen to analyse individually, Burgman said he loved the vinyl versions of those songs the best. He attributes this to its association with nostalgia, inherent noise, a sense of time and place, a comfortable resolution and a pre-existing process. As he writes;

> I love the sound of vinyl, with its warmth, with all the ritual of getting a record out, putting it on the turntable, dropping the

> needle in the groove, and then remembering to not walk across the floor in front of the stereo so you don't make the needle jump. They are what we were trying to achieve back then. Records were it. We wanted to make a good record and I think we did. The 2014 CD is really good too, but it's more transparent than the record, more revealing, more of us are laid bare. In a way the warm blanket that is vinyl encourages a listener to use their imagination when it comes to lyrics, chords, which guitar is doing what, as opposed to a well mastered CD through a decent stereo which shines a more fluorescent light onto the performances, and reveals more internal detail. That warm light was what we used to listen to everything back then. A bit like old school PAL TV versus 1080p, there is a charm, magic, a measure of comfort, in not having everything so crystal clear. It encourages the imagination to participate and allows for mistakes, it means each person's appreciation, understanding and enjoyment of the music will be different and it will be their own (Richard Burgman, personal communication, 29 March 2017).

Burgman suggests that the extra detail evident through the 2014 remastered CD in terms of clarity and performance comes at a cost when compared to the original vinyl release. For him, there appears to be an associated charm, sense of cultural heritage and meaning associated with the vinyl release and the process surrounding consuming vinyl that is lost through the more detailed 2014 digital replica, in particular the meaning and imagination of the audience through its consumption [5].

For Oxley, the 1981 vinyl release was a great representation of the band, particularly for fans, as that was the way the band sounded live, the record had a nice sense of separation of the instruments and a warm bottom end. Although Oxley was initially excited about the 1991 CD release in terms of *Sunnyboys* being made available on the considered new 'superior' format, he was disappointed as it sounded like it had been badly cut from the original vinyl master, was a bit muffled, and he refers to it as not a very thoughtful release from Mushroom Records. However, Oxley said he was very pleased with the 2014 remastered CD release, citing he perceived a beautiful quality of sound of the vocals and instruments, fantastic separation and overall it sounded amazing. For the track 'I Can't Talk To You', Oxley said he perceived the 1981 version to have an 'edgier' sound and the 2014 copy portrayed a 'heavier and fuller sustain on the guitar part and better separation overall'. For the song 'Happy Man', Oxley said he believed the 1981 original version to be 'happier and lighter' when directly compared against the 2014 release. Oxley depicted the 1981 original version of 'I'm Shakin'' in a positive manner as 'wild' and also thoroughly enjoyed the better 'separation of instruments' on the 2014 copy. Oxley's brief descriptions of the sonic elements he associated with the vinyl version of these songs as being 'happier and lighter' and 'edgier' may suggest, when compared to the 2014 copy, that there is less amplitude overall across the frequency ranges and perhaps sharper transients and peaks in the frequency range, roughly around the 2 kHz mark that is associated with 'edge' [26]. Oxley's perception of 'greater separation overall' of instrument parts and a 'heavier and fuller sustain' associated with the 2014 version may also indicate a greater use of equalisation and dynamics (limiting and compression) signal processing employed in its creation.

CULTURAL HERITAGE

The 1981 *Sunnyboys* album forms part of the cultural heritage surrounding the Sydney 'Guitar-Pop' sound and fashion associated with this period, and consequently I, along with many others, attributed a significant level of meaning to this release over a long period of time [20]. The published newspaper review below by well-respected Australian music critic Mengel is a reflection of the popularity, significance, perception of cultural heritage and meaning associated with the original 1981 recording musical artefact *Sunnyboys* by Australian society.

> There are other Australian albums as great as Sunnyboys' 1981 debut *Sunnyboys*. There aren't any that are better. Hearing it again thirty-two years later, it still feels like one of the best rock'n'roll debut albums ever, one of those collisions of bright-eyed youthful energy, lyrical intensity and razor-sharp rock'n'-roll instincts that couldn't have been scripted. Like singer and chief songwriter Jeremy Oxley's assured, expressive guitar playing, something about this music hits you in the guts, the heart and the head. It says all you need to know about being young and alive, with all those doubts, joys, heartache and hopes. And it rocks – Noel Mengel, The Courier Mail [27].

I was therefore initially concerned upon learning of the decision to create the 2014 digital replica as to whether perceived cultural heritage and meaning attached to the original musical artefact would be maintained [28,29]. When I interviewed Oxley regarding cultural heritage considerations, he was quick to point out that because they were not remixing from the multi-track or adding anything new (a new guitar part for example), and that they were only working from the original stereo master tapes and not the multi-track session tape for *Sunnyboys*, the overall intention was to only make the recordings sound better, not different and hence any sense of cultural heritage was not diminished. As he writes;

> If we had gone back and put a couple of guitars on there, then that would have been a whole different thing. But all that was

> being done was using the technology today to get it sounding better. You go, okay, that sounded good there, it's the same band, same instruments, same sessions, so you get it to that point and you change it slightly or you do whatever you need to do (Peter Oxley, personal communication, 5 October 2015).

This then led me to question Oxley on whether he believed if the Sunnyboys' existing fan base, primarily those original fans from the 1980s, wanted to hear these modern and updated digital replicas. In particular, whether these perceived 'better' sounding digital cultural representations resulted in fans experiencing a diminished perception of cultural heritage and meaning towards the original recordings and vinyl release. As he writes;

> I think fans do. I know myself that I've rebought records when they've been rereleased and I probably didn't need to, but I did. For example, the Nick Cave and the Bad Seeds album Push The Sky Away. Twenty year olds in 1980, they still want to be 20, well, you don't want to be 20 but the music that you loved when you were 20, if you can go and see it played by the same people that played it then and they can play it well, that's really tight, good, powerful, then it's going to be just as exciting as it was when you were 20. I mean, the worst thing I think is when a band plays and they are not very good, and you think, oh god that band was good when you were 20 or something. So remastering and re-releasing, it just part of that and if you're playing, to have a remastered recording out that you could sell them at the shows, it's part of the marketing. And, I think they go hand-in-hand (Peter Oxley, personal communication, 5 October 2015).

Oxley's response appears to indicate that despite the digital replica sounding in his opinion 'better', the perceived original meaning and cultural heritage associated with the original recordings and live performances from the band and fans is not diminished. In fact, Oxley viewed the remastered digital replica as helping to sustain and potentially build a commercial market and to give the band a reason to resume playing live. Burgman said that without the creation of digital cultural representations much of the music recorded before 1990 would be lost from tape degradation and that the ability to take old tape masters and convert them to digital has saved many recordings. With regards to remixing and remastering, Burgman shared a similar view to Oxley in that 'we're not trying to change anything we did back then, we're just much more in control of what comes out now that has our name on it, and that's a good thing' (Richard Burgman, personal communication, 29 March 2017). However, Burgman cited the recent example of digital intervention involving remixing, remastering and preservation being of tremendous value in the 2017 remastered CD release of *Shy Imposters* (a self-titled band release Burgman and Peter Oxley were in together shortly before Sunnyboys) in 2017, from a recording session in 1980. As he writes;

> In 2016 we looked into finding our original tape and seeing if we could remix it, remaster it for release on CD. The tape was dug up, and was unusable. It had to be baked and transferred to digital to be of any use. This was done in Sydney, the tracks were rescued, the songs remixed and remastered, and it's now available – cool huh (Richard Burgman, personal communication, 29 March 2017).

Burgman's depiction of the original music artefact having to be transferred to digital to be of any use is an interesting insight, particularly when we consider cultural heritage and meaning. The fact that the original recording was of little use in its present condition would suggest that any attached cultural heritage and meaning associated with the original recordings were in danger of disappearing altogether. To salvage the musical artefact and create a digital replica for fan consumption, Burgman was not only happy for remastering to be used, but also for the recording to be remixed, which implies a greater level of digital manipulation and intervention on the original as opposed to the practice of remastering.

CONCLUSION

The digital audio analysis revealed perhaps predictably that the 2014 CD remastered version of *Sunnyboys* was generally louder, possessed a wider frequency spectrum and a reduced dynamic range as compared to both the 1981 digitised vinyl transfer and the 1991 CD. However, although the frequency spectrum analysis appeared to show a general consistency of shape across the three releases, mostly between the 2014 and 1991 remasters, the 2014 CD version seemed to display distinct differences in some frequency ranges (80–100 Hz in Happy Man, 800Hz–1 kHz, and 10–12 kHz in I'm Shakin') as compared to the other two releases. This would suggest that the modern remastering practice used on *Sunnyboys* involved greater manipulation, particularly concerning EQ adjustment, than on previous releases. It is therefore plausible to propose that this change in remastering practice has evolved in an effort to make digital cultural representations of older recordings sit comfortably sonically alongside new recordings that incorporate modern mastering practice and technologies. Furthermore, these measured 'differences' appeared to support the sonic perceptions of both Oxley and Burgman in their written descriptions of sonic alterations they identified which suggests that such manipulations are audible and noticeable. These noticeable differences between the digital cultural representation and the musical artefact therefore bring into question whether perceptions of cultural heritage associated with the original artefact may be impacted by this change. Oxley was adamant that because they were not remixing from the multi-track or adding anything new (a new guitar part for example), and that they were only working from the original stereo master tape, the overall objective of the remastering process was to make the recordings sound 'better' as opposed to 'different' so therefore the cultural impact was minimal. Burgman reiterated Oxley's view that the band's motivation and intent was for 'improvement' as opposed to 'change' and viewed digital cultural representation in general as promoting the perception of cultural heritage through the preservation of degraded musical artefacts that would ultimately be destroyed. Additionally, Burgman believed digital remastering made cultural musical artefacts more accessible through modern playback devices which further fostered and supported perceptions of cultural heritage.

NOTE

1. Although discussion about Loudness Unit Full Scale (LUFS) is fashionable in literature lately, it is more of a broadcast standard and hence RMS was used. It is acknowledged that LUFS is similar to RMS in that it uses a more sophisticated statistical methodology and a frequency-based weighting system.

REFERENCES

1. Shelvock, M. *Audio Mastering as Musical Practice. (Master of Arts)*, The University of Western Ontario, London, Ontario, Canada (2012).
2. Rumsey, F. 'Pound of Cure or Ounce of Prevention? Archiving and Preservation in Action'. *Journal of the Audio Engineering Society*, Vol. *60*, No. 1/2, (2012), pp. 79–82.
3. Benzine, A. *Court Hears Apple Corps Plan for Beatles Digital Release.* Music Week, (2006, April 15), p. 4.
4. Richardson, K. 'Still Remastering, Still Dreaming'. *Stereo Review*, Vol. *62*, No. 9, (1997), p. 96.
5. Vickers, E. The loudness war: Background, speculation and recommendations. In *Audio Engineering Society Convention 129*, Audio Engineering Society, (2010).
6. Hjortkjaer, J. and Walther-Hansen, M. 'Perceptual Effects of Dynamic Range Compression in Popular Music Recordings'. *Journal of the Audio Engineering Society*, Vol. *62*, No. 1/2, (2014), pp. 37–41.
7. Nielsen, S. H. and Lund, T. Level control in digital mastering. In *Audio Engineering Society Convention 107*, Audio Engineering Society, (1999).
8. Walsh, M., Stein, E. and Jot, J.-M. Adaptive dynamics enhancement. In *Audio Engineering Society Convention 130*, Audio Engineering Society, (2011).

9. Deruty, E. and Tardieu, D. 'About Dynamic Processing in Mainstream Music'. *Journal of the Audio Engineering Society*, Vol. *62*, No. 1/2, (2014), pp. 42–510.
10. Nardi, C. Gateway of Sound: Reassessing the Role of Audio Mastering in the Art of Record Production. *Dancecult*, Vol. *6*, No. 1, (2014), pp. 8–210. doi:10.12801/1947-5403.2014.06.01.01
11. Hodgson, J. *Understanding Records: A Field Guide to Recording Practice*. 1st edn. Continuum International Publishing Group Ltd, New York, (2010).
12. Cousins, M. and Hepworth-Sawyer, R. *Practical Mastering: A Guide to Mastering in the Modern Studio*. CRC Press, Abingdon, Oxon, (2013).
13. Sexton, P. Repaving 'Abbey Road'. *Billboard, 121*, (2009, September 12), p. 24.
14. Bennett, A. '"Heritage Rock": Rock Music, Representation and Heritage Discourse'. *Poetics*, Vol. *37*, No. 5-6, (2009), pp. 474–489.
15. Silverman, H. and Fairchild Ruggles, D. *Cultural Heritage and Human Rights*. Springer, New York, (2007).
16. Roberts, L. and Cohen, S. Unauthorising Popular Music Heritage: Outline of a Critical Framework. *International Journal of Heritage Studies*, Vol. *20*, No. 3, (2013), pp. 241–261. doi:10.1080/13527258.2012.750619
17. Baker, S., Doyle, P. and Homan, S. Historical Records, National Constructions: The Contemporary Popular Music Archive. *Popular Music and Society*, Vol. *39*, No. 1, (2016), pp. 8–27. doi:10.1080/03007766.20110.1061336
18. Hennessy, K. *From Intangible Expression to Digital Cultural Heritage Safeguarding Intangible Cultural Heritage* (Vol. 8, pp. 33–46). Boydell & Brewer, Brewer, Suffolk, (2012)
19. Clayton-Lea, T. *The Magical Remastering Tour*. Irish Times, (2009, August 25), p. 12.
20. Easton, M. *Down the Lane: Sydney's DIY Music Scene*. The Lifted Brow, No. 17, (2013), p. 42.
21. Barham, J. '"Not Necessarily Mahler": Remix, Samples and Borrowing in the Age of Wiki'. *Contemporary Music Review*, Vol. *33*, No. 2, (2014), pp. 128–147. doi:10.1080/07494467.2014.959273
22. O'Malley, M. The Definitive Edition (Digitally Remastered). *Journal on the Art of Record Production*, Vol. *10*, (2015).
23. Barry, B. (Re)leasing the Beatles. In *Audio Engineering Society Convention 135*, Audio Engineering Society, (2013).
24. Owsinski, B. *The Mastering Engineer's Handbook: The Audio Mastering Handbook*. Cengage Learning, Boston, MA, (2008).
25. Shepherd, I. *7 Crucial EQ Bands to Help Balance Your Mix*, (2010, October 27). Retrieved from http://productionadvice.co.uk/using-eq/
26. Katz, R. A. *Mastering Audio: The Art and the Science*. Elsevier/Focal Press, Amsterdam; Boston, (2014).
27. Pittman, T. *Liner Notes. Sunnyboys*. Warner Music Australia Pty Ltd, Sydney, (2014).

29. Grainge, P. 'Reclaiming Heritage: Colourization, Culture Wars and the Politics of Nostalgia'. *Cultural Studies, Vol, 13*, No. 4, (1999), pp. 621–638. doi:10.1080/095023899335077
28. Cartwright, A., Besson, E. and Maubisson, L. 'Nostalgia and Technology Innovation Driving Retro Music Consumption'. *European Journal of Innovation Management*, Vol. *16*, No. 4, (2013), pp. 459–494. doi:10.1108/EJIM-06-2012-0062

10

Mastering success

A ROUNDTABLE DISCUSSION WITH EARLY TO MID-CAREER MASTERING ENGINEERS

Russ Hepworth-Sawyer

INTRODUCTION

The predominate interest for many readers of music production in the past 20 years or so has, rightly, been those interviews with engineers and producers who hands have touched the microphones and consoles that managed the signals that recorded the classics. We're still coming to grips with the phenomenon that was the boom in popular music in the latter part of the twentieth century. Academically speaking, within the music production field, we're still processing!

To read a Ken Scott interview, is now quite commonplace, such as ours in 2012 for the *Journal of the Art of Record Production* [1]. Since Howard Massey's success in raising considerable interest in the views of engineers and producers, the likes of Ken Scott and several other seminal colleagues, too numerous to list here, have been interviewed frequently and deeply, providing the researcher with so much material to discuss and interpret.

During the preparations for Producing Music, another book in the Perspectives on Music Production series (POMP), I considered the voices so far neglected in the discourse from producers in a 'Behind The Glass' style. There are few interviews with Ken Scott and his peers at the time they were doing their seminal work, or even before their first crowning achievements. I considered what it might be like to be able to research an engineer's view perhaps before they'd done their most popular and defining work. In Producing Music, I interviewed three colleagues and friends that are what I considered 'mid career'. This can be read with interest. Since that chapter, British engineer and producer Adrian Breakspear, now working out of Sydney, Australia, for example, has won more awards and done more defining work. Equally, Mike Cave has since worked on the hugely successful album launching Lewis Capaldi's career. Their interview contributions to that chapter, which stand before further, recent successes, will stand as a useful resource for future researchers looking into their work at a defining period.

Once again, in preparations for this POMP series book, Mastering In Music, I felt it important to continue the theme. This time looking at early to mid-career engineers to capture their views on their education.

THE QUESTIONS

In the preparation for our roundtable discussion, held mainly over email, but also direct Skype interview for some with a later transcription, I prepared nine questions relating to their start in audio mastering, their training journey and views on current situations within the field. This is a current field of interest for me presently and it is interesting to see the views and consideration of the engineers included here who have not received a 'standard' apprenticeship route in audio mastering.

The questions are listed below.

1. How did you get into audio, and was mastering your first work in the area?
2. If not, how did you come to know mastering existed and how did you fall into it?
3. Would you describe yourself as self-taught, or has some form of 'teaching/mentoring/apprenticeship' been at the heart of some of your training?
4. What are you views of the routes into mastering?
5. If you're self-taught, how have you trained yourself. Has it been organic, figuring stuff out, or Internet or other methods I've not listed?
6. What do you think you've missed out on within a apprenticeship or higher education context? One example might be the passing down of vinyl cutting for example.
7. What do you think are the advantages of being self-taught, if there are any?
8. If you were asked to write a higher education course based on your experience, what might you include?
9. What do you see as the future of audio mastering and how you might advise those coming up in the field?

Not all questions were answered sequentially and, as perhaps a poor interviewer, conversations naturally ventured elsewhere. I have tried to edit the work back to its original path where necessary.

PARTICIPANTS

The participants are all known to the author personally and have been chosen due to their early to mid-career stage in their careers.

Nick Cooke attended Leeds College of Music and from there toured both as a musician and soon landed a job with a production music company in their mastering department. Nick is fascinating as his experience was being dropped in the field without preparation. He's since gone on to set up his own mastering business and exceptionally renowned White Mark designed studio.

Katie Tavini, known for her contributions to Audio Media Europe and as a keen supporter of the AES UK Mastering Group that is co chaired by JP Braddock and the author. Katie's business is growing exponentially and it is wonderful to capture her thinking at this pivotal growth in her career.

Jay Hodgson started life as a performer in several bands in his native Canada. Later in life he became one of the first academics to study popular music production and be awarded the Governor General's Award for his contribution. Jay came to mastering from a purely creative angle and has enjoyed several successes and Juno nominations for his work internationally.

Jeremy Graham started working in the live audio industry in his native New Zealand. Later he came to the UK to study audio mastering and became a valued assistant in MOTTOsound with the author. Successfully Graham now runs his own mastering business, Transfer Lounge AMS and continues to gain success as a music producer. Jeremy's insight here is exceptionally valuable, as we predict his career to be one to watch.

QUESTIONS

How did you get into audio, and was mastering your first work in the field?

Nick Cooke – Prior to my degree I was already interested in audio, as I have been playing in bands since I was 13 and I continue to work as a musician. My parents and grandparents had a huge influence on me from a young age with their interest in music and Hi-fi. My dad also had a PA system for his band, so I started off by doing live sound for them and other local events. I was lucky to go to a school with a strong music department and I was involved in a range of audio related activities – both in and outside school – which naturally led to me doing music technology at A-level and then to continue on to higher education. Throughout and after completing my Music Production degree, I continued to gain work experience in different areas in the audio world.

Mastering was definitely not my first work in audio. I initially wanted to work in foley for TV and film and managed to gain some time at both Aardman Animations and the BBC in Bristol as well as North One TV and Nickelodeon in London. I also assisted on various location recordings for a Classical music label to gain as much experience as possible, whilst working in my local pub to fund it all. Eventually this all led to paid work in Outside Broadcasting for televised tennis, horse racing, and football. This covered rigging, maintenance, comms, and simple mixing tasks. I also continued to obtain work as a session musician, both live and recording, as well as some live sound work at a few small festivals around the country.

Katie Tavini – I got into audio because I was super into music, played the violin since I was a kid, and was fascinated by computer music. I used to borrow books from the library on midi sequencing and super nerd out. I got a copy of Cubasis when I was about 16 and started making MIDI music. I studied both music and music technology in college but didn't really know what career options there were. A career officer told me the only job in music was to become a performer but I knew that I never wanted to do that. So I applied for an acoustics based course at university, but then changed to a BA Music course when I realised I was the only girl on the course. There was a tiny 'recording studies' module which I loved, and so I spent all my time in the studio – I used to buy my friends studio 'credits' off them so I could have longer in the studio. One day my tutor asked if I'd like a job working in a studio which was ace. I really wanted to be a studio engineer at that point. The job was amazing and the producer I was working for taught me so much. Eventually I went freelance as a recording engineer and started doing a bit of production work too. It wasn't really my thing because I was expected to mix, and hated it. Too many choices.

Jay Hodgson – I got into audio by participating in recording sessions as far back as 1992 or 1993. I was lucky enough that the first recording session I ever participated in was for John McDermott's platinum selling 'Love Is A Voyage' album, so I got to see an elite operation for my very first session. McDermott was somewhat mean to me, though. He would often introduce the band, at live shows, and tell everyone that I was only there because my father was a very expensive lawyer. So needless to say, I didn't stick around for very long. I soon travelled down to Boston, and worked as a session guitarist a fair bit while attending Berklee. I also wrote my own songs and had a band. Slowly I became increasingly interested in how some producers were able to get particular sounds and translate my songs into particular arrangements that I found super compelling. Once I put out my first record, in 1997 or so, I was completely hooked on recording, and utterly bored with performing and touring. But it wasn't until 2002 or so, when I bought a laptop with Garageband on it, that I got really interested in audio. I didn't start mastering in earnest until a few years after that.

Jeremy Graham – I started working professionally in the music industry towards the end of 2007 where I began a career as a live sound engineer, initially starting out as part of an internship programme during my bachelors degree. Although my strengths leaned more towards music production, there weren't too many opportunities to be involved in the recording/mixing side of things, and I didn't view myself as a studio mix engineer at the time.

Live sound and working in the events side of the industry became a real solid foundation that would provide financial security over the following decade and really trained my ears to be able to listen critically, both sonically as well as socially, in terms of communicating and working alongside a diverse range of people, events and a variety of spaces.

If not, how did you come to know mastering existed and how did you fall into it?

Nick Cooke – My first mastering job was at a small family run music production library, De Wolfe Music, that have been going since 1909 when they produced music scores for silent film before sound recording was available. I didn't know I was applying to be a mastering engineer: the role had been advertised as 'Music Editor' and my interview went so well chatting about mainly Glastonbury Festival that I forgot to ask about the job!

On my first day I was put in a small room with a tape machine, a pile of tapes, a couple of Weiss units and a SADiE system. I was asked to compile a CD out of a few CDRs from various composers and to make any slight EQ adjustments. I later realised this was mastering. Although I had learnt a bit about mastering at university, and had the opportunity to meet mastering engineer and audio education guru Bob Katz at a University conference, I hadn't really thought about mastering until accidentally getting that job.

De Wolfe have a vast and varied back catalogue on tape and a big part of my job there was remastering past vinyl releases from the original master tapes. This was an amazing opportunity to hear the music from a time when budgets were big and large studios still existed. The audio and production quality was eye-opening. I really enjoyed this work and started to consider mastering as a career.

Katie Tavini – Someone on Gearslutz (I think) told me that if I wanted to improve my mixing I should learn how to master. So I started stealing tracks from studios, trying to 'master' them on crappy speakers, and then when the pro masters came back, I used to compare and try and work out what the mastering engineer had done. I didn't know anything about the technical side or formats, I just enjoyed having a go.

Jeremy Graham – I was exposed to post-production and mastering during the final year of my bachelors degree. Although I excelled at the post-production side of things (in terms of syncing audio & video), Mastering was honestly still kind of a mystery to me.

It wasn't until a few years after I completed my honours degree, alongside continuing to produce music that I slowly began to understand what mastering was all about sonically, and out of necessity more than anything else (in terms of releasing music and getting tracks ready to perform live). I decided to move to the UK at the start of 2017 and consciously stepped away from live sound, which is when I really started to hone in and understand the full breadth and craft of audio mastering from a more professional context; building on the framework I had already developed.

Would you describe yourself as self-taught, or has some form of 'teaching/mentoring/apprenticeship' been at the heart of some of your training?

Nick Cooke – I haven't had direct training in mastering like an apprenticeship or a higher education course focussed on mastering, so I could be described as essentially self-taught. However, in learning on the job, I have always been picking up bits from engineers, both mastering and other types of audio engineer, and other people around me. You could also say that I have had indirect mentoring through keeping bosses and clients happy. They would tell me what they wanted changing and I would work out how do it. They feel like mentors to me.

When first starting at De Wolfe I was shown the basics of SADiE and how to use a tape machine, and then left to my own devices. Luckily, as most of the time I was working with amazing sounding tape recordings, I was a bit scared to alter things too much and was encouraged not to. I quickly learnt that a little can go a long way. Later moving to Extreme Music at Sony/ATV I was working with of all sorts of producers/engineers of various different calibres, some just starting out in their careers and some iconic big names in the business. These were often all on the same album that I needed to make sound coherent as well as creating a specific sound to meet a given brief. Working with producers, engineers and artists with very clear ideas of the way they want their material to sound was a quick learning curve and showed me a very different side to mastering. I gained a lot of experience from working to briefs and meeting their goals.

I think I was lucky to have these experiences of such different styles of mastering so early on in my career. Throughout all of this, however, it has always been really important to me to read a variety of audio and music related books not just aimed at mastering, and of course a huge amount of listening to masses of music and recordings.

Katie Tavini – I'm totally self taught and still learning!

Jay Hodgson – I am completely self-taught. I mastered for friends and family for years, and figured everything out by myself – Totally self-taught.

Jeremy Graham – I'd consider myself an observant person, willing to put in the time to research varied opinions and processes in order to draw my own conclusions, so yes, fundamentally self taught.

Having said that, I've been fortunate enough to have had some enlightening and extremely educational conversations with some well respected mastering engineers in the industry very early on in my career, and that advice, along with my own investigations, trials, errors, and overall persistence, has really paid off more than I could have hoped for.

It's probably not the most traditional route, but I'm not convinced there really is one.

What are you views of the perceived routes to becoming a mastering engineer?

Nick Cooke – It is hard for me to comment on apprenticeships as I have technically not been through one but I certainly believe they are valuable. Probably the biggest challenge with that route is finding an opportunity in the first place. It seems to be the traditional way to move into mastering as a career, and may well be a more straight forward route into it. Certainly assisting and observing a notably great engineer is a wonderful opportunity that I would have jumped at the chance if one came along. My route into mastering however, feels like I have snuck in through the fire exit: it couldn't really have been planned and was in a large part down to chance. And even after 12 years it still feels like I'm only just getting started.

I am pleased to have gone through higher education and although I wasn't focussing on mastering at the time it certainly helped open my mind to possibilities and cemented the idea that I wanted to work in audio. I personally feel that it is better to keep a broader focus at degree level, rather than specialising too soon, then you make the most of whatever opportunities arise. Plus I don't see mastering as a 'black art': it is part of music production as a whole, perhaps one from a different perspective, but it is necessary to understand all the parts to know how mastering fits in.

In my opinion building a career in mastering requires a lot of self-teaching even with apprenticeships or higher education. I would also be bold enough to say that a mastering engineer cannot be fully self-taught. It relies on working with and learning from other people throughout the whole career. That is certainly what I have found so far anyway. Rather than three distinct routes, I would argue they can often intertwine.

Katie Tavini – Self taught is cool because it's the only way I know, and it's enabled me to develop a very clear style and tastes. It's also given me the freedom to work on the types of music I enjoy the most. Apprenticeship I guess gets your sense of the actual business side of things good fast because you're watching someone deal with stuff on a daily basis, but I would struggle to sit there and shut up when someone else is mastering. I can't learn that way. Higher education seems like a great way of learning the technical skills, but you won't really learn the business side of things and honestly I'm really not an academic person so it's not an option I would personally consider.

Jay Hodgson – I can't speak to anything other than being self-taught. I imagine it is a tonne of extra work, though the results are very personalised. I would have liked an apprenticeship, and even higher education, but unfortunately I couldn't afford either, as I worked my way through to a PhD in music which left little time for anything else professionally speaking. Being self-taught does allow you to have a certain confidence in your own tastes and artistry, I think, that other routes perhaps wouldn't emphasise. I truly don't believe in a right or wrong, or a good or bad, beyond client approval, when it comes to mastering. When I hear people slagging other people's work I just roll my eyes, and move on. We're all just getting client approval how we see fit. It's that simple. Anything more is just reassuring ourselves that our job matters. It's so loud its distorting? Cool. That sounds good sometimes. Any more words on the matter, beyond 'the client approved', is just pretentious, insecure bullshit to my mind.

Jeremy Graham – I suppose I'm biased in that I really enjoyed and valued the five years of full time music study I completed straight out of high school, but I'm not sure what else I would have done; it was the only logical option given the nature of opportunities that existed for me at the time, or that I was aware of.

Tertiary study is definitely not for everyone (and I suppose neither is an apprenticeship), but regardless of whichever path you might choose, only you know the right path for you, and whatever that path is, there is no evading the abundance of hours that are required to figure out and develop your skills and overall craft.

If you're self-taught, how have you trained yourself? Has it been organic figuring stuff out, or internet or other methods I've not listed?

Nick Cooke – In terms of self-training, it's been mainly listening, over and over, and trial and error. I have read a lot of books on mastering, which has been essential to learning the technical bits, but also wider reading around audio in general, as well as how to run a business etc. I have learnt bits at university, conferences, networking events, YouTube, podcasts and internet forums. I also learn from reading magazines and equipment manuals, even old ones from the 1950s, 1960s, and 1970s, I have a brilliant set of manuals from Neumann and Scully cutting lathes for example. But where I've learnt most is through talking to and collaborating with people. Plus having the opportunity to work in an acoustically accurate room from pretty much the beginning of my career has definitely made a huge difference to my understanding of mastering processes.

Being a musician and knowing the music business from other angles has been key for me for working with artists and producers and understanding their needs and processes. This has really helped me with building relationships with clients and having clear communication in both directions. I know that some of the top engineers have also come from a musical background so it obviously helps.

Katie Tavini – Figuring stuff out, having a go, comparing work to pros work (like I said before, I used to nick tracks and try and master them, and then compare them with professional masters that came back). My background in classical music gave me really good critical listening skills I think.

Jay Hodgson – I trained myself by working. I just worked, worked, worked. I must have cleared 1,000 masters before I had any sense of certainty about what I was doing. I read a book to figure out what meter numbers to peg records to so what I produced would sound roughly the same on my monitors as they did once distributed.

Jeremy Graham – In terms of my own mastering development, I'd say it's been a fairly organic process.

It's really been a matter of first figuring out the specific tools I wanted to work with, knowing their strengths and weaknesses (including my room/monitors), picking up bits of advice here and there (whether directly from other engineers or via the Internet, books, interviews, etc.), and then developing my own approach to problem solving issues/problems (where they exist) and tuning in my ears/mind-set somewhere between the expectations of the client and my own intuition based on my personal tastes and references.

Aside from that, I'd say trying to remain consistently self-critical over the long term has been valuable for me personally, but not being too hard on myself either; understanding my own shortcomings, expectations, and seeking to improve and refine my abilities.

What do you think you've missed out on within a apprenticeship or higher education context? One example might be the passing down of vinyl cutting for example

Nick Cooke – I wouldn't say missing out as I wasn't aiming to be a mastering engineer, but aspects of an apprenticeship such as on-the-job guidance and reassurance of my work is something that I have had to manage without. The main things I feel I missed out on in my degree was working within a professional studio and a work-placement scheme within the degree. I am aware of some courses that do offer these, which I would say is definitely beneficial. As work placement opportunities were not available on the course I was on, I had to find my own way to build experience.

Yes, vinyl cutting. I would love to learn to cut vinyl, it's still a big dream of mine but unless you are already working in a facility with cutting engineers it is a very daunting and expensive skill to obtain, although it is not impossible. Saying that I have learnt a lot from various conversations with cutting engineers either casually or as part of a project. And there is still time!

Katie Tavini – Cutting vinyl is the main one, but no one's ever complained that I don't cut. I just give them the details of some good cutting engineers I know. But I had a good background in analogue formats from the first studio job I worked at.

Jay Hodgson – Well I've mastered a lot for vinyl, but I don't cut. The whole 'cutting vinyl' thing is a stodgy last ditch effort to gatekeep, IMHO, anyway. If you master for vinyl, you're usually preparing masters for cutting now. And that's fine. It's kind of like complaining about how kids nowadays have always had a screen to stare at while working, so everything sounds too much 'on the grid'. So what? Time moves on. I don't feel like I missed out on anything, really.

Jeremy Graham – If I were starting all over again from scratch, I'd definitely consider going back into higher education, but it's certainly not the only route to take, and as mentioned earlier, it's such an individual thing, down to your own determination, available opportunities, and what you're willing to sacrifice in order to follow that specific path.

Aside from the obvious musical/technical knowledge I gained during my years studying, I would say that I also had instilled in me a sort of lifelong camaraderie for those involved in the music industry, right from the get go. I don't believe this has anything to do with something you're necessarily taught per se, but more to do with an attitude that develops over time while you're immersed in that kind of environment. I'd like to think the same applies in the traditional apprenticeship route but I'm not really in a position to speculate in that area, as it just wasn't the path I went down.

From an individual standpoint, I've consistently been put in positions throughout my life where others around me are either older, wiser, and more experienced than myself, and it's forced me to raise my own bar in order to keep up so to speak, which I think has a lot to do with the route I went down, but that's just me, and I don't feel like I've missed out on anything really, it's just a different path.

What do you think are the advantages of being self-taught, if there are any?

Nick Cooke – Possibly not inheriting bad habits, although of course it's absolutely just as likely to develop your own. Higher education offers a good starting point, particularly in discovering and learning about the basics of the industry, but if someone knows they absolutely want to do mastering, perhaps getting stuck in is more efficient than completing a degree course first. As far as I'm aware, the basic aspects of mastering haven't changed much so can be self-taught through the many mediums available. Hopefully the continued exploration, experimentation and creative side of mastering will also develop naturally though engineers that are self-taught.

Katie Tavini – Freedom to experiment, make mistakes, get a good solid network of people who I can trust to skill share with, the ability to work on the types of music I'm most interested in (particularly the DIY scene and classical music, I'm not really interested in big pop artists for example).

Jay Hodgson – It is 100% your creative art, then. You were led to it, or drawn to the craft, and you have created a process from a void.

Thus, your work is entirely creative. You are not beholden to notions of right or wrong, or good sounding vs. bad sounding. That means you can just do the work. That's an advantage, to my mind. But then, I wouldn't know the obverse. So who can say?

Jeremy Graham – There are certainly pros and cons with either route. If you're able to critically assess yourself alongside the information you absorb, I think being self-taught can be a real advantage. However just like higher education, it's not a path for everyone. We each learn and pick up skills differently so it really depends on the type of approach you want to utilise along with your own learning style.

There's always value in learning first hand from others while you're in the same room, where you might have those light bulb moments far earlier on than you might have otherwise, and that can apply to both higher education and apprenticeship routes, but that also doesn't necessarily mean you'll gain more or less, it depends on the environment, your own situation and the attitudes of the person/people you're learning from or with.

Either way, you have to be able to communicate effectively with your peers, colleagues, and ultimately your clients; the right attitude has a big part to play in my opinion and experience.

If you were asked to write a higher education course in mastering what would it include?

Nick Cooke – Conferences and meeting mastering engineers would be my top priority. Plus chances to really listen to music in a proper mastering facility. I don't think teaching specific methods are that important, as mentioned earlier, these can be pretty much self-taught. More focus should be on the overall goals of mastering and creating differences in feel. Therefore, providing space to experiment with different methods would be important.

The biggest skills required I think are people skills. Mastering is often not just doing what you think sounds best but making something how someone else wants it to be, which can often be difficult to pinpoint. My main advice would be to not to over complicate your approach. I find simple is usually best, and its ok to go back to the beginning and start again.

Katie Tavini – Focus on critical listening (musical dictation although not enjoyable has really helped my work, as it's enabled me to pick up on areas which can be improved in the mix super quickly). Ditto for music theory – it's great to be able to talk to clients in their language and makes communication so much quicker and easier. The importance of metadata! Developing your own tastes and learning how to enjoy music you wouldn't normally listen to (because I don't know any mastering engineers who only work in one genre). Business skills – good communication, being nice to people, tax returns, how to find work. Understanding formats such as disc description protocol (DDP) and MFiT, now Apple Digital Masters and so on.

Jay Hodgson – I've actually done this. I teach them my process, from track arriving to the final emailed goodbyes to the client at the end of the process. I take them through what the goals are, and how I achieve those goals. And then I really push them to develop their own processes. In this way, I create the same conditions I learned under, but in a controlled 'safe' way. I honestly don't believe there's any other way to learn this craft.

Jeremy Graham – If I really had to, and in a very broad sense, I'd likely focus the majority of the course time on ear training, whether that was musically/sonically, micro/macro, attentive/passive, space/depth, height/width, etc. in order to help engrain more readily usable listening habits earlier on, despite the intended role or application; I think it would benefit all involved across the music industry to have a fundamental foundation of critical listening skills.

I'd also factor in some kind of critical thinking element as well (specifically tailored to the music industry), as this can really open up your mind to other possibilities and allow yourself to have more unbiased opinions, especially while researching new topics or information; it was certainly valuable for me during my studies and something I still use everyday.

What do you see as the future of audio mastering and how you might advise those coming up in the field?

Nick Cooke – I think there is a big future for mastering although it is constantly evolving as it always has done. I'd advise people to keep up with the different formats and industries developing: we already have streaming, gaming, apps, film, and TV as well as the standard music business and the syncing and production music areas are stronger than ever. There will always be new formats and mastering applications arising.

Advice that I have been given that I would pass on is to not undersell your services. As an engineer, look at how you fit in with other engineers, keep prices reasonable but don't undercut people or you'll be undercutting the industry as a whole. We need to keep quality high and value that quality.

Katie Tavini – Over the past year I've seen more artists and mix engineers wanting detailed feedback in order to perfect a mix. I've not really mastered long enough to see trends come and go, but it really seems like clients are valuing mastering engineers as a fresh set of ears this year rather than 'just making stuff louder'. I've also had to learn as much about the rest of the music industry as possible too, as clients often ask me which distributor they think would be the best for them, can I put them in touch with any music blogs, what different types of royalties can they claim etc. So even though I'm just doing mastering, I get asked a lot of other stuff too which is becoming more of a thing lately.

Jay Hodgson – AI-based algorithms doing the transferring work, and the rest of us doing the creative work. I recommend being as much of an artist as possible, and shaping what you receive to suit your ears. Everything else is being done by machines now. And will continue to be for cheaper and cheaper… my creativity is the only aspect of my mastering that can't be automated. So it's the only thing I really care to develop and invest in. Of course, that means I'll lose some clients. But they didn't want my mastering anyway; they just wanted their work to be mastered. Big difference.

Jeremy Graham – As technology improves for everyone who's involved with the creation or process of making music, I'd like to think more emphasis would be focussed on the core of the song or production i.e. how to best translate the intended feeling or emotion across to the audience, whether that's at the recording, mixing, production, or final mastering stages, even how the visual elements are presented from cover art, through to artist presentation. Although this concept is far from new, I feel like technology is becoming less of a focus and more of a tool or foundation to extend expression, however one may choose to use it in order to do so, musically or otherwise.

In terms of mastering specifically, and from what I can infer from at least my own experience, is that the right mastering engineer can offer a lot more than just the final creative decisions and format deliverables for a project, and can be a real asset to the artist and their projects over the long term. As budgets appear to becoming smaller (or perhaps just consciously tighter), I personally think it's more vital than ever before to retain real lasting connections throughout your immediate team and become invested with a mastering engineer who can work alongside you as a source for not only technical knowledge, but who can also assist the artist in collating and expressing their vision from a trusted and committed outside perspective.

Some advice I try to remind myself of from time to time is as follows; listen to and develop your intuition, practice patience with yourself and others, be prepared to play the long game (not just the innings), take your time (but hurry), and you're only as good as your last gig. Finally, talk with as many professionals and peers in the industry that you can; those interactions are invaluable, however fleeting they may appear at the time.

CONCLUSION

The pool of early to mid-career mastering engineers selected here have all launched their careers in a somewhat self-taught paradigm. This is not to promote this method, or route to professional mastering is by any means the norm, but there does seem, to the author in his research, to be more acceptance in this route. As many of the interviewers note, they do not cut to vinyl, which is often seen as Jay Hodgson notes as the 'gatekeeper' to the dark art of the field. To many in other spheres or generations would uphold this mantle. To others this is a fairly insignificant feature of the tonal adaptation and improvement that occurs before the cut is made, or the 'bounce' is done.

Had time permitted, further questions beckon around what each engineer has learnt and how it has been achieved. There have been discussions of 'trial and error' but also Jeremy Graham being brutally realistic about being honest at the same time as not being too hard on oneself. These appear to lead to further discussion at a later point around the soft-skills and personnel factors in audio mastering, at a time when non-attended sessions are most definitely the norm.

Of particular note is Jay Hodgson's striking comment that he believes mastering engineers are potentially all too critical, and that the approach of 'what the client approves is what the client gets' is his rule no. 1. This suggests a question for further research that is 'when do we know when the mastering is done'? There also seems to be a sophisticated psychological dynamic at play here, which is sadly not addressed in these interviews, but warrants further research.

A concurrent theme through the interviews is the clear professionalism demonstrated. Each interviewee highlights the importance of being able to listen and to adjust music for the improvement that can be offered by a positive experience in mastering.

As a snapshot of early to mid-career mastering engineers in the new decade of the 2020s, the aim of this interview has been to provide interesting insights into their training and their opinions – just at the point when audio mastering may change its identity forever.

REFERENCE

1. Golding, C. R. Rosen, D. Hodgson, J. Hepworth-Sawyer R. 2012 *Interview With Ken Scott. Journal of The Art of Record Production,* Issue 7.

11

Peak-Fi

HAVE WE REACHED THE ULTIMATE REPRODUCTION STANDARDS NEEDED TO MAKE HUMANS HAPPY?

Crispin Herrod-Taylor

INTRODUCTION

I've spent most of my life trying to advance the quality of audio recordings, and reproduction, through the design and sale of effective electronic devices. Over the last few years, however, I've slowly begun to look at what people in the industry are actually doing, separate from what they are saying. This essay reflects my experience and current thoughts on the subject of audio quality, and how it's being used today as a sales technique, rather than a pure or necessary aim. A quick caveat: this is a pure opinion piece. Feel free to disagree with me, but I think you'll find that I'm right... ☺

IN THE OLDEN DAYS

Let's go back to the late 60's and early 70's, when the young author was stuck on a farm in the middle of nowhere.

My first real memories of music were recordings. Off my old man's collection of 50 & 60's Jazz and Big Band records. They were played from this piece of furniture, like a sideboard, but with a turntable in the middle. Perhaps because I grew up with it, it seemed really normal, not at all magical, that music could be released from these black discs of plastic. This was just how music happened.

And what music it was! Faced with unchallenging telly or music, music won hands down. In a 1970s world, there was nothing else to compete with. No video, no PCs, no arcade games. There was alcohol, sex and drugs of course, but even these were allied with music. From where I stood, I couldn't get enough of the music from the weird sideboard, it stirred up emotions I didn't know I had. It moved me, it motivated me, I wanted to dance, sing and groove. In an otherwise dull 70's existence in the middle of England, music was transforming.

The only "real" music I heard was from choirs in church or (again) my Dad's so –so trumpet playing. Now it was very clear that the trumpet sounded completely different from those on the recordings, but I somehow just accepted this. The recordings had a lot going on, lots of instruments and voices, so perhaps this was how things were. The gulf between real and recorded didn't bother me, I could get energy and emotion from the recordings, no problem. A bit later I got a transistor radio, which I would listen to pirate stations and radio 1 on. It was pathetically tinny, but again you could easily extract the music from it, and the music was amazing, so different from Jazz. It was wild. Man.

Eventually my big brothers came back from University, bring with them loads of cool 60's and 70's records and Hi-Fi's. Music no longer came out of sideboards, instead it came out of separate boxes, all joined together with wires and it looked really high tech. It sounded good, really good, much more bass, more level and top end than the sideboard. By this time (about 10 years old) I had already shown an interest in breaking things apart and putting them back together, so I started to investigate what was causing this change in sound. Was it the new bands, just playing better, or something in the Hi-Fi? I remember playing a Jethro Tull album on the sideboard and then in comparison on the Hi-Fi. For reference I then played some Acker Bilk on the Hi-Fi too. It was obvious that it was the Hi-Fi that was releasing more music. I wanted one. Now. So started a career in trying to make things, to make music sound as awesome as possible.

I won't bore you with my career path, but I bet it was common to a lot of you. I mended, modded, borrowed and built gear, working with bands and then finally with recording technology. My route went via University to study Physics, 'cos I wanted to know how things worked, via top end HiFI manufacture, ending up at the incredible Solid State Logic (SSL). My last stop before Crookwood was with Focusrite. All in all, I've spent some 45 years making music gear, most of it at the very highest level.

At all stages of this path, I could or thought I could manipulate technology to improve audio quality. The more experience I had, and the more budget, the better I could do it. There was a clear relationship between engineering and audio quality. It all seemed so straight forward.

AND THEN IT ALL WENT WRONG

I can't point to a definitive point in time, and I guess in reality I knew there was a problem looming. But these are the pointers I started to get worried about.

Analog consoles

Moving from Hi-Fi to SSL, I was excited to see what the pro guys were doing. I mean this was the gear that was making all the records that my Hi-Fi gear had been tweaking. I was amazed. Well horrified actually! In Hi-Fi we were removing as much circuitry as possible, but in pro audio everything was littered with capacitors, Op-Amps Integrated circuits (ICs), and miles and miles of cable. Not even Oxygen Free cable. Why didn't it sound like effluent?

The simple truth was that the electronics needed to be there to create the user experience which enabled the engineers to produce the amazing music I was tweaking. But if the sound degradation after going through perhaps 30 ICs, 60 electrolytic capacitors and 200M of cable wasn't actually that bad (it was in fact very good on bypass on the small faders on an SSL), what had I been worried about in the Hi-Fi environment?

Lesson 1: Don't assume audio quality is determined by the electronics. A little knowledge is a bad thing.

NE5534s

In the 80's, any respected HiFI buff was using the latest Burr Brown OPA something Op-Amps. Really fast, low offsets, precision devices with impressive looking spec sheets. Pro audio used loads of NE5534s – an "inferior" 20 year old design (Edit: they're now nearly 50). Surely if we replaced these with the OPAs it would sound better? Err No. I tried. It became apparent that the pro engineers knew what they were doing. They were actually engineers with experience. Stability, suitability and reliability beat spec sheets. OPAs were often unstable (too fast) and couldn't always drive the required loads. Sonically I couldn't hear a difference in a controlled environment when I got them working. Prior to that point, I just accepted the logic that faster was better.

Lesson 2: Experience beats excited imaginations. Focusing on one area of technology does not make for a better whole product.

Digital and the Media

I know my maths, and it was clear that digital systems were going to take over the planet. They were essentially a perfect method of moving audio around and storing it. Far superior to analogue tape, vinyl, FM and all the old analogue technologies. Let's be succinct here - if we engineers did out job properly, digital audio transmission and storage should be perfect. QED.

However, I remember hearing one of the first CDs in a Hi-Fi showroom (Dire Straits from memory) and it was certainly different, but stupidly shrill and slightly unpleasant. Subsequently I see that there wasn't the experience out there to master the new format at that point, and some of the converters were pants, but again at SSL, I saw and heard some great digital stuff, so it wasn't inherent in the format.

I also saw a media backlash. Punters loved the convenience and the high-tech nature. I believe on the whole the audio quality wasn't really a factor, other than with bragging rights. The immediate sonic benefit over vinyl was easy to demonstrate however: background noise. But the Hi-Fi media started a purist stance and a whole load of urban myths arose. Even today these still haunt us – pictures of bar graph samples under perfect sinewaves = it's obvious that digital is missing signals isn't it? And so on.

The reality of course is very different. All of the purists were listening to stuff that had at the very least been digitally multitracked, fed through digital delays and reverbs, composed and created by engineers and artists, not extracted by magic from the artists soul. But these myths continue and have become mainstream. Today there is a strong market for "special" digital SPDIF, optical, USB and HDMI cables, all extolling the virtues of audio improvements from these magical devices. Heads up: digital cannot be corrupted over 1–2M of cable. Jitter and RFI performance of this length of cable can be measured, and probably ignored. If it truly does sounds different, your DAC is inept, replace this first. Thing is, people review these cables and comment on them. Other people respect these reviews. People buy the cables and believe. Science is not allowed to have a voice.

Lesson 3: Maths is pure and true, but difficult to understand. Humans trust other humans and want to believe. There are too many enthusiasts in the process.

Selling Crookwood gear

Finally running my own company, I have to sell stuff, as well as design it. This has given me a much wider understanding of how people think and what motivates them to purchase. I think we make some good gear, and usually we do very well in unseen shootouts. But I'm not the best marketer and I spend most of my time on the electronics, rather than promotion. Consequently, I lose sales to other competitors, that I really should win. Why is this? Well, it could be because I'm kidding myself and our gear is rubbish, but I don't think so. It's more likely to be because people buy with their eyes, they buy based on fear of making the wrong choice, they buy under peer pressure, they buy with short term views and lots of imagination. In short, they buy because they've been marketed to. Oh and if they buy something so-so, they aren't ever going to admit they made a mistake.

Lesson 4: People don't buy on technical audio quality, they are influenced by other people's opinions. If you build it, they will not come.

Summary

The result of my audio experiences over the last 45 years or so, can be summarised thus: There is a huge gulf between what people believe is happening, and what is actually happening.

HERESY

The culmination of all these factors meant that around 2005, I started to get concerned about what people were saying, versus what they were actually doing. So here I am, about to put this on the record, about to be labelled an audio heretic. Despite the fact that I've got a solid track record, I'm sweating slightly as I write this – will anybody ever buy from me again? We're not very good at digesting disruptive views.

Bear in mind, I'm not talking about processors, bits of gear that are built to add a sound, nor am I talking about transducers (speakers, microphones etc) that are flawed electromechanical beasts. Rather I'm talking about the glue in the studio- amps, routers, converters etc, the stuff that should be invisible.

So here goes.

- I believe that science and knowledge is being dismissed, and people are being manipulated emotionally by marketing for commercial gains. Fake audio news is real and prevalent.
- I believe that we already know how to make audio gear that is within a percent or two of perfection. Indeed, this gear already exists in the world today.
- Further, I believe pursuing audio perfection to the nth degree is more about engineers being bored and wanting to mentally challenge themselves, than progressing audio science. This manifests itself in our current gear lust, believing that technology purchases will give us a shortcut to the top, an edge, bypassing the tedious process of spending time learning your craft.
- I believe that the end consumer of music values convenience above everything else. They are not fussed about absolute sound quality, rather look for experiences and ease of access. The quality they experience now is more than sufficient for their needs. They have no drive, no need for finer reproduction. They can however extract emotions and energy from the music they listen to on their MP3s quite easily.
- Finally, I believe that if you spend the time examining your own studio, understanding all of the elements and practices you use to make music, you will reap the biggest gains in audio quality. These elements get ignored because they are harder to achieve over just buying a new bit of gear.

Phew. So that's out in the open, and I'm still here. So, let's be clear, I'm not saying that audio quality doesn't matter, just that if we look around and choose carefully, essentially perfect gear is already available. To make better music, we need to focus on the whole production system and our own abilities instead.

THE CASE FOR THE PROSECUTION, SO WHAT'S THE PROBLEM WITH AUDIO QUALITY PER SE?

The first issue is that it is relative. Audio quality has to have a ceiling where we can say it is perfect. However we treat it as if it is infinite. We measure relative improvements rather than absolutes, and as shown further below, these relative changes are so susceptible to influence. We end up believing that A is better than B, when in fact they are the same.

Some of this is down to the lack of scientific processes in choosing. Science has had a long time to work out how to prove things, removing variables and influences, so you can say with certainly whether A is better than B. It's just that we don't use these processes in audio. Hearsay is stronger than fact. And if we're honest, audio is full of enthusiastic amateurs, who totally outnumber the seasoned pros. Lack of rigidity is good for creativity, but bad for evaluation of gear.

There's also the question of what is better? My preferences may be different from yours, and any recording or recording process is a piece of complete fiction. A musical story. Some are based on a true story, but we create the sound, and manipulate it in the studio. It has no reference to an absolute rendition. Even classical recordings are engineered to produce a view, a sound, by choice of mics, amps, location and balance. What are we actually comparing things to? And despite knowing a lot about stuff, we still can't weigh the elements of reproduction to say with any certainty, what elements matter more than others. For instance, is a modern MP3 better than my Dad's vinyl sideboard? They certainly sound different, and I think the MP3 would win, but how can we quantify this?

Finally, who are we making this music for? Ourselves or the people who pay us? Or the end consumer who pays us all? At what point do we let our work ship rather than constantly reworking it? What standard does it need to pass?

WHAT DO WE KNOW

Quantifying audio quality really starts with understanding how people's hearing works. Over the years there's been an enormous amount of work done on humans and animals to understand how we hear and what we can hear. While a lot of the original work was done early last century, it's been repeated and verified, and with every new bit of medical scanning tech, new understanding is reached about how we hear.

It's interesting to note that we evaluate audio very differently from video or photography. With audio we generally use a human to evaluate it. With visual stuff, we use machines. This arises because audio is utterly transient – it only exists while it's being played, and it ceases to be musical when snippets are played on loop. In contrast, pictures and images are constant, and only change when the light on them changes. We have machines that can measure the light intensity and spectrum from every section of an image and then repeat the same scan on a reproduction, even if it's been recorded and transmitted to a screen. It can tell you with pinpoint accuracy if any pixel has been corrupted in terms of level or colour to a far higher degree of accuracy than the human eye. This absolute mechanism means that it's really easy to characterise and prove the accuracy of any device in the video chain - lens, camera, lights, transmission, and monitor. Subsequently we don't have a video industry telling us that this monitor reveals subtle highlights when used with this HDMI cable, floated on a bed of seaweed. It's just too easy to disprove it.

Contrast this with how we qualify audio quality. Our distortion figures are pretty useless in labelling quality, as are our frequency response charts. They can indicate certain things, but provide a very limited view of the perceived quality of a device. Indeed, in the early 70's there was a rush to make amplifiers with vanishingly small levels of THD, as measured in the lab. They all sounded pretty average. To my knowledge there is not a definitive set of measurements you can use to determine the audio quality of a device (although each vendor may have their own secret sauce). Hence all quantification is ultimately done on human test subjects.

It's further interesting to think about MP3s, the hated format that "ruins" sound. When the MPEG people invented it, they had a very good understanding of what humans can hear. It wasn't knocked up after a bet in the pub, they really looked at what elements of sound most humans **can't** hear, and removed this data to compress the audio. It's shouldn't then come as a surprise that most consumers can't tell the difference between a CD and the same source as a MP3.

But we should note that we're talking about most humans here, not all. Guess what, our abilities are all different, and are probably Gaussian distributed (bell curve). This means that there are always going to be some folk who can hear something that the average and 1 standard deviation's worth of human can't hear. And we know that we can train humans, exercising their ear muscles so they become ear fit and can discriminate better than the average. There will always be a limit however.

The last part of the puzzle is that the ear as a device, is a transducer (i.e. flawed), and it's a pretty crap one at that. Fair play to evolution, it only had flesh and bones to make an ear, but our mics and electronics walk all over the flesh things bolted onto the sides of our head. If you monitored the electrical impulses coming out of the cochlea, you'd wonder how we can actually hear anything.

The real secret to our hearing is the amazing ASP (analogue signal processor) that is part of our brain. This translates and extracts the lovely sounds that we hear from the poor incoming data stream that our ears produce. It is utterly remarkable. The main takeaway here is that, what we actually perceive as hearing is controlled by our head. Our hearing changes if we're tired, drunk, stressed or in love. Our current emotional state and environment influences our hearing far more than the use of a particular capacitor in an active feedback loop.

HOW TO WIN FRIENDS AND INFLUENCE PEOPLE

So, knowing all this, how can ensure that we make accurate decisions re the ability of audio gear? Or conversely, being naïve or evil, how can we make people believe something sounds better than it really is? Let's make a list.

We can divide this into two main areas, things that we know we can actually hear, and things that influence our ASP to make false decisions. The real stuff is all related to how our hearing evolved to let us communicate and protect ourselves from danger. The second thing is all to do with our insecurities and our need to fit in with fellow human beings.

Condition #1 – audio level intensity

That's loudness measured in dBs to you and me, and any fule kno (Molesworth) that comparing devices with different audio levels is the most obvious way to make poor decisions. The general population can resolve about 1dB of difference. Mastering engineers reckon they can easily do ½ a dB, so for safety, we should calibrate stuff to within ¼ dB or finer. Many people trim to 0.1 dB, which seems like a good engineering aim. If you've ever tried to trim stuff to 0.1 dB, you'll know that this is not always as easy as it sounds, and 0.1 dB is only an accuracy of 1%, pretty rubbish for most scientific measurements. With audio however, the problems come thick and fast.

Firstly, we normally trim at 1 KHz. There's probably a good reason for this, but I think it's now habit, when you bear in mind our peak hearing sensitivity is around 3 KHz. But what if the device doesn't have a ruler flat frequency response? Most electronics should be flat, but don't bank on it. Transducers will have response variations well over 1 dB across the frequency spectrum, but compressors and EQs when set to flat (not relay bypassed) are almost certainly going to wobble. For example, if you compare two EQs set to flat, and calibrated at 1 KHz, and hear a difference, what are you actually hearing?

Secondly, don't expect the environment or your setup to be stable. Any change can uncalibrated your nicely setup levels. A fun example from the field was an EQ with a high output impedance, but with a relay bypass that bypassed the output impedance... The setup had two EQs in series, and the idea was you'd bypass each one in turn to hear the "sound" of the other. But all you heard was EQ #2 rising in level by 1 dB when EQ #1 was bypassed.

We then have all the issues that let the loudness wars exist. Any compression can alter the perceived loudness of a device. Calibrating with a static tone may not reveal issues. Comparing compressors, or processors is fraught and even at a simplest level, running small speakers loud for a long time will cause voice coil compression, so they will drop in level compared to a larger, better vented speaker. It would be very interesting to also measure overall loudness levels of room correction systems. Again, do we really know what change we are listening to?

Finally, we need to think about absolute as well as relative levels. We all know our equal loudness or Fletcher-Munson curves (or IEC 226:2003 for the most accurate ones), but we still don't work coherently across the industry with monitor levels. If I compare something at one level, but my colleague Crispina Horrid-Tyler does the same test at another level in another room, I'm pretty sure we'd have different conclusions on the merits of certain pieces of gear. In this day and age, we really should all be working at an industry standard level when making production decisions. Sure, ramp up the level for fun, but let's start working together at a standard given level.

One more thing (so it wasn't finally above was it?), our perception of loudness re frequency alters with training, age, exposure and fatigue. So even if I have a perfect setup, somebody may prefer a coloured device that fits their hearing curves better to a more accurate flatter device.

Oh yeah, I forgot to mention phase. I don't think you can hear phase (others disagree), but you can hear its effects sometimes. Make sure everything you check is in absolute phase to each other.

Condition #2: – clicks and gaps

In order to avoid getting eaten, our hearing is very good at spotting sudden changes in loudness, like twigs breaking etc. You know that experience of lying awake at night, then suddenly hearing a noise? You head feels like it's on fire, you're examining every little sound you hear with a heightened sense of awareness. Well, odd clicks and sudden transients can cause the same high sensitivity during listening tests.

I made a quite sophisticated A/B/X box to help me decide what I was really hearing. Unfortunately, the way I designed it, I could hear the clicks the relays made when switching between devices, and one direction's relay click sounded different from the other. I ended making false preferences based on me subliminally listening to the relay clicks! I corrected the issue, but I became aware of other possibilities – if I switched at certain dynamic points in a song, I could feel myself over analysing the sound. I resolved to switch during quieter passages, or on the beat to minimise the musical jarring, and not to make decisions based on audio around the switch point. I also redesigned the box.

It turns out there are many more audible clues that feed create subliminal drives. We're very good at judging timing with music, and we can easily identify different gaps of silence between tracks or switches if A/B ing. If these gap differences are regular, we subconsciously and automatically label the A and the B devices depending on the gap.

Similarly, if you're comparing stuff and you experience background noise, this will affect your ability to judge. Computer fans, doors opening, traffic human or motorized, all will affect your judgement. To be at your best, you need to be relaxed, and comfortable in your environment.

Condition #3 – listening environment

All of the above works fine in mono, and in some ways mono is a good way to listen to stuff and evaluate it because of its presentational simplicity. However let's be honest, apart from circuit designer geeks like me, everybody else will only listen and judge stuff on a stereo system.

You can listen with a set of cans (headphones), but for me they give a very artificial presentation, that make it difficult to judge. I'm always listening to a side effect rather than the music. They're great for low level detail however due to isolation, and I guess if you spend a lot of time with them you'll get used to the presentation and you can use them to do anything. But for most people, we're going to be listening in a room. With speakers, imperfect ones.

If it's your room, you will have got a very good mental ASP preset as how things sound. This helps you detect changes pretty well. But strange rooms, or your room with strange speakers or listening out of your normal position, will mess up your ability to relate to what you're hearing. It is really easy to come to false conclusions in a strange room, as anybody who's been to a Hi-Fi show in a hotel can testify.

Lastly beware the nearfield. Because you're listening in the nearfield, they can act like a set of cans except that you can change the sound by just moving your head. A set of speakers working in the farfield is much better for judging absolute quality issues.

Condition #4 – What we see

We're now entering the realms of our ASP being reprogramed by the rest of our brain. This is the domain of behavioural economics. Normal economics works on the basis that all humans are rational and will evaluate choices appropriately. Behavioural economics however says we're all a load of children whose behaviour can be modified by the presence of sweeties (candies). Slight over-simplification, but you get the point.

Sight is our dominant sense, and if it's working overtime, our ears get to work in the background instead. This is really easy to illustrate. Watch a snippet of a movie, then play it again, but with your eyes closed. You know what's happening, but you become super critical about what you're hearing, as you put all of your processing power into your ASP, and the two feelings are very different.

So, having flashing lights, bargraphs, edit screens etc. in our eye line while working, will and does lower our ability to judge audio. They distract us. Screens also interfere with the direct sound from our speakers, but that's another story. This is one reason why old boys and girls with hardware kit can perform so well in the engineer's seat. They're not looking at the audio on a screen, and trying to use a mouse to turn a virtual knob, they're listening to the music, ASP on full, and they're adjusting a real hardware knob with their out-stretched hand, without even looking at the gear. Their ASP has all the processing power it needs, eyes are turned off.

The second problem with our eyes is that they load up our learnt preconceptions from what they see. More below.

Condition #5 – What we know

If we know something, it's very hard to un-know it and be objective. For instance, if I see the devices in an A/B situation, I will form an opinion, based on their looks as to how they sound. If I know which is A and which is B, the games completely over.

Similarly, if you A/B things with price tags on them, you will expect the more expensive item to sound better. You can't help it, your upbringing has loaded this bias into you. This follows with sizes – bigger boxes are better, colours – yellow coloured gear is frightening, finishes – how we perceive quality (think Apple), and so much more. Even reading a spec sheet, webpage or brochure will skew your opinion if you know this relates to what you're listening to.

In short, the more information we have about a thing, the more we will try and prejudge its abilities and qualities. We can't unlearn these prejudices, and in many cases they benefit us, saving us time and money from having to start from scratch each time. The problem is that short term salespeople know this, so pig's ears can be made up to look like lovely sow's purses. Certainly it completely messes us the A/B process.

Condition #6 – what others think

Finally we rarely make decisions alone. We often doubt ourselves, and want confirmation that we are right. Most of us don't want to stand out and be different. Conversely others enjoy being trailblazers and being first at any cost. All of these emotional elements affect our judgement, even when our ASP is saying one thing, our brain will override it. We buy status and position.

The most obvious place this goes wrong is with peer pressure. In a group, you may not hear a difference, but if everybody else says they do, you're unlikely to disagree with them. You may also be carried away with the group enthusiasm, and really believe you can hear a difference. In both cases, your need to fit in has trumped what your ASP is telling you. This is also where all the audio urban myths original from: you know digital is flawed, cables are as important as amps etc. A story, told well to the right people will spread and spread. And once it's big enough, it takes a lot of confidence to disagree in public with the majority.

Another place it occurs is when faced with an unknown competitor and a brand, you chose the brand. Nobody ever got fired for buying an IBM (a 1970s slogan). So even if the unknown bit of gear sounds better, you go for the brand because what would your friends say?

And lastly if you've bought something, unless it's really bad, you are never going to publically denounce it. So audio forums are full of people saying their last purchase is the best, even though they can't all be right.

Getting it right

So if you really want to make a judgement on audio merit between two things, how do you do it? Here's my guide to evaluating audio gear.

1. Try and work with only two things at a time. The more variables you add, the more difficult it becomes to know what's what.
2. Make sure you are relaxed and comfortable. This means you're doing it in your own room, when you're fresh, not when you're in a strange demo room, and not when you're up against a deadline. No extraneous noises, no phones etc.
3. Ideally use a proper A/B/X box. This will switch the signals and report the results properly. You need a number of results to know the truth, one or two won't cut it unless the differences are really obvious. There aren't a lot of A/B/X boxes around, but see me after school if you want to find one. Alternatively take great care with your setup.
4. Adjust levels at 3KHz to around 0.1dB with sinewaves, or you can use pink noise with a really slow averaging meter. Measure at the level you're going to monitor at.

5. Make sure there's nothing else in the chain that may mask the result, like limiters, weird cables. Draw out a picture of what's in the signal path before you start, to make sure.
6. Listen by yourself, not with others. If they want to judge, they can do it themselves after you've had a go.
7. Try to have as little information as possible in advance about what you're hearing, so you don't go looking for audio clues. Unless you want to use the audio clues you've learnt.
8. Don't know what A or B are. If you know option A is brand A and option B is brand B, you won't be able to make unbiased decisions. A good way to do this is with a colleague who connects the items up randomly without telling you.
9. Again, don't look at screens, meters, or the devices while A/B ing. You will get fooled.
10. Try and switch segments of music in the gaps or on the beat, listen for a few seconds before judging, don't base your decision on the switch.

It's not easy hey? And without becoming paranoid, it also makes you wonder what you're doing as you're working, A/B ing levels and processes. Thankfully, this process is individual to you and the job in hand, not an absolute test of your work versus somebody else. So, if you're happy with your work, that's all that matters.

SELLING OUR WARES

At the heart of all of this is money. Audio is a mature business, it has a wide penetration, and the technology to make it happen is mostly low tech and cheap. However, the scale is enormous, there are lots of people selling stuff, and everybody wants a slice of the pie.

The problem is that stuff today lasts a good length of time, it doesn't break like the old stuff. So in order to sell more things, we need to invent "new" things to replace the near identical old thing – we need to convince our customers that the old thing they have is deficient. Digital gave audio a boost, re-making analogue gear into plug ins, and adding new things you couldn't realistically do in analogue. But now there's not a lot more we can do to stereo in order to sell more stuff, other than making things cheaper. Basically, we've done it all several times over.

One area which we haven't yet got to, is where the consumers have the same tech as the pro's. In the olden days there was no way a consumer could afford or maintain an analogue setup, but digital changed that. It's now possible for a consumer to have the exact same production format as the studio. Digital wasn't made for audio folks, rather we piggy backed onto a much larger beast, aimed at consumers and pro-sumers – the Personal Computer (PC). The ever lowering cost of entry together with the ability to duplicate product or programs without cost, meant that a consumer could own the same tools as we professionals do. This clearly benefited us too, a cutting lathe, the prerequisite to mastering a record would cost about $750,000 in today's money. I can get shareware and a $500 PC to master an album today.

This has obviously created a few issues however. If you splashed out ¾ of a million dollars on a bit of kit, you were going to staff it properly. You'd train the engineers, invest in the surrounding gear, be productive and manage your investment with a long term view. Today, training and experience are not valued, certainly in terms of recompense. Why invest in hardware if you can do everything with a PC? Suddenly your labour bill becomes the dominant cost of doing business and we all know what this means. Musicians historically never were rich. Nor were roadies or musical helpers. For most of history they were quite poor, perhaps we're just going back to the historical trend lines? Certainly, as the end cost of music goes down, there's less available for everybody in the chain below the customer. My Spotify family account for instance, allows me in theory to listen to a track for less than 0.5p. People now expect music to be virtually free, and this puts pressure on everybody to reduce the amount of time spent on production (labour is the dominant cost), and with this, quality will start to take a back seat. The consumer doesn't care do they? They're listening to MP3's.

All the large corporations who've risen on the back of audio however need to keep the churn going. The need to get people to buy gear. Video successfully managed to get people to constantly upgrade their TVs to higher and higher res (or bigger and bigger), to buy Blu Ray over DVD, over VHS. And digital means the consumer can have the same formats as the pros, and this makes for a really powerful "them and us" marketing tool to make people shell out with their cash. But DVD-A and SACD just didn't resonate with the pubic, who in fact just went backwards and chose MP3 over CDs. And the audio public are buying smaller and smaller speakers to replay music. Completely in the opposite direction to their TVs. What's a manufacturer to do?

I KNOW, LET'S RELEASE HI-RES AUDIO. AGAIN

New formats create employment for everybody in the chain, but the public seem really against investing in new audio formats. Surround sound was an obvious area where the public could invest, and unlike Hi-res audio, the differences are really, really obvious. Yet it just hasn't taken off, despite every DVD since the year dot having 5.1 streams on them.

The reason seems to be to again due to convenience. It was too fussy, too complicated, too expensive to add extra speakers and amplifiers to the telly. The sound from the TV was (unbelievably) good enough. Contrast this with buying a bigger TV. This was a no brainer, one box, same connectors, no learning required, instant gratification. Today we might buy a soundbar for the telly, one box, not too big, HDMI in, easy. The upgrade is again easy to demonstrate, more level, more bass, more impressive. But we're not buying 6 speakers and amps. The hope with hi-res audio this time, is that it will be more convenient, and people will buy in their droves. But there is a problem. Hi-res will not provide a better experience for consumers. It has low bragging rights, unlike a large TV and it's really difficult to A/B differences from CD to Hi-res audio. This is an area where the comparison with video tech doesn't hold up.

Video has always suffered from bandwidth issue, specifically the amount of data that had to be moved around in a given time. The tech was more expensive because it was faster, and you had to effectively downsample the original data to get it to be transmittable across the airwaves in analogue. But the bigger the screen, the more the original image is magnified, so the need to capture and then distribute Hi-res video became really easy to quantify and understand. Even video is starting to hit the limits now, with 4 and 8 K needing very big screens to show a difference over 1080p. Our equivalent in audio is loudness – if you're close to a speaker it doesn't have to be very loud, but if you're 200M away in a field, the speakers need to be much louder. However, adding more bits or increasing the sample rate doesn't make the far away speakers sound better.

It turns out that since CD, we've been working at higher resolution that humans can process, so increasing this resolution really is driven by a commercial and engineering lust, rather than science. This subject is very marmite, because we've all heard stuff and want to believe, but see above for why you can't always trust what your head is saying about your ASP. I think a lot of us are living in echo chambers, only talking to people who believe in Hi-res.

Let's look at the issues.

Bit depth

Because we engineers manipulate audio a lot during the production process, we need more bits to play with than the production release, and 24 is a good number. In the box (ITB), we may even work in 48 bits plus or floating point equivalent. However once it's processed, releasing the final effort in 16 bits is more than adequate, and if our SRC's (Sample Rate Converters) are good, we shouldn't be able to tell any difference.

Not convinced? OK, so at the start of the chain somewhere there's an analogue source, perhaps a vocal, or heaven forbid a real band. The noise floor of most mics is about −116 to −120 dBu. We typically add 40–60 dB of gain to them, so our noise floor is now worse than −80 dBu. Our mic amps and ADCs clip at about +24 dBu, so that's a

S/N of about 104 dBu, or about 17 bits. And that's before we degrade the noise figure every time we mix tracks together.

These figures fit very well with a human's ability to hear sounds of different levels (the best we can get is about 120 dB dynamic range as long as they haven't got hearing damage, or tinnitus). It's also well above the most dynamic music compositions. As an aside our best DACs and ADCs have a noise floor around the 20 bit level, once you get rid of the flattering A weighing figures, and 24 bit pro-sumer DACs have noise floors around 17 bits. And yes, if properly dithered you can hear music at a level well below the LSB in the noise, just like analogue.

24 bits are enough to capture everything. When processed and released to the public, 16 bits are enough to replicate human music and distribute the finished production file.

Sample rates

Sampling theory is only a theory, but given that in every digital system that uses it, from washing machines to spacecraft, the theory works perfectly, can I just ask us to just accept it now? You don't have to understand it, the same way you can't understand exactly how your phone works, but you have to accept that it, just like your phone, works. Unless you've just sat on your smartphone of course...

We are also pretty sure that we can't hear above 20 KHz, and the quality of music at 20 KHz has limited value. It's a really easy test to do with your DAW's oscillator, just see how high you can hear. Now try it on a teenager, who's been wearing cans all his/her teen life. While you're at it, try it on a 20 year old raver, your elder peers and any musicians who've been cranking up the volume all their lives. I reckon with modern life, most people have a fairly hard limit of 12 to 15 KHz, and some a lot less. Also because of the slope of the cutoff, raising the volume of the HF tone has little effect on our ability to hear it. High frequency hearing amongst humans will follow the bell curve, and some will hear a 20 KHz tone clearly, but nobody can hear a 30 KHz tone, and these HF tones don't sound nice.

Again there are engineering reasons why it's easier to double the sample rate to say 88.2 or 96 K, so when we're working we don't have to worry about analogue filtering, but we don't need to extend it any further. Nor do we need to take up twice or four times the bandwidth to transmit the same music.

But I can hear it/the public want it

Yup, I've heard it too, but then another time, I've also not heard it. And under my A/B/X setup using my current generation of ADCs and DACs back to back, I can't hear the difference with any statistical certainly. So what's happening? The first point is that it's subtle. It's not like comparing a CD to Vinyl. I think most of it might be covered by the "how to win friends" section above, but there may also be some abnormalities with digital filtering approximation's which might explain some of the differences. It's worth remembering that most/all converter chips sample analogue at about 4–6 MHz using a sigma/delta approach (like DSD), then apply internal digital filtering to give you the PCM or DSD output of your choice. Or vice versa for DACs. That is, the fundamental sampling mechanism is unaffected by the input or output sample rate or format DSD/PCM. Anything we can hear, is then down to the application of the maths in the converter chips, rather than the sample rate. Perhaps some filters sound "nicer" than others, perhaps some are "louder", but if we want to work out what's happening, we need to start here.

That the public want it, is clearly another half-truth. There is no drive for hi-res audio. They chose MP3 over CD, Flac has been around forever, but isn't widely used, a ubiquitous YouTube video stream isn't far off a CD quality audio stream rate, yet every attempt to date to mass market hi-res audio has failed.

There's often reported interest from polls and focus groups to illustrate the public's want for high quality audio, but these again are influenced by our known biases. There's a classic example of a manufacturer employing a market research group to see if there was any mileage in a new product. The results came back really glowing, everybody wanted it, what a great idea it was etc. Just to be sure, they commissioned another group to check. This group got the same results, but at the end, they offered the participants a $20 note or a more valuable discount voucher against the wonder product they had just been talking about. Everybody took the money, nobody took the discount voucher. Go figure. And question the methods focus groups and market research use to gather data.

WHERE DO WE GO FROM HERE?

I see the drive for widespread consumer hi-res to be a complete red herring. We don't need it, and it won't make the world sound better. It's a commercial drive to try and create more revenue from a base of audio consumers. It's the equivalent of a CES vendor showing off a robot to fold napkins for your dinner table – just because we can do it, doesn't mean we need it. So what will make a difference to the quality of audio we sell and make?

A return to CD quality

CD format wasn't a wild guess. It's the optimal audio data rate for humans. There is no reason not to move back towards this being the default rate for streaming. We can easily losslessly compress the data if we need to, and storage is not a problem. Can we go back to this as a standard please? Rename it if you need to, in order to sell it, I don't care. You can even make it 24 bits if you must, but let's make this the standard bearer of audio rather than MP3s. While we're at it, let's also try to use the dynamic range it offers, rather than just playing loudness wars with it.

Love your hardware

There's constant pressure for everybody to lower costs, and as digital tech can emulate hardware well, you can do an excellent job ITB (In The Box, that is inside a DAW without any outboard hardware). I have three issues with this route: Your focus on a screen, your efficiency and the ability to replace you with a machine. I've dealt with the ideas that your eyes need to rest while listening above, and at the very least, you should get a hardware control surface to be efficient. Imagine typing a novel using a streaming remote on a TV instead of a keyboard! While a mouse is clever, it's no substitution for a real fader or knob. Real hardware or a surface will increase your efficiency hugely and help you be more creative.

The machine replacement issue is real. If you're making music by numbers, a computer can do it better and cheaper. There are so many companies that want to get rid of those pesky audio engineers, throwing them out onto the street and trousering the profit. The more you work to a formula, ITB, the easier it is to replace you. Hardware gives you an edge and like a vinyl record, it's not easy to replicate. Consider what you can do to legitimately avoid being replaced.

Niches for everybody

The world's a big place. Just as there are a few people who've invested in kick ass surround systems, there are a few people who would love 192K/24 bit sounds. And others who love vinyl, ½ tape or even cassettes. Niches are good, you can often charge for exclusivity, and scarcity and everybody is happy. Just don't try and force it on everybody. If you believe and you can find others that do, great, focus on your niche, and get good at it.

Experiences

Live concerts, VR, interactive and immersive sound are just a few of the areas where advances in audio are being made, and where the results are exciting. Not at home, but in the company of others. The buzz you can get from these events, well executed is huge and restores your belief in music as a power for good. It doesn't matter that you can't (and shouldn't) replicate these at home, it's important that people identify great sound with effort and expense. We need to establish value and experience. Good music is not free.

More audio design engineers

We're a dying breed. Most of the best engineers out there are starting to look at retirement brochures. Sadly, the money just isn't in boring, slow speed, analogue electronics. Graduates want to work in DSP or communications, where they'll get paid a fortune. Those who do end up in audio work, often don't understand audio, and do things by the numbers. If you look at most analogue circuits these days, they're straight out of the manufacturer's application notes. To be a good engineer takes a lot of experience, an understanding that all engineering is an approximation of the true maths, and you need to think holistically. A printed circuit board is just as important a component as that fancy Op-Amp. Great audio is about choosing the right compromises, but without the new generation of design engineers, in the future, you'll be buying heritage gear instead of new.

It's in the air tonight

The air is full of RFI. Nasty stuff that wants to get into everything and mess it up. PCs, Phones, Wi-Fi, comms, everything electronic is polluting the air and your gear. You can't see it, but it alters the noise floor of your kit and can subtly change how some circuits work. As you can't get rid of it, you need to filter it out by being careful with wires and metal in your room. As an extreme example, British Grove studios (Mark Knopfler's) has Faraday cages around all the rooms and shielded mains cables. The result is that guitar pickups sound really quiet and amazing. If you look at most production rooms, a lot seem hastily put together. A bit of planning, cable management and grounding will work wonders.

Back to school

We should never stop learning, and we should listen to those with experience. Not blindly, but again in a learning mode. Your skill as an engineer is what makes records, not the gear per se. I have lots of old timer clients, who produce stunning work, despite not having all the latest gear, indeed sometimes they only have a few pieces of outboard. But they know how music works, how to get the best out of it. Their knowledge and experience is priceless. Their gear however, can just be bought.

Monitoring

I saved the best till last. Having seen a lot of rooms, I can testify that the standard of monitoring across the world is, err, different. I accept that some rooms are historic and can't be changed, and others are constrained by their location, but if you want night and day experiences, travel between rooms. This to me, together with listening level standards is where the biggest gains can be made. It encompasses acoustics, monitors, amps, ergonomics and sensory feedback. When you hear the "truth" in a tailored environment, you will produce your best work. Speak to people, learn and ask professionals for help. You'll be amazed at what can happen.

THIS IS THE END

Well it's taken me some 9000 words to say what I stated succinctly with at the start of this essay. Audio quality matters, but we just need to use our heads and ears to sort it. We don't need new tech or formats. There is no audio revolution about to happen. We've reached the summit, we've just got to figure what to do about it.

What there is however, is a large body of evidence, knowledge and people, who can help you make better records. We can use our brains to see through the distractions and we can start to invest in our humanity, using our skills to produce great music and adapt to our customers accordingly. We're humans, not machines, let's embrace that, using science, not magic as our framework.

Part Three

Mastering: future

12

The shifting discourse on audio mastering

Carlo Nardi

INTRODUCTION: WHAT IS MASTERING?

One of the striking things when visiting websites that offer audio mastering services is that they habitually feature a paragraph or a section with a definition of mastering. Similarly, music production magazines, which in the past years have dedicated several cover features and articles to the topic, almost inescapably feel the need to 'demystify' mastering before starting to tackle it more or less in depth.

Why is there a perceived necessity to define audio mastering again and again? This chapter explores the following non-mutually exclusive hypotheses: mastering as a 'dark art' that potential customers and general readers are somewhat still unable to fully grasp; mastering as a shifting object that changes along with audio formats, music distribution and developments in musical styles; mastering as a discourse that needs to be constantly reinstated to counteract actual or potential threats posed by digitalisation, AI, and changes in the organisation of the music industry.

As a matter of fact, since most musical productions nowadays are mixed and released in the digital domain, one of the original purposes of mastering – namely, the transferal between different physical formats – has virtually become unnecessary, with the notable exception of the resilient, albeit niche, market of vinyl. Nonetheless, mastering has never been as visible as in the past two or three decades. According to Birtchnell [1], when in the 1980s record companies started to release their back catalogues on the new medium, CD remasters allowed the wider public to finally acknowledge the figure of the audio mastering engineer (p. 3). At the time, however, they could be still understood to perform 'anonymous work', a term that Zweig [2] uses to describe those workers whose contributions become visible only when something goes wrong, as the controversy surrounding the so-called 'loudness war' well-illustrates. Around the same period, a deeper change was affecting the music industry, with the diffusion of personal computers and the establishment of smaller, more flexible independent studios: 'As the technology changed over the 1980s and various technical skills became industry standards independent "freelancers" began to dominate the market' [1, p. 3]. This trend has continued in the following decades, nurturing a profitable market for a wider gamut of consumers. In Nardi [3] I describe the apparent increase in media coverage of audio mastering in music production magazines at the dawn of the new century. In an age marked by increased commodification of cultural practices, there is an undeniable connection between this newly gained visibility and the expansion of the market for music equipment (DAWs, plugins, monitors, sound treatment products, etc.).

In order to shed light on the shifts in the discourse on mastering, other interrelated facts have to be taken into account: the blurring of the distinction between production and consumption with musicians having progressively become consumers of technology [4, p. 4]; the digitalisation of music production, distribution, consumption, and business followed by the breaking of price barriers for software and recording equipment; the shift from waged work to mobile and flexible forms of self-employment; the merging of private space and work space; and, more in general, a diffuse process of individualisation that charges social and financial risk on the individual while disintegrating pre-existing networks and safety nets [5,6].

More recently, automated mastering services have insinuated the idea that machines will eventually substitute humans in this line of work. At first, this idea has crept into the public arena as a remote possibility received with either scepticism or enthusiasm. The most famous of these services, LANDR, has already attracted scholarly attention [1,7,8]. Like an elephant in the room, even when not mentioned explicitly, the mere existence of 'an automated online service which is being misrepresented very strongly by commercial interests' – citing the words of a mastering engineer in an interview with Hepworth-Sawyer and Hodgson [9] – has the power to condition the way we think about mastering.

In this chapter I will critically discuss the discourse about mastering as it promotes a more or less balanced relationship between creativity and technical knowledge. If it is true that 'the aesthetic ideal now inheres in the sonic materiality of the recording itself' [10], then it can be assumed that also mastering engineers, as manipulators of sound, perform, at least to some degree, a creative task. By impressing their 'sonic watermark' on a record, mastering engineers' 'capacity to interpret sound has just as much potential to impact on a recording' [11, p. 13]. Yet, the creative aspect of mastering has to be harmonised with other fundamental functions, so that from time to time either creativity or technical expertise will be emphasised to avoid that the engineers' work obfuscates the creative agency of their clients.

While mastering engineers might actually not consider AI as a threat, it is still worth inquiring the role and reception of platforms, such as LANDR, that offer automated mastering services to see how they impact our understanding of the functions of mastering (see [8]). In this regard, reputation is crucial both in emphasising the relational aspects of the profession and in counteracting the use of automation by competing forces in the 'New Economy'. In relation to AI, other questions can be raised: are we facing a particular shift in the often-conflictual relationship between labour and automation? Where does this place mastering in comparison to other professions in the culture industry? If we consider the particular role of mastering within the music industry as a bridge [12] or gateway [3] between production and consumption, what are the implications of highlighting the creative factor at this stage of production?

STUDYING MASTERING AS DISCOURSE

AI and creativity

Until recently, it has been common belief that machines could not emulate human creativity. Advances in AI, however, have led to artificially generated artworks that are displayed or performed for an audience and in some cases also sold in the market. The branch of AI called computational creativity deals precisely with how computers can simulate artistic behaviour. Within this framework of thought, also machines can operate creatively, replicating human thinking and reasoning: '[C]reativity is not some mystical gift that is beyond scientific study but rather something that can be investigated, simulated, and harnessed for' [13, p. 102]. Du Sautoy [14] applies Margaret Boden's idea that there are three kinds of creativity: exploratory creativity, which remains bound by the rules of the current paradigm while exploring its edges; combinational creativity, which establishes new connections between two constructs or cultural domains formerly separate; and transformational creativity, which performs a paradigmatic shift by breaking the rules or abandoning a long-held assumption (pp. 7–10). Of the three types, apparently the third is the hardest to link to the capabilities of AI, as far as it requires machines to disregard a programmer's inputs. Nonetheless, du Sautoy insists that nowadays AI can achieve transformational creativity since 'the new approach to AI allows for meta-algorithms designed to break the rules and see what happens' (p. 280).

While these breakthroughs in AI may be exciting for some artists, scientists, and companies, for others they can increase the fear that machines would eventually replace humans in the activities that mostly define what being human means. Birtchnell [1] investigates if and how audio mastering engineers are aware of emergent algorithmic cultures involving AI. He also stresses that, as these cultures may be a cause of human obsolescence and redundancies in occupations, there is as well evidence 'showing that in other cases they enable, and even revivify, forms of expertise and pose alternative business models and ways of performing creativity and labour' [1, p. 3]. Sterne and Razlogova [8] come to analogous conclusions, affirming 'that although LANDR suggests itself as an alternative to working with a mastering engineer, it actually has not eliminated jobs, instead leading to a reassessment of what mastering "is"' (p. 3). Similarly, Collins et al. [7] argue that, since automated mastering services cannot perform the same tasks as a human, it is 'extremely unlikely that these technologies will entirely replace and eliminate professional roles such as that of the mastering engineer' (p. 13).

Technology, anxiety, and reputation

A narrow focus on the dichotomy automation/human labour, however, might understate the role of technology – AI included – within capitalism, as far as 'the "effects" of AI are heavily shaped by the social fields and contexts in which it is allowed to operate' [8, p. 3]. In this regard, Rosati [15] argues that 'the speculations about AI also take the present social context for granted, as the natural – rather than political – environment in which AI and all machines are developed and utilized' (p. 97). Tension between optimistic and pessimistic views about AI 'presents us with a moment to interrogate not technology but the politics of autonomous objects and alienation in capitalist life' (p. 97). As a matter of fact, '[t]he central struggle of digital capitalism is not (yet) between machines and humans but between social life and its forms of mediation, which *already* – and have for so long – subjugate humans as they provide for their liberation' (p. 97).

Whether mastering engineers see AI as competitor or not, there are other costs linked to technology that can generate anxiety, such as the perceived need to constantly having to update equipment or to learn how to operate new software. Most importantly, technological anxiety among cultural workers is framed within the structure of the New Economy, which is based among other things on market concentration and individualised freelance labour. Competitors organised around platforms, in particular, raise the standard regarding delivery times (automated mastering is virtually instantaneous) and pricing. Insecurity is structural to the functioning of the cultural industry, where labour is often based on flexibility and project work (the so called 'gig economy'):

> 'Insecurity is a feature of a great deal of working life, but is arguably worse in the cultural industries than many other sectors, because of the uncertain and short-term nature of many cultural-industry job contracts, and the high level of subjective investment that many creative workers have in what they do' [16, p. 113].

In this context, workers have to rely especially on their social capital to find or maintain a job. Gandini [17] stresses that existing research, while recognising the increasing relevance of social capital in the New Economy, has not sufficiently examined the features of this phenomenon. Since social capital is now predominantly based on self-branding, flexible work arrangements, and digital interaction and networking, Gandini suggests that 'one's personal reputation within the knowledge economy is today a newly determinant element for career success' (p. 8). He then argues that reputation is the social capital of this economy:

> 'In such a system, reputation functions as a networked asset that intermediates unequal transactions in the allocation of resources, information and goods. Most importantly, it also means that reputation represents the source for the establishment of trust among the actors involved in the system, as it is instrumental in entertaining economic transactions among "quasi-strangers" who interact in newly mediated ways across the offline-online spectrum' (p. 28).

Creativity, states Eikhof [18], is important to establish reputation in the world of arts: 'A person's or an organisation's artistic reputation is intangible and hard to measure. It consists of the collectively shared perception of this person's or organisation's artistic capability and capacity for artistic innovation' (p. 248). In order to achieve reputation, hence, a cultural worker has to contribute creatively to a product: 'Artistic reputation is earned by producing, but not simply reproducing art. The kind of artistic reputation gained depends on the degree of innovation inherent in the work' (p. 249).

The argument that audio mastering, rather than merely performing technical tasks, contributes artistically to the final product has to be sustained by an acknowledgement of the entire production process – mastering included – as a constitutive part of the artistic product itself. In regard to the link between creativity, human agency, and product, de Mántaras [13], stresses that the Turing test, which is aimed at distinguishing computer generated from human generated artefacts, is inappropriate for creative software – and, we might add, for creative work at large – since it denies the fact 'that the production process, and not just the outcome of it, is taken into account when assessing artworks' (p. 103).

As long as the general public accepts the dichotomy humans as creative versus machines as imitative, it is predictable that engineers will emphasise the creative aspect of mastering in their self-presentations. In their study, Collins et al. [7] establish that, for mastering engineers, 'despite lacking expertise, the agency of the musician in DIY mastering appears acceptable to some extent, whereas automated systems are strictly criticised for the lack of human involvement. Thus, a key theme here is the presence of a human agent in the mastering process' (p. 12).

The data collected for this study will illustrate mastering engineers' attitude towards technology in general and AI in particular as well as the role of creativity in building their reputation.

METHODOLOGY

This study is based on data collected through two methods, thematic analysis of media content and a questionnaire for sound engineers. The first method targeted communication on mastering engineer's websites and music production print magazines. For the website analysis, I identified a set of relevant themes and examined the emphasis given to them: equipment, knowledge/competence, client portfolio, services offered, integration with online platforms and services, delivery time, facility and/or location, and price. I also looked if websites feature a definition of mastering, an explicit reference to AI, and the engineer/s' biography.

The data, collected between November and December 2019, concern 115 websites with services based in or targeted explicitly to the UK market. More precisely, these websites feature a facility located in the UK or at least a UK contact address (in at least two cases, the engineer is based abroad, which is not apparent on the website). 39 of these websites offer only mastering services, 22 offer mastering, mixing and other services (except for tracking), and 52 offer tracking alongside mixing and mastering. For several of the latter, more emphasis is placed on the facility whereas mastering has a marginal role. The remaining two websites are specialised on other kinds of services, namely CD duplication and post-production for games and films. Websites promote either a company – ranging from large establishments such as Abbey Road Studios to smaller ventures – or an individual engineer or a team of freelancers, although there may be overlaps between these categories. There are also a few occurrences of websites that promote a group of engineers or companies (hubs) that also have their own personal website. Finally, acknowledging that a significant sector of the cultural economy, similarly to transportation or food delivery, takes the form of a '"gig economy", in which work is increasingly precarious, insecure, and yet highly optimized for both firms and users' [19, p. 178], this web analysis included British mastering engineers' profiles on SoundBetter, which is a platform based in the US and acquired by Spotify in fall 2019 where freelance music performers, songwriters, and sound engineers can be hired for piecework on a global or local basis.

I then examined four music production magazines aimed at a mixed readership including both professionals and (possibly on a larger scale) amateurs. Three of these magazines are based in the UK – *MusicTech* (of BandLab UK Limited, which is part of BandLab Technologies based in Singapore), *Computer Music*, and *Future Music* – and one – *Electronic Musician* – is based in the US. All have online subscription and are available as digital files. I took into consideration all the monthly issues published between January 2015 and December 2019 and looked for cover features dedicated to audio mastering. For two of these magazines – *MusicTech* and *Electronic Musician* – I also analysed all the other articles or mentions of audio mastering over the same time span. In those five years, *MusicTech* had 4 cover features, 1 special insert and 24 articles about mastering, while *Electronic Musician* had 9 articles. *Computer Music* had three cover features and *Future Music* two. Thematic analysis of magazines focused on some of the topics cited above: AI, definitions of what mastering is, knowledge related to mastering, perception of changes in the profession.

The questionnaire was sent between 18 October and 3 December 2019 to 132 freelance engineers and studios. Their contacts had been obtained through the websites included in this study. There were 33 valid responses. The questionnaire consisted of a first section with sociodemographic items and labour-related issues. This was followed by a set of multiple-choice items investigating technology, AI, social capital, relationships with customers, propensity for risk taking, and the functions of mastering. These items used various five-point Likert scales, where only the extreme category responses were specified. The last item was an open-ended question inviting respondents to express their thoughts about the topic at hand.

Due to the sampling technique, which was bases on search engine results, and to the response rate (25%), these data cannot be considered as a representative sample of the population of mastering engineers working in the UK. Moreover, in some instances, especially in bigger companies, there was an administrative gatekeeper who precluded me from getting in touch with individual sound engineers. On the other hand, the data collected provide valuable insights on how actual mastering engineers perceive their work and the challenges they are facing. Data from the questionnaire were complemented with informal interviews and email exchanges with some of the respondents, who agreed to further discuss with me some aspects of their interest.

DATA ANALYSIS

Sociodemographic data and labour situation

There were 33 respondents between 18 and 74 years old. 29 are British (one of which with also Italian citizenship), one Irish, two of other European nationalities and one Canadian; 30 have their main place of work in the UK, with a prevalence of businesses situated in Greater London (11), North West England (6) and Scotland (5); 3 work prevalently with UK clients but live abroad (3). 19 have a bachelor or higher education title.

Of the 26 respondents that are self-employed, 1 started doing audio mastering professionally before 1975, 2 in the 1980s, 7 in the 1990s, 10 in the 2000s, and 6 in the last decade. Of the 7 respondents that are employed for wages, 2 started in the 1980s, 1 in the 2000s, and 4 in the last decade. For 17 respondents, audio mastering has been the main or only source of income in 2018. For 12 of these, audio mastering was already the main source of income ten years before that. This information is consistent when compared to the date engineers started doing mastering professionally, with only 1, among those who started in the 1990s or before, for whom mastering in 2018 has not been the main source of income as compared to ten years earlier.

Data do not seem to suggest significant changes in the type of employment. We must however consider that data collection targeted only engineers that are currently working. Moreover, I did not inquire their type of employment in the past. For these reasons and due to the sampling method, response rate and sample size, these data do not allow for a longitudinal analysis.

Attitude towards technology and AI

Eight items measure attitude towards technology and/or AI (by all means, a particular development of technology in itself). In particular, three items measure trust towards technology in general and two items concern AI. The three remaining items are about automation. Since automation is a central aspect of AI as well as of other instances of technology, these items can be used to measure attitude towards both technology and AI. Nonetheless, mentioning technology in general has unsurprisingly brought different results as compared to emphasising a specific aspect such as automation. Using a scale from 1 to 5, where 1 expresses the lowest level of trust and 5 the highest, trust in technology, based on the three specific items plus the three about automation, registers a mean of 2.8 ($\sigma = 0.61$) and a median of 2.7. Interestingly, if the items about automation are ignored, the mean rises to 3.6 ($\sigma = 0.63$; median = 3.7), revealing a more positive attitude towards technology in general. On the other hand, the average for the three items about automation is 2.1 ($\sigma = 0.86$; median = 1.7). Also trust in AI is low, with a mean of 2.0 ($\sigma = 0.63$; median = 2.0).

Automation is a double-sided feature of technology in that it can be seen positively as a way to save time and resources or, negatively, as a threat to human agency. Automation has economic costs – software and hardware have a market price that firms and professionals can or cannot afford depending on their capital and business turnover – and social costs, due to the risk of obsolescence of cognitive labour and human expertise. The perception of risk, as well as its factual dimension, can be higher in certain situations and for certain social categories, bringing about anxiety. Anxiety can arise not necessarily from the fear of job loss but also from the quality and conditions of labour. Bauman [5] and Beck and Beck-Gernsheim [20], among others, argue that the organisation of labour in late modernity exacerbates the feeling of insecurity: risk is lived as a natural, inevitable component of many professions.

The questionnaire indicates that mastering engineers are not worried that AI might endanger their work. In this regard, Sterne and

Razlogova [8] argue that there is no evidence that companies such as LANDR are causing job losses: 'LANDR's claims to automate a process through AI hide the ways in which LANDR actually creates labour for its users' (p. 11). They also mention that none of the mastering engineers contacted for their study 'expressed any concern about LANDR as a threat to their business' (p. 16). Collins et al. [7], on the other hand, registered a mixed response among the mastering engineers they interviewed through music forums and social network communities: 'At one extreme, these technologies and practices are perceived as a threat, and, at the other, the notion of effective mastering outside of a professional studio is inconceivable' (p. 12).

In the present survey, indicators of anxiety towards the future do not reveal significant information (mean = 2.95; $\sigma = 0.70$; median = 2.50). There is also no significant difference between employed ($\mu = 3.00$; $\sigma = 0.58$; median = 3) and self-employed engineers ($\mu = 2.93$; $\sigma = 0.74$; median = 2.88). The higher dispersion of data for self-employed engineers, however, can be explained with the high variety of working situations among individuals. The judgement of freelance workers possibly depends more closely upon contingent business performance. As a consequence, their perception of the future appears to be less stable.

Four questionnaire items measure the propensity for risk taking as an indicator inversely correlated with anxiety for the future: the higher someone is worried about the future, the less they will be willing to take risks. On a scale where 5 corresponds to a high propensity and 1 to a low propensity for risk, the mean is 2.74 ($\sigma = 0.42$; median = 2.75). Overall, employed engineers ($\mu = 3.00$; $\sigma = 0.35$) register a higher propensity for risk taking than self-employed engineers ($\mu = 2.63$; $\sigma = 0.41$). On the one hand, these results seem well-founded, since employed workers have to worry less about their future. On the other, this observation seems to contradict common sense: if the stakes for someone are higher (i.e. if someone might lose a stable job), then it should be less feasible that they take risks. In the culture industry, however, we might assume that a certain degree of risk taking is intrinsic to creative roles. In other words, employed mastering engineers might have more headroom for creativity thanks to the higher security of their job.

Social capital, reputation, and the functions of mastering

Reputation in the economy of knowledge is an asset that defines the workers' position in the market and their capacity to increase or maintain their social capital and get jobs [17]. The digitalisation of most aspects of labour – means of production, communication and promotion, networking, economic transactions, etc. – means, among other things, that reputation 'is now increasingly tangible, visible and, to some extent, also measurable via the activity of individual users on social media platforms. This measurability extends its effects over the whole labour market' (p. 38).

Five items in the questionnaire concern the acknowledgement of professional reputation, measuring the importance for mastering engineers of having connections in the business, keeping customers satisfied, welcoming feedback from customers, keeping existing customers updated with news and offers, and promoting their business to find new customers. The combined responses to these items indicate a marked recognition of the importance of reputation for this kind of activity (on a scale from 1 to 5, $\mu = 3.68$; $\sigma = 0.44$; median = 3.8). Interestingly, the mean does not change significantly between employed engineers ($\mu = 3.79$; $\sigma = 0.27$; median = 3.83) and self-employed engineers ($\mu = 3.90$; $\sigma = 0.39$; median = 4.00), although the slightly higher standard deviation for self-employed engineers suggests that, among this category, there is higher disagreement about the relevance of reputation. A positive thinking attitude is probably necessary for self-employed engineers to succeed in their business. This, however, might induce some of them to underestimate the efforts they normally make to obtain and maintain reputation.

Websites normally (but not always) highlight human resources. The name of the engineer or engineers is featured on 70% of studio websites, 86% of freelancers' websites, and on all hubs (i.e. independent networks of mastering engineers) and SoundBetter profiles. Here, rather, the interesting information is that some freelancers do not publicise their real name and show instead an impersonal interface to their potential clients. This strategy of communication often employs the plural form, implying that there are more people and competencies involved. The assumption that there is some mediation between the professional and the client adds to the prestige of a company. Moreover, it might be advantageous to distinguish the business side from the creative side of an enterprise to preserve the engineer's artistic independence and highlight their cultural capital.

Thematic content analysis of websites confirms the relevance of reputation. For this purpose, I took the following eight themes into consideration: gear (outboard processors, software, microphones, etc.), knowledge (skills, competences, and experience), portfolio (past clients and works), services offered, integration with online services and platforms, delivery time of the final master, facility and/or location, price. Then, for each website, I identified the three most prominent themes based on a qualitative judgement of the emphasis given. The judgement criteria consisted of the location of a theme in the website navigation, the extensiveness of the respective content, and the presence of pictures or videos in a prominent position.

The results show that studios place, as expected, emphasis on facility or location (67% of the websites) and services (67%). Secondarily, they also highlight the engineers' knowledge (56%) and the gear they make available to clients (55%). Freelancers, on the other hand, emphasise first of all their knowledge (77%), then the kind of services they offer (64%) and their past clients and works (60%). This seems to confirm

Gandini's [17] conclusions about the importance of reputation for freelance workers in the knowledge economy. Here, reputation is obtained mainly though association with artists and through experiences gained in the field. Since reputation is an acquired rather than an inherited trait, it makes sense that engineers highlight their accomplishments and that, in some cases, they do so by using impersonal communication. Since they are selling their own labour, they can benefit from an external entity vouching for their value. Other websites, however, chose a more direct and approachable style, where engineers reveal some peculiar aspect about their way of work or even their personality. In other cases, freelancers employ irony to establish complicity with the potential client or, more rarely, a more aggressive tone, as if to imply that doubting their professionality is out of question.

Music production magazines voice the concern for AI with titles such as 'Automated Mastering: Can a Robot Replace an Engineer?' [21] and editorials and article discussing the pros and cons of automated mastering services. Articles normally include practical tips, tutorials and suggest, either directly or indirectly, which plugins and DAWs to use – we must remember that these magazines normally depend on their advertisers and, hence, are outlets for promoting music equipment.

In the past five years, *Future Music* has dedicated two cover features to mastering, one in April 2018 and another in October 2019. Comparing the two issues, there seems to have been a shift of perspective. While the first issue is mainly focused on how to master in-the-box, with tips about workflow, procedures, and proprietary plug-ins to suggest that music producers can do mastering themselves, the more recent issue draws a different picture, focusing on the role of the mastering engineer as a professional figure that demands recognition. This issue, while also including tips and workshops on more advanced topics, features two interviews with two mastering engineers working in the UK. Whereas both issues mention AI-based mastering, the more recent one concludes that with LANDR 'there's no getting past the fact that you're handing your mixes over for processing to a "mastering engineer" without ears, in the conventional sense' [22, p. 30].

Both websites and magazines draw a description of the work of the mastering engineer that highlights a great variety of tasks and the capacity to cope with a wide range of situations. They list some recurring functions implying that audio mastering is still relevant today: performing objective listening techniques during a stage of music production that should be kept distinct from mixing, correcting defects in the mix (removing noise, taming unwanted resonances, smoothing out abrupt endings, etc.), evaluating if a premaster is ready or requires to go back to the mix stage, giving consistency to a multitrack release, ensuring that a music release sounds well across different listening situations, and many others. It is important to note that these functions are not fixed once and for all as they concur to a particular definition of mastering that is historically embedded and that establishes the role of the mastering engineer in relation to other roles in record production accordingly.

The ability to respond creatively to different situations in order to obtain standard results in terms of audio characteristics might actually be the most distinctive aspect of audio mastering. Again, as Hepworth-Sawyer and Hodgson [9] stress, 'every mastering engineer works differently, and often using different tools, even if they pursue the exact same aesthetic goal, namely, producing the best record possible from the mixes they are given' (p. 2).

CONCLUSIONS: MASTERING AS A SYNTHESIS OF TECHNICAL, AESTHETIC, AND SOCIAL ASPECTS

In this chapter I have discussed the role of creativity in audio mastering as opposed to a discourse in which it is considered a technical task that does not require creativity or even human intervention. One of the biggest supporters of the latter view, LANDR, supposedly relies on machine learning (ML) to deliver its automated mastering service – although Sterne and Razlogova [8] express some doubts about this: 'Its operational sequence cannot be known to users, between corporate secrecy, ever-changing back ends, and the status of algorithms as golem-like assemblages' (p. 14). This has several implications: firstly, automated mastering can only replicate human-made mastering, applying fixed and repeatable procedures to given input data and, at best, maximising economic performance. As a corollary, any deviation from a given set of conceivable situations is seen as an error that requires fixing rather than as a chance for innovation. Secondly, ML, implemented through big data collection and analysis, appropriates the work and knowledge that is the product of human labour, alienating their original creators from any economic gain that can be generated in the process.

Mastering engineers often emphasise the creative and relational (social capital) aspects of their craft that inherently resist mechanisation and automation: the ability to work on a case-by-case scenario that treats each premaster as a unique piece of art, tailoring services to meet very specific customer's needs or requests, providing clients with feedback about their mix ('a second pair of ears') and allowing for the resubmission of a revised premaster, maintaining links with particular music scenes and creative communities, demonstrating aesthetic competence in different genres and styles, being able to take complex decisions in situations where priorities cannot be taken for granted (very often, an improvement in one aspect of a track may go along with a deterioration in another), being able to look at the bigger picture (the artist's signature sound, an album, etc.), and so on. The aim of this emphasis is not necessarily caused by worry of job loss but more probably by an attempt to reclaim the definition of their work in spite of competing discourses, such as that promoted by LANDR.

Indeed, the work of the mastering engineer falls between the somewhat fuzzy – and historically determined – border line separating creativity and technique. In this regard, Shelvock [23] discusses dynamic range compression (DRC) in mastering as an essentially creative task, involving a balance between loudness requirements and musical taste mediated by industry standards and genre aesthetics. In fact, audio mastering requires a continuous negotiation of objective tasks, foreseeable goals, and creative solutions. For this reason, if computational creativity were to successfully emulate human behaviour, it should be able not only to consider how creativity can respond to problematic situations but also and simultaneously to apply several criteria of different nature to establish when it is appropriate to use creative thinking as a response to a task and when not.

If for a moment we remove a recording from the context in which it is produced and consumed, then du Sautoy's view of AI as capable of performing the three different kinds of creativity above mentioned – exploratory, combinational, and transformational – appears plausible. A complication, however, arises when we challenge abstract definitions of the term 'creativity' and place it in a larger context. Du Sautoy summarises creativity as 'something that is new, that is surprising, and that has value' (p. 3). This apparently unpretentious definition conceals some problematic issues, as far as any judgement about creativity is contextual. In other words, the attribution of novelty to a certain artefact, its value, and its capacity to surprise should matter to someone specific. As McIntyre [24] argues, creativity applies to ideas, products or processes that are unique in reference to a certain body of knowledge or domain. Only those who hold that body of knowledge are entitled to judge whether an idea, product or process is unique: '[N]o judgement ever occurs in a vacuum' (p. 151).

These observations imply that, if we stand by du Sautoy's definition of creativity, judgements about the creativity (or derivativity) of an idea, act or object cannot be isolated from the social context and the actors involved. Furthermore, cultural products and performances will be judged as creative, or not, by different groups in the same historical moment.

A/B referencing is a common practice for mastering engineers, who employ an array of listening techniques as they compare their masters to tracks of their choice. It might seem that algorithms operate very like human engineers as they collect and analyse data from multiple sources to obtain similarly objective results. However, it is not just a matter of scale or procedure, since the criteria for mastering – whether through algorithms or A/B techniques – should respond to judgements that are only in part technical. Whereas the procedures of

AI are typically standardised, mastering engineers often follow a more subjective and extemporaneous approach. In fact, they adopt ad-hoc solutions while managing simultaneously a large number of variables, which are not always technical or even related to sound in the first place. Complexity does not prevent mastering engineers from taking prompt and effective decisions to achieve their goals while synthetising technical, aesthetic, and social aspects.

A final consideration concerns not so much what technology can achieve, but what we allow it to achieve and, more importantly, to whose advantage. In a strictly economic sense, if we follow Marx, only human labour can add (surplus) value to goods. This means that, by definition, AI by itself does not have the capacity to create new value. In the case of LANDR, surplus value is generated by quality control, which, in the end, is performed by customers. Whether platformisation is actually destroying human jobs or not, the vast concentration of power and capital that it entails and its aspiration to centralise and homogenise some crucial steps of music production should at least raise some eyebrows. At this point, it seems apparent that this trend has already affected the music industry at large by redefining the definition and conditions of work of cultural workers and by changing the relationship between mastering engineers and their clients and, possibly, even between music makers and their public.

REFERENCES

1. Birtchnell, T. 'Listening Without Ears: Artificial Intelligence in Audio Mastering'. *Big Data & Society*, Vol. 5, No. 2, (2018), pp. 1–15. https://doi.org/10.1177/2053951718808553.
2. Zweig, D. *Invisibles: The Power of Anonymous Work in an Age of Relentless Self-Promotion*. Penguin, New York, (2014).
3. Nardi, C. 'Gateway of Sound: Reassessing the Role of Audio Mastering in the Art of Record Production'. *Dancecult*, Vol. 6, No. 1, (2014), pp. 8–25.
4. Théberge, P. *Any Sound You Can Imagine: Making Music/Consuming Technology*. Wesleyan University Press, Hanover, NH and London, (1997).
5. Bauman, Z. *Globalization: The Human Consequences*. Polity Press, Cambridge, MA and Oxford, (1998).
6. Bauman, Z. *Liquid Modernity*. Polity Press, Cambridge and Malden, MA, (2000).
7. Collins, S., Renzo, A., Keith, S. and Mesker, A. 'Mastering 2.0: The Real or Perceived Threat of DIY Mastering and Automated Mastering Systems'. *Popular Music and Society*, (2019), pp. 1–16. https://doi.org/10.1080/03007766.2019.1699339.
8. Sterne, J. and Razlogova, E. 'Machine Learning in Context, or Learning from LANDR: Artificial Intelligence and the Platformization of Music Mastering'. *Social Media + Society*, Vol. 5, No. 2, (2019). https://doi.org/10.1177/2056305119847525.

9. Hepworth-Sawyer, R. and Hodgson, J., editors. *Audio Mastering: The Artists. Discussions from Pre-Production to Mastering*. Routledge, New York and Abingdon, (2019).
10. Greene, P. D. Introduction: Wired Sound and Sonic Cultures. In: Greene, P. D. and Porcello, T., editors. *Wired for Sound: Engineering and Technologies in Sonic Cultures*. Wesleyan University Press, Middletown, CT, pp. 1–22, (2005).
11. Hinksman, A. 'The Mastering Engineer: Manipulator of Feeling and Time'. *Riffs*, Vol. 1, No. 1, (2017), pp. 11–18.
12. Katz, B. *Mastering Audio: The Art and the Science*. 2nd edn. Focal Press, Orlando, FL, (2007[2002]).
13. de Mántaras, R. L. 'Artificial Intelligence and the Arts: Toward Computational Creativity'. In: González, F., editor. *The Next Step: Exponential Life*. OpenMind, Bilbao, (2017), pp. 99–123.
14. du Sautoy, M. *The Creativity Code: Art and Innovation in the Age of AI*. Belknap Press, Cambridge, MA, (2019).
15. Rosati, C. Spectacle and the Singularity: Debord and the Autonomous Movement of Non-Life in Digital Capitalism. In: Briziarelli, M. and Armano, E., editors. *The Spectacle 2.0: Reading Debord in the Context of Digital Capitalism*. University of Westminster Press, London, (2017). pp. 95–117.
16. Hesmondhalgh, D. and Baker, S. *Creative Labour: Media Work in Three Cultural Industries*. Routledge, Abingdon and New York, (2011).
17. Gandini, A. *The Reputation Economy: Understanding Knowledge Work in Digital Society*. Palgrave Macmillan, Basingstoke, (2016).
18. Eikhof, D. R. Does Hamlet Have to Be Naked?: Art Between Tradition and Innovation in German Theatres. In: Pratt, A. C. and Jeffcutt, P., editors. *Creativity, Innovation and the Cultural Economy*. Routledge, Abingdon and New York, (2009), pp. 241–261.
19. Rahman, K. S. and Thelen, K. 'The Rise of the Platform Business Model and the Transformation of Twenty-First-Century Capitalism'. *Politics & Society*, Vol. 47, No. 2, (2019), pp. 177–204.
20. Beck, U. and Beck-Gernsheim, E. *Individualization: Institutionalized Individualism and its Social and Political Consequences*. SAGE, London, (2002).
21. Cooper, M. *'Automated Mastering: Can a Robot Replace an Engineer?' Electronic Musician*, (2015), July 2015, pp. 70–74.
22. n.a. *Mastering: The 2019 Guide,* (2019). Future Music 348, October 2019, pp. 28–39.
23. Shelvock, M. T. *Audio Mastering as a Musical Competency*. Ph.D. Dissertation. Western University, London, ON, (2017).
24. McIntyre, P. Rethinking Creativity: Record Production and the Systems Model. In: Frith, S. and Zagorski-Thomas, S., editors. *The Art of Record Production: An Introductory Reader for a New Academic Field*. Ashgate, Farnham and Burlington, VT, (2012), pp. 149–161.

13

Loudness – the sales gimmick

Jay 'Jedi' Hodgson

Mastering is an art. It's definitely not a science. There is no single 'correct' master for a track, just as there is no single 'correct' mix. One master is not 'better' than another. One mastering engineer is not 'better' than another mastering engineer. The only 'correct' master – the 'best' master – is the one the client approves. Mastering isn't *solving for x*, in other words. If it were, why not automate the process?

A number of digital entrepreneurs are trying to do this very thing. In fact, it wouldn't be too much of a stretch to say that the current industry is *devoted* to automating audio mastering. Everywhere you look online, some automated audio mastering service or other is soliciting clientele. I'd say the algorithms haven't yet succeeded the work of human hands and ears, but they are trying.

Browsing Instagram over the last few months – the only social media I bother with – I'm sent sponsored stories from LANDR, eMastered, and other automated mastering services, soliciting my business, almost every hour. These services promise better mastering through science, as it were. 'Artificial intelligence' gets thrown around in their ad-copy a fair bit. It's all a bit disconcerting for today's mastering engineer. Are we the next agency to be automated from existence? And do we really live in an era where artists want their productions partially automated? How did *that* happen?

'*Master your track, instantly*', eMastered.com beckons (see Figure 13.1). '*An online audio mastering engine that's easy to use, fast, and sounds incredible. Made by Grammy-winning engineers, powered by AI. Upload your track below and hear it now for free*'. Does anyone believe that the algorithm was really programmed by '*Grammy-winning engineers*'? And if so, did they win a Grammy for computer programming? Of course not! I suspect they consulted on the algorithm and that's it. Nonetheless, some engineers seem to have indeed

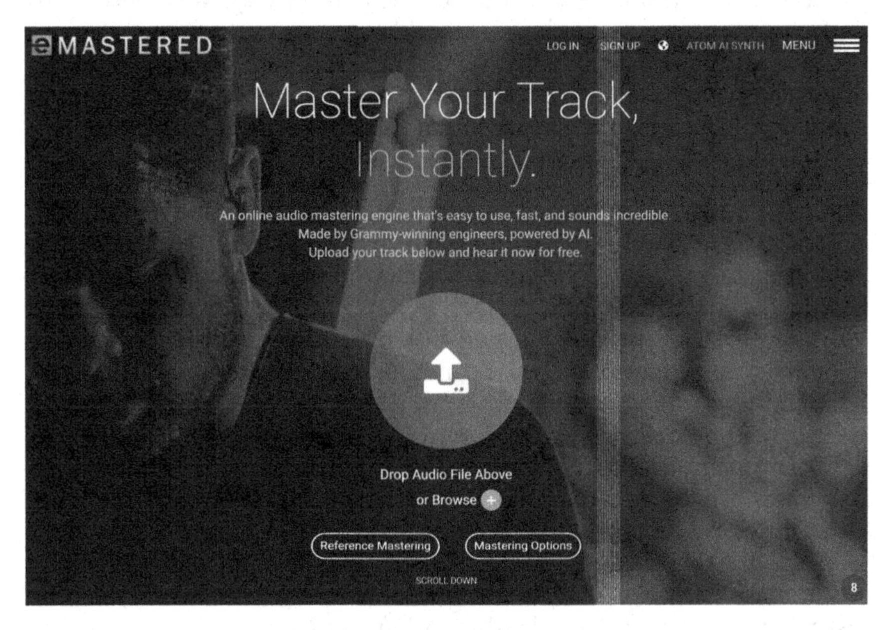

Figure 13.1 Advertising for emastered.com on Instagram. Accessed 23 January 2020.

'bought in' completely. Clinton Sparks, who has a Grammy nomination for his work with Lady Gaga, offers the following testimonial lower down on the site: '*eMastered has become an essential element in my music making process. Simple, fast, and most importantly, incredible quality. This will become standard for anyone making music*'. Incredible quality? Really? Tae Best, another Grammy nominated engineer, most famous for his work with Kendrick Lamar, fawns just as unashamedly. '*I tried eMastered on an instrumental that I did some light mixing on, and the engine did some great EQ'ing and compression*', he explains. '*It really enhanced the track*'. Carlos 'CID' Cid, probably best known for his work with Lana Del Ray, tells us he was '*blown away*' when he '*heard eMastered for the first time.... It did such an incredible job*'.

What exactly is being offered here? What are these artists responding to, besides the likely kickbacks for testimonials they've received? Do they realise they are praising machines to replace them? More importantly, at least to my mind: what is the model of audio mastering they're applauding? What, in other words, do eMastered.com and their ilk sell? I'm not 100% sure, to be honest. But I'm fairly certain it's going to play a major role in audio mastering's future. More than 40,000 tracks are now uploaded to Spotify daily, and many are run through one or another of these audio mastering algorithms before they see the light of day. Whether or not that means mastering engineers should get computer science degrees remains up for debate. I'd say 'no', of course, but I've been wrong as often as anyone else in the last three decades about things like this so I just don't dare prognosticate anymore.

Maybe it's true. Maybe audio mastering *can* be digitally modelled. I'm sure some aspects of it can be programmed and coded. But is audio mastering a *mathematical* pursuit, something where the numbers are all that really matter?

In fact, I don't find this idea all that threatening. Maybe what I do as a mastering engineer *can* be numerically modelled and coded, rendered as some computer program designed to sequence 'musical' bits into a predetermined and numerically fixed mould. *Why not* upload some music, and pay a digital entrepreneur very little to make it louder .... Who cares about artisanal discretion and oversight? Does this sound like the sort of thing Pharrell would go for? I'm sure George Martin would've had no issue telling 'the boys' that their mono master for *Sgt. Pepper's* was almost ready for delivery, that he just needed a second to upload it to LANDR.

If you've only ever read textbooks on mastering, and never engaged in the act itself, I'll forgive you for believing such nonsense. And I'd blame mastering engineers themselves for helping you to reach such a conclusion. It was, after all, mastering engineers who marketed their services by insisting that they alone had some technical 'dark' knowledge which clients absolutely needed if they wanted their mixes to ever be 'ready for market'. 'It's such a dark art', explains M.C. Schmidt, half of the electronic duo Matmos. And his partner, Drew Daniel, agrees: 'It's this mysterious process that a lot of musicians don't understand, including us'. (https://pitchfork.com/features/article/9894-the-dark-art-of-mastering-music/).

'The mastering engineer's job breaks down into roughly two parts–a final edit of the music, and preparation of the files for release. Much of the work is so technical, so reliant on an obsessive knowledge of gear and acoustics and physics, that to talk about it with any level of detail essentially requires speaking in another language altogether', writes Jordan Kisner, in an another article published by Pitchfork (https://pitchfork.com/features/article/9894-the-dark-art-of-mastering-music/). 'If rock stars are the sex gods of music, mastering engineers are its druids, the ones who work methodically and meticulously, and to whom people come for mystical wisdom and blessing'.

Tunecore takes an even more scienifistic approach in its description of the process:

> The original mastering engineers were literal technicians who were highly trained in electricity and mechanical engineering. These engineers knew a ton about how electricity works, how

> mechanical things work, and how frequencies impact each other. They also had an understanding of everything that involves vinyl, tape, and any other mediums the audio might need to be transferred to. From the start, mastering has been a bit hard to understand because it requires such specific knowledge. (https://www.tunecore.com/blog/2018/10/mastering-isnt-a-dark-art-its-just-misunderstood.html)

But mastering is not mystical. It's definitely not science, nor is it math. Nothing in music is really all that mysterious or esoteric. You don't need to know what a 'LUFS' is, nor the math behind calculating 'LUFS', so long as your master at about –12 LUFS integrated when you run it through a meter you trust. Make it sound as best you can, even while you peg it to that reading, and you've done your job, at least inre: loudness normalisation for streaming. For goodness' sake, drop every other pretense – *please*! My master will be 'better' than yours only if it sounds 'better' to the client, regardless of who the client is or what they think sounds 'better'. Our job isn't to educate anyone on anything. Our job is to entertain, and our market is recording artists. So educate yourself on all the technical details you need to learn to produce viable masters, but don't ever think that there are numerically or quantifiably 'more correct' or 'better sounding' mastering moves beyond what artistic discretion dictates. As Jeff Carey puts it:

> *You're trying to make this impossible match between what the artist has done in a fantastic studio situation–or even a shitty basement–and something that will sound great in someone's earbuds or a shopping mall. What's the meat of this music and how can I make it so that it still sounds good in all these impossible situations, like something the artist was really thinking about?*

'I try to achieve what an artist, engineer or producer can't do on his or her own', explains Emily Lazar. 'Sometimes that's technical, but just as often it's conceptual'.

Carey and Lazar don't describe a scientific – or, even, really all that technical of an – endeavour. It's an artistic pursuit they describe, a practice of informed discretion. It's as much about listening and shaping as it is about forcing sounds into some technical mould.

I've cleared thousands of masters over the years – in fact, a few of those masters have gone up for Juno awards here in Canada (in the category of Best Electronic Recording, to be precise) – and never once did I need to know, let alone broadly understand, the math used to determine a 'LUFS' measure. All I needed to know was that my master sounded good, and that it measured at whichever LUFS reading on my meter I was matching it to. You don't need to be able to calculate the RMS (Root Mean Square) of a track to know, for example, that when compressing programme with lots of sustain, switching the detection circuitry from 'peak' to 'RMS' will optimise the compressor's responsiveness to those musical areas you hope to address. And yet, despite lacking this mathematical skill, I compressed many mixes, and expanded them, and equalised them, and performed other kinds of processing, with no more knowledge of the algorithms I deployed in so doing than a guitarist has of the psychoacoustic difference tones produced by strumming an E major chord while they noodle by a bonfire.

When a mastering service tells you there are hard and fast empirical destinations where your mixes need to go, and that they alone know how to get them there, they are marketing to you. Mastering engineers are artists, not scientists. They should live and die by their creativity, like everyone else who takes the risk and devotes themselves to artistic pursuits. There is only sound. Either it sounds good to the mastering engineer, whether they be a PhD in the subject or a total and complete neophyte, or it doesn't. It is just that simple, as far as I'm concerned. I have always loved what Claude Debussy had to say on the matter of artistic discretion:

> *There is no theory. You have only to listen. Pleasure is the law. I love music passionately. And because I love it, I try to free it from barren traditions that stifle it. It is a free art gushing forth – an open-air art, boundless as the elements, the wind, the sky, the sea. It must never be shut in and become an academic art.*

And now we are being told we can automate this creative process, which I love so much. And I think we mastering engineers ourselves, with all of our pretensions to empiricism and science, are to blame. Some of the best books on mastering read, to my eyes at least, as psychoacoustic summaries of signal processing, with little or no information about how those processes are deployed to make music. The psychoacoustic procedure behind compression, for instance, is already readily available from psychoacoustics textbooks. Why do we need a page rehashing the math when a basic explanation of what the knobs do will suffice? As if it matters at all what scientific principles undergird any aesthetic pursuit anyway! How many guitarists are thinking about Hertz and kiloHertz, let alone deciBels full-scale or SPL, as they rip and shred? Or how many laptop musicians care at all what the numbers on a meter read, so long as a track sounds good and makes them want to move?

LUFS and RMS ranges and targets are likewise regularly thrown at readers, as though they mean anything. These authors usually market some golden age of dynamic listening, and a direct passage back to that era. As though quantifying dynamic ranges really matters all that much beyond knowing roughly where a master should hit so as not to sound too quiet or distorted through earbuds. Seriously, who are we kidding?

The advertisements that have recently made their way onto the various social media where today's laptop musicians lurk aren't so much introducing a new idea as they are amplifying the 'scientifistic' discourse of audio mastering to its logical conclusion. To verify this, just google 'loudness wars' and watch mastering engineers pull out maths and numbers to explain why they are 'right' to say that dynamic range decreases as loudness in the full-scale realm increases, thanks to brick-wall limiting. Of course they're right about that – it's really not that complicated a notion to understand. In fact, I usually explain it to students and assistants in about five minutes, and then we move on. But the really important question is: *so what? Who cares* if things are getting louder, or dynamic range is collapsing? Flying Lotus sounds fantastic to my ears, and he's not exactly known for dynamic range. I like a lot of obnoxiously loud records. I also like a lot of records that are so dynamic they could easily cause a car crash if listened to without prior experience. Music, and recording, is a vast terrain. As a vestibule of human experience, record production is too large not to occasionally contradict itself.

I'm personally so sick of the 'loudness wars' conversation that, rather than rehearse all of the tired old arguments about dynamic range, I'll simply say that loud, even distorted, masters sometimes make mixes sound 'better'. What do I mean by 'better'? Who knows .... And that's my point! I make loud masters. It's my specialty. I work mostly in electronic genres, and hip hop's many modern variants (i.e. chillhop, lofi, etc), and I send back masters that dominate the streaming output without much notable clipping. In fact, one of my favourite compliments came from a client working in pop: 'It's offensively loud. We love it!' I sent them a very loud master. The artists, the label client, the mix engineer, and the mastering engineer all approved it. All of the complaints against 'loudness' remind me of the complaints against rock 'n' roll. It's a social hysteria .... But an extremely conservative and uptight one.

So is the hype against automated mastering. It may be that automation sometimes produces a master that the client prefers to one produced by our artisanal hands. That's okay. If that client approves another master than mine, they didn't want me to master their record. They just couldn't know that until they asked me to try my hand at it.

Two of my loudest masters were nominated for Juno Awards in the category of Best Electronic Recording. The loudest master I ever cleared was for a client currently sitting at more than almost a billion streams on Spotify alone, numerous gold and platinum records, a Juno Award for Dance Recording of the Year, and two songs that so dominate the airwaves that my relatives and friends can only roll their eyes when their songs come on the radio in some public place – which happens at least twice a day to me – because I can't help myself, and I have to yell: '*Those are my boys!*' I produced the masters that way for the same reason anyone does anything musical: because I thought it sounded best that way! Again, as Debussy said, '*Pleasure is the law!*' Or as the company I work for puts it, '*whatever it takes to make it sound right*'.

Rules are meant to be bent. Theory is for weaklings who need reassurance that their work is 'good'. Musical creativity is all about one thing: being brave enough to say, '*I think this sounds good*', and then withstanding the response without rationalising our emotional choices behind 'theory'. Personally, I think theory is *all* just a lot of emotional reassurance. If we know our master peaks at the pre-ordained meter register, thanks to streaming's new loudness normalisation (which is anything but standard, by the way), and we know we have some quantifiable amount of dynamic range (as if anyone wants a number rather than something that just 'sounds good' to them), we have some tickable box to reassure us that the master we produced is 'good'. If you like something else, say Metallica's *Death Magnetic*, well then you must be 'wrong'. What a boring world we live in!

Yes, there are standards we as mastering engineers must ethically peg our masters to. No, it's no different than saying guitarists who want to play in tune with a band should tune their A-string to 440 Hz. It's really not *that* complicated. We're artists. We learn to produce a master pegged to required levels for optimal playback on the world's ever proliferating streaming services; we learn to create masters for vinyl, and tape, and DDPi for CD. We educate ourselves and obsess about our instruments. But even this education is an artistic pursuit, a way to optimise our tools for the craftsmanship we use those tools to engage. Like guitar players experimenting with various pickups, and amps, pedals, and so on, we mastering engineers obsess about, say, op amps, and nickel-wound transformers, and the best I/Os. I've known some drummers who will only play drums made of maple wood; and you won't catch me dead running an analogue master on a rock track that hasn't at least touched my elyssia stereo-comp. But all our gear are just tools, 'sciency-sounding' musical instruments. What matters, at least in any empirical sense, are the ears we engage while we use those instruments; our discretion, our decades of expertise, and the judgement with which we deploy that expertise, are the final markers of our mastering abilities. Artists should come to a mastering engineer because they like the sound of their masters. That's it.

Mastering engineers: stop advertising your technical expertise. Nevermind your gear! Forget about advertising your environments. Advertise your work. Let the work do the talking for you. Be an artist. Just link records you've mastered, let clients listen to those masters, and if they like them, they will likely want to work with you. Yes, they'll never hear the progression from mix to master, but they'll have heard a lot of your masters! And that's all they need to hear. Algorithms and AI have taken over every other aspect of our work, in any event. The artistry of audio mastering is all we have left. And it's the best part of our job, as far as I'm concerned.

14

Audio mastering facing automation

THE EMBRACEMENT OF THE HUMAN

Thomas Birtchnell and Andrew Whelan

INTRODUCTION

Once a musical composition has been recorded and mixed there is still some way to go before it can be released to a listening audience. The status quo dictates that before listeners can digest a sound recording comfortably, or as it was intended to be listened to, it must be scrutinised by someone who was not involved in the creative process: an objective set of ears. Yet, since 2014 a wave of digital platforms have arisen that promise automated signal processing via machine learning, big data analytics and artificial intelligence (AI), notably the Montreal-based company LANDR [1]. While it is not entirely clear how these platforms process music, commentators suggest the companies deploy algorithms to compare uploaded content to a vast database of existing music organised by genres, and then applies effects and corrects for errors with no human being involved in the process.

A key innovation in these digital platforms is to push the responsibility for mastering onto the artists themselves. LANDR, for instance, invites artists to upload the same song multiple times and correct issues they detect or remix the song until it reaches their satisfaction. Indeed, the tool is marketed as a new form of instrument or tool for the artist to learn and experiment with.

This certainly bucks the trend of how audio mastering was done in the past. Traditionally, the responsibility for tailoring a recording to an audience's expectations lay with the audio mastering engineer. This expert is technically trained in the science of sound and appreciative of conventional listening standards. Audio mastering engineers have access to dedicated studio spaces, technical equipment, and assemblages of software and hardware. Through training and intuition their craft involves critical listening and an awareness of aesthetic expectations. They are also usually musically trained or at least able to comprehend musical theory.

On the surface, audio mastering engineers correct for audible clicks, pops, distortion, and phasing. Another routine task is to encode digital files with metadata: song title, length, artist name, and so on. They also, however, apply 'sweetening' and make the recording 'louder' – a disputed term meaning sonically more powerful or energetic – through carefully applied audio effects. Once the recording is ready, engineers adjust and compress the quality of the waveform to a range of media formats including CD, vinyl, MP3 and WAV – an aspect of the role that has diminished in importance with the shift to online content.

A standard list of equipment in the audio mastering engineer's repertoire includes monitors, spectral analysers, multiband compressors, limiters, maximisers, equalisers, and audio editing software. Audio mastering engineers are also familiar with a sited listening environment that has been 'treated' for sonic biases with foam and other materials and learned through careful listening over time. A room (or at the least high-end headphones) that the engineer trusts is balanced to deliver reliable quality of sound across a range of playback situations is critical. While some audio mastering engineers train at a tertiary institution, the majority gain their skills through a combination of industry experience and a term as an apprentice with an established mastering engineer.

The research question that drives this chapter inquires: how is audio mastering moving from being a craft limited to experts who operate separately from the artistic process to being a key part of the artistic workflow? The trend is occurring due to a wider movement towards the adoption of new technologies including samplers, synthesisers, drum machines, and digital audio workstations (DAWs) across diverse music genres ranging from metal to hip-hop [2]. Artistic familiarity with these technologies enables competence in mastering tools (for instance multiband compressors or spectral analysers), some of which are now stock features within DAWs such as Ableton Live or Apple Logic. Yet, audio mastering engineers will often contend that the artist is not in a position to master their own production due to their subjective bias.

In this chapter, we address the introduction and possible implications of AI to audio mastering from a social science perspective. AI often conjures sentient robots in the public imagination, but the phrase – sometimes defined as 'narrow AI' – more pragmatically refers to algorithmic processes, involving machine learning, which automate specific workflow tasks.

The chapter is set out as follows. After a description of the changing nature of the role in the next section, we then present a brief overview of the relations between music and technology as these have been understood and described by social scientists. We highlight two key themes which social researchers have addressed in investigating contemporary music industries and the role of technology within them. These are work or labour (in the contemporary idiom, precarity, and the gig economy in the cultural industries), and the aesthetic and cultural implications of technological disruption. To provide an empirical grounding to this topic, the chapter draws on qualitative interviews conducted in 20 recording studios with audio mastering engineers in Australia. In the penultimate section, we document a set of core issues facing audio mastering engineers, and offer conclusions in the final section.

BACKGROUND

A brief history of mastering

In the twenty-first century it is acknowledged that the job of the mastering engineer is more than simply managing the intermediate step between taking the audio from the mixdown to replication and distribution. After Bobby Owsinski, 'It's an art form that, when done conscientiously in its highest form, mostly relies on an individual's skill, experience with various genres of music, and good taste' [3, p. 3]. However, this individual, at least in industry circles, is understood to not be the artist who produced the artwork, but a trained audio mastering engineer availing of objectivity and distance from the creative process.

In the early days of the craft, audio mastering engineers were indeed separate from artists, as functionaries within music labels. With origins in the emergence of radio transmission and recording media in the mid-twentieth century, the dedicated role of the audio mastering engineer arose to chaperone sound productions from artists to audiences with industry-agreed standards that held across diverse playback technologies and environments. In the early days of this profession, the role was tied to the physical manipulation of production technologies, principally magnetic tapes and lathe cutters, to create a 'master' recording disc for mass reproduction in vinyl. Here expertise was grounded in a combination of prior experience, muscle memory, technical competence and an awareness of sonic standards in relation to specific room dynamics and playback equipment. There was no perceived creative element to audio mastering in this period, instead it was a set stage in the process of audio post-production and media manufacturing – in short, akin to a 'blue collar' profession of manual labour.

With the advent of digital recording and transmission technologies in the 1980s the role became less visceral and more intellectual through competing 'styles' of mastering, the demand for 'loudness' [4] through the manipulation of the recording waveform's RMS ('root mean square', or the average loudness of the signal) and careful addition of harmonic distortion. In this period arose a generation of celebrity mastering engineers with signature techniques and even aesthetically unique 'sounds' derived from assemblages of technologies and techniques. Here audio mastering became closer to a 'white collar' job, with engineers holding positions in mastering houses, and operating in virtual environments on computers.

Once online and cloud-based systems took hold in the 2000s, a wave of freelancing occurred, and audio mastering engineers joined the ranks of software start-ups, creative professionals, and entrepreneurs: 'no-collar' workers. The role reached a tension point between the provision of routine craft labour and intellectual and creative input, partly inspired by automated audio mastering systems competing through cost and speed reductions, notably LANDR and competitors [5].

Of chief importance in this movement towards automation is the inclusion of the artist in the workflow and the recasting of audio mastering as an artistic or creative tool with aesthetic influence and appeal [6]). Dissimilar to the pre-production role of the audio mixing engineer [7], a profession long recognised as creative and directly engaged with by the artist, the role of the audio mastering engineer remained (until now) distant from the artistic process with an emphasis on objectivity and the unbiased and scientific analysis of sound. We contend in this chapter that such transparency or neutrality in audio post-production is contested by the social nature of music and its embedding in human culture, morality, and meaning [8].

At present audio mastering is still a 'no-collar' profession, based in a form of artisanal technical expertise with a relatively high degree of professional autonomy; however, with competition from AI and automation, the industry faces a perceived threat to job security and succession [9]. As with other forms of skilled cultural labour, audio mastering as a profession is predicated on the use of particular technologies within networks of diverse actors. As in any professional field, the development and introduction of new technologies presents the possibility of disruption to how labour is conducted and what it consists of. The prevalence of the 'DIY' music career [10] is due to the proliferation of audio production and distribution systems, many made available to audiences illegally, such as Napster [11]; for free, such as Youtube; for subscription, such as Spotify; or for donation, such as Soundcloud [12]. Another aspect here is the diminishing financial returns for audio recordings via major music labels, and the saturation of the market by genres and recordings, resulting in self-management by artists [13]. As a result, mastering is no longer held to be 'mysterious' [14] or a 'black art' [1], out of the reach of understanding or competence of most audio artists.

Music and technology: an overview

The history of modern music is a history of technological innovation and disruption, and AI in music must be understood relative to this history. Technological innovation rationalises music, in the sense that it leads to new forms of standardisation. Such standardisation increases the reach of music. Perhaps the most famous example treated in sociology is the Western chord-scale system, as discussed by Max Weber [15]. In its dissemination and uniformity, this system permitted the development and transmission of repertoires and gestures across space and time. This standardisation was and remains in constant tension with affective expression, and this tension can be understood as a driver of further technological, and aesthetic, innovation.

As we now understand and experience it, music is inextricably bound up with technologies of recording and distribution. For example, the length of popular music (songs and albums) is an artefact of the phonograph. The very word 'album' is derived from the book within which to house a set of 78s, much like a photo album. Nearly one hundred years ago, media formats, notably vinyl, and burgeoning industries and networks for recorded music distribution, played fundamental roles in the establishment of international recorded music markets and the development and proliferation of genre forms [16]. Broadcast media, and technologies at the site of reception (for example, the car radio), worked together to further refine music taste, genres, and markets [17].

The broader political and cultural ramifications of the relationships between music and technology cannot be overstated [18]. Technologies of distribution and consumption, such as MP3 format, CD-R, and cassette, have directly and indirectly shaped the development of copyright law [19]. Music markets and technologies of recording and distribution have been fundamental to cultural globalisation, both reproducing and disrupting forms of colonial exploitation [20,21].

Remixing and file-sharing – repurposing 'end products' into new networks and new compositions – have led to new styles of music and new forms of sociability around music. The possibilities of (re) production at the site of consumption have thus been a flashpoint for both cultural creativity and innovation [22,23], and concerns about the political economy of the music industry [24,25]. Digital mediation or the dematerialisation of music has been understood as enhancing professional autonomy for musicians, as the basis for creative engagement from fan communities, and as the source of new revenue streams.

End-user consumer technologies have become inextricable from the everyday culture of music. Format design and audio production and programming processes at the back end have been fundamental to this development. For example, the algorithms by which Spotify playlists are generated are receiving increasing interest. AI can be conceptualised as the next incoming 'wave' in the technological mediation of music, shaping how music is produced and sequenced – getting further 'in' to the music – rather than the means by which it is circulated and consumed.

The social science interest in music is not accidental. Social scientists since Weber, including, notably, Adorno and Attali, have long suspected that developments in music often foreshadow developments in society at large. It is not a coincidence that the phrase 'gig economy' is in common parlance, and next we review the shift towards precarity in the creative industries. We situate the development and introduction of AI into audio mastering alongside other technologies which have had disruptive effects. Specifically, we discuss how electronic dance music evolved in conjunction with the consumer electronics market that afforded artists the ability to mix, spectrally alter and even remix their sound productions. We consider how technologies disrupt notions of audio production standards and processes.

Labour and aesthetics

We now discuss evidence of disruptions both in terms of labour and aesthetics. Two central and related themes warrant attention in terms of how to understand the role AI is coming to play in music production. These are its implications for labour, and for culture and aesthetics. By 'labour', we mean the work involved, and who (or what) does it. AI in audio engineering can be understood as a set of shifts in labour allocation in the production of music: artists become (more) involved in mastering, software engineers, technically working in an anterior market, produce AI services which enter the market for mastering services, and the work and practice of professional audio mastering engineers is thereby effected.

By 'culture and aesthetics', we mean the effects a given technology has on the sound of music, understood and experienced as aesthetic effects. All musical styles have characteristics derived wholly or in part from the kinds of technologies used in production and recording. Some genres, for example, black metal, have signature lo-fi and low budget aesthetics (features such as tape saturation and harmonic distortion), derived from the techniques and technologies through which seminal releases in the genre were recorded. It is important to keep in mind that this consistent encroachment, as it were, of aesthetics into and through recording technologies (which in audiophilic terms are designed with the goal of serving as a clear window on the musical product) is usually not foreseen by those who originally develop that technology.

One convenient way to think about these two themes and their relationship is by reference to Hesmondhalgh and Baker's work on creative labour in the cultural industries [26]. They present a kind of heuristic in the shape of 'good and bad work', where 'work' includes both (labour) process and product. As process, good work has attributes such as higher wages, fewer working hours, and higher levels of safety, autonomy, interest and involvement, sociality, self-esteem, self-realisation, work–life balance, and security. Bad work is characterised by poorer wages, more working hours, and lower levels of safety combined with powerlessness, boredom, isolation, low self-esteem and shame, frustrated development, overwork, and risk. Some of these features inhere to the work itself, and some to the conditions under which it is conducted. 'Good' work is that which produces products to a standard of excellence, and/or products that contribute to the commons. Conversely, 'bad' work yields low quality products, and/or products that fail to contribute to the well-being of others. Hesmondhalgh and Baker are focussed on the experience of creative labour, and they situate the 'good' or quality of the output in that context (rather than something we might independently value irrespective of the circumstances of its production). They acknowledge that the good of the process and the product could be untethered: a poor-quality product could be produced through good work, and vice versa.

Not all effects of all technologies are uniform, and a given technology could have a positive outcome on work as a process and/or product in one context, and a negative effect on work conducted elsewhere, as when the introduction of word processing, over time and alongside other developments, eliminates the job of the typist and intensifies labour for those who used to depend on their services. A longstanding concern with technology, including AI, has been around how such technology might variously automate and de-skill or replace human labour.

Hesmondhalgh and Baker's heuristic gives us a reference point from which to discuss the effects of new technologies. The terms process and product map roughly to labour, and culture and aesthetics, as described above. One important distinction is that culture and aesthetics is intended here to refer to a whole range of cultural products (pieces of music), rather than a particular instance. It is at this supra-individual level that the effects of technologies on music are most evident.

Take, as an example, the 'bad work' that has resulted from the ubiquity of sampling technologies. The amen breakbeat is undoubtedly the most sampled recorded sound in human history [27]. It is a drum loop, just under seven seconds long, which has featured in countless tracks across a range of genres. It was first sampled in hip-hop; it became ubiquitous in jungle and drum 'n' bass, it has been deconstructed in breakcore and gabber, and it continues to be used in all of these genres. It has been sampled by Nine Inch Nails, Oasis, David Bowie, and Janet Jackson, and features in the *Futurama* theme song. Used alongside the synthetic drum sounds derived from the Roland TR-808, the amen has an unmistakable and unrepeatable combination of features: swing from the drummer's skill [28]; air and crunch derived from the recording studio; and unique sounds from the drum-kit's nuances and use of magnetic tape machines and tube powered desk. It was six seconds that has shaped thousands of songs [29].

The amen originates in a funk track, 'Amen Brother', itself a version of a gospel number, released as a B-side in 1969 by a band called The Winstons. The drummer on the track was a man called Gregory C. Coleman. He was a session musician, who also played for Otis Redding and Curtis Mayfield, among others. The track was popular among hip-hop block party DJs in the late 1970s in New York. It resurfaced on a bootleg *Ultimate Beats and Breaks* compilation, directed to the hip-hop DJ market, in 1986. From there it began to appear in hip-hop recordings, including in tracks by Salt-N-Pepa, N.W.A., Schoolly D, 2 Live Crew, and Eric B and Rakim. From U.S. hip-hop, it began to appear in U.K. 'hardcore' rave music in the early 1990s, which mutated into jungle and then drum 'n' bass. Rave culture continued 'underground' as that decade ended, but the amen came to prominence again in breakcore in the early 2000s. Breakcore is a genre of 'intelligent dance music' devoted to tight and aggressive edits, largely centred on the amen and notable for being one of the first genres to coalesce online. Neither Coleman, nor Richard Spencer, another former member of The Winstons and the copyright holder for 'Amen Brother', received any royalties for this reproduction of their work. Coleman spent the last few years of his life homeless and died destitute in 2006.

The trajectory of the amen break is an instructive example of how technology effects work as process and product in music. Coleman was essentially a cultural worker of the kind Hesmondhalgh and Baker are interested in. It is likely that much of his time as a working musician would have accorded with their description of 'bad work process': irregular and unpredictable hours, poor wages, uncertainty, precarity. The amen, though, the output of his labour on one of his working days, was certainly 'good work product'. His playing was so remarkable and the sound so singular that, starting about seventeen years after that day's work, it developed and maintained wide appeal for nearly forty years and running. It exhibits excellence and has contributed vastly to the common good.

We cannot tell the whole story of how the amen break became canonical. One critical feature in this story, however, was the emergence of samplers like the Akai [Linn] MPC60, S950 to the S3000 and the E-MU SP-12, 1200 and eventually E6400 Ultra. Sampling and sequencing technologies, analogue and digital, afforded looping, re-sequencing and experimenting with the individual hits of the amen, and as such played a central role in making the break ubiquitous. Producers wanted the amen and breaks like them (for both their sonic and cultural properties) and samplers preconfigured their creative process. The popularity of breakbeats led to the use of more samplers, not the production of more drums and the employment of more drummers (or even the remuneration of the original breakbeat producers). Culture and aesthetics as described above, specifically, breakbeat culture, is not precisely the result of The Winstons' singular 'good work product' that day in 1969. Breakbeat culture and aesthetics evidences instead how that work was taken up in a growing mediated network, predicated on and reproduced through access to sampling technologies. None of this mitigates the brute injustice of the hardship of Coleman's life, a tragic demonstration of the more profound structural problem about how rarely cultural value is meaningfully rendered as economic sustainability for creators. The history of the amen highlights how technological developments change work as a process for musicians, and how other workers elsewhere (including the people who usually do not get a mention, like the workers on the assembly line at Akai) are involved in these changes. With an appreciation of how disruptive technologies have previously interacted with both labour and aesthetics, we can turn to see how these dynamics are likely to unfold with the introduction of AI to audio engineering.

FINDINGS: MITIGATION STRATEGIES

From the interviews we conducted we flag two strategies human mastering engineers are enacting in order to mitigate competition from emergent automation systems. In each case, we present an anonymised aggregated vignette assembled from the interviews to provide a tangible example.

Informal networks in creative infrastructure

Meeting in Tyson's waiting room, his presence precedes him through the golden discs adorning the wall, the signed posters, and photos with both famous and up-and-coming bands. The strategically placed instruments and stacks of amplifier rigs illustrate that this is a working studio and that he does not lack for work. A booming voice and laughter herald his arrival as he emerges from the studio with a clutch of young men in their early 2000s clutching guitars and drumsticks. Sweat in their long hair illustrates their recent creative outburst.

Relocating to a nearby café, where Tyson bear-hugs the waitress and high fives the manager who greets him as a friend, he explains the importance of getting to know his clients and of engaging in the local creative music and art scenes. Tyson explains that he is eager to get mentioned in the liner notes or credits on his clients' artistic productions, a departure from convention in traditional audio mastering. A lion's share of his work derives from opaque, all too 'human', unpredictable forms of advertising: a conversation with a client in a pub, a recommendation from another audio mastering engineer who has too much work, social media exchanges, word of mouth. There is no algorithm to the works he accepts, or which goes on to find success; he notes that his clients are diverse and their music not always to his taste, but that should not affect his work, he assures.

A key shift in recent times is the need for what Tyson terms a 'funky' meeting space close to where the clients live, work, rehearse, or spend their leisure time. While inner city studios are expensive, spaces in the inner city where students and artist spend time does provide a sense of authenticity and convenience for his clients. Interestingly, Tyson has a small studio space he hires in the city for client liaison and recording, alongside a more extensive working studio at his home out of the city, where he lives and spends most of his time, and does not meet clients.

Given his work spans the world due to his online presence, he is able to balance his time between the urban core and such remote work in a leafy suburb that offers quietude and access to nature. Yet, being tuned into the local creative scene is vital, since reputations spread in complex and unpredictable ways, and overseas clients often hear of his expertise through local artists' touring ventures or 'underground' networks in creative genres that do not map to geographical borders. On leaving, another youthful, trendily-dressed client arrives at the door, guitar in hand. 'Next' Tyson jokes, ushering him into the studio, after the obligatory high fives end our meeting.

The first strategy discovered from the research recognises the importance of place to music and how digital platforms and technologies enhance rather than diminish locality [30]. Pertinent is the growing importance of networking in providing human engineers with an advantage against automation and delocalised services. Paradoxically, the affordance of working remotely through utilising cloud-based and online services has heightened the demand for localised and spatially embedded services. A key idea here is of 'creative infrastructure' [31] pertaining to the cultural resources that communities create together [32], tied to certain spaces within cities or communities: galleries, bars, pubs, installations, museums, libraries, venues, parks, and so on. The reasons a certain site is incorporated within creative infrastructure are unpredictable, illogical and nuanced, often dictated by nostalgia or irony, rather than quality of service or location [33]. Therefore, the ability to recognise and exploit creative infrastructure is a distinctly human trait offering an advantage over automated services which can only mimic cultural significance and relational significance [34].

Visceral praxis and the human edge

Jerome is running late. Meeting in the car park of his inner-city studio space, he appears harried as he parks his car while talking on his mobile phone. Catastrophe has occurred for him: the plumbing in the gym space next door has sprung a leak, flooding his studio, and causing irreversible damage to key pieces of equipment. While he is insured, some of the vintage equipment is irreplaceable and vital to his signature workflow and style of mastering.

Inspecting his studio for damage, Jerome explains how the equipment works. He runs the summed 'mix-down' digital file – that is, the pre-mastering, post-mixing version of the aggregated 'stems' within each music track (each stem roughly representing an individually recorded instrument) – through an analogue mastering console and vintage 1960s mixing desk to capture the unique warmth of its transistors and tubes, adding effects from his range of outboard equipment racks. These include standard audio mastering tools, such as multiband equalisers, maximisers, limiters, and compressors, to rare pieces affording 'coloration' through harmonic generation and saturation processing. Some of these tools have to be manipulated solely by hand according to each musical track's distinct needs. Most of the analogue equipment does not include digital components that allow automation through a computer; however, Jerome notes that there is an emerging market for boutique devices that provide digital connectivity and an analogue signal chain. He also sometimes incorporates magnetic reel-to-reel tape into his workflow to add further 'warmth', a throwback to the past where all mastering was done from this format.

Using analogue equipment is not merely an aesthetic choice, although Jerome notes that clients who visit the studio are impressed by the appearance of the many racks of outboards with coloured lights, faders, and knobs. More importantly, it is a strategy that gives humans an advantage over automation systems, since analogue circuit-based devices require ears and hands as well as the critical listening that occurs in studio environments carefully designed and calibrated. Jerome compares the use of software products to AI systems where no human actually listens to the recording. Algorithms that analyse the peaks and frequencies of a track to apply effects according to templates of other people's music face many issues with quality control. Mastering using analogue equipment requires an engineer to get to know a musical work, and repeatedly listen to it, imagining how it could be improved, and experimenting with different treatments and combinations of devices. We finish our meeting with Jerome suddenly discovering, as he moves a heavy rug from the wall, an infestation of fungus that he worries could be a health hazard to us.

What the vignette shows is that the foregrounding of human intuition beyond the capabilities of machine learning represents a method for audio mastering engineers to stand out from the crowd, so to speak. What this point highlights is that audio mastering is shifting closer towards artistic and creative endeavour through the privileging of the performativity and improvisation of mastering with tools that require haptic control and audible perception, akin to musical instruments within digital technology-mediated environments [35].

To be sure, the adoption of both legacy and new analogue mastering equipment is a resistance to the ubiquity of digital mastering tools and indeed automation systems with no human elements in their workflows, such as LANDR.

CONCLUSION

Audio mastering engineers consistently describe their professional practice in 'scientific' terms; however, they face a counter-narrative privileging the creative possibilities of their practice in response to competition from automation. A discursive shift occurs in the development of AI, and this is evidenced in its marketing discourse: mastering becomes not only available to artists, but available to artists as part of their creative process. It is taken out of the realm of objective expertise and 'aestheticised', rendered a feature of the music amenable to compositional modification for aesthetic or creative reasons, rather than as a technical process of enhancement.

The artist is thus put in the driver's seat by AI. To offset this phenomenon, audio mastering engineers turn to the 'human' aspects of their craft, through heightening their links to local creative scenes and genres and offering charisma and an 'experience'. This turn is contrary to how audio mastering was traditionally viewed. Paradoxically, audio mastering comes to be part of the artist's remit, but this aestheticisation of the mastering job draws on 'sciencey' mastering for authenticity, through a second strategy: utilising human intuition, performance and technical prowess derived from analogue and boutique equipment unavailable to most musical artists.

AI engages with 'big data' sets, such that an AI music mastering service could adjust a particular piece of music relative to all the other music it has encountered. AI learns, and knows more than a human could. According to the marketing, AI has a formal, objective, quantitative understanding of how to get your music sounding the best, relative to (all) other music. The quality of the mastering becomes simultaneously aesthetic, in the hands of the creative artist, and more technical.

As such, this discourse imagines musical artists as the principal stakeholders or beneficiaries of changes in the creative/industrial process. Artists are positioned as insiders, experts, and astute consumers, caring for their output and their creative process by seeking out automated mastering. Audio mastering engineers are offering a counter-discourse, highlighting their own place in the creative process and positioning themselves as foundational to humanly artistic processes. We conclude by suggesting that the ongoing development, expansion and accessibility of AI audio engineering may well accord with the broader history of the relationship between music and technology, and the effects of technology on forms of skilled labour and artisan technical work in other industries. Yet we offer a beacon of hope, that through facing competition from automation, humans will turn towards creativity to continue their engagement with art and music.

REFERENCES

1. Birtchnell, T. and Elliott, A. 'Automating the Black Art: Creative Places for Artificial Intelligence in Audio Mastering'. *Geoforum*, Vol. *96*, (2018), pp. 77–86.
2. Marrington, M. 'From DJ to Djent-step: Technology and the Re-coding of Metal Music Since the 1980s'. *Metal Music Studies*, Vol. *3*, No. 2, (2017), pp. 251–268.
3. Owsinski, B. *Mastering Engineer's Handbook: The Audio Mastering Handbook*. Course Technology, Boston, (2007).
4. Hove, M. J., Vuust, P. and Stupacher, J. 'Increased Levels of Bass in Popular Music Recordings 1955–2016 and Their Relation to Loudness'. *The Journal of the Acoustical Society of America*, Vol. *145*, No. 4, (2019), pp. 2247–2253.
5. Birtchnell, T. 'Listening Without Ears: Artificial Intelligence in Audio Mastering'. *Big Data & Society*, Vol. *5*, No. 2, (2018), pp. 1–16.
6. Sterne, J. and Razlogova, E. 'Machine Learning in Context, or Learning from LANDR: Artificial Intelligence and the Platformization of Music Mastering'. *Social Media + Society*, Vol. *5*, No. 2, (2019), pp. 1–18.
7. Hepworth-Sawyer, R. and Hodgson, J., editors. *Mixing Music*. Routledge, New York, (2016).
8. Whelan, A. 'The Morality of the Social in Critical Accounts of Popular Music'. *Sociological Research Online*, Vol. *19*, No. 2, (2014), pp. 1–11.
9. Collins, S., Renzo, A., Keith, S. and Mesker, A. *'Mastering 2.0: The Real or Perceived Threat of DIY Mastering and Automated Mastering Systems'. Popular Music and Society*, (2019) 10.1080/03007766.2019.1699339.
10. Bennett, A. 'Youth, Music and DIY Careers'. *Cultural Sociology*, Vol. *12*, No. 2, (2018), pp. 133–139.
11. Nowak, R. and Whelan, A. *'Editorial: A Special Issue of First Monday on the 15-Year Anniversary of Napster — Digital Music as Boundary Object'. First Monday*, Vol. *19*, No. 10, (2014), https://firstmonday.org/ojs/index.php/fm/article/view/5542/4121.
12. Hesmondhalgh, D., Jones, E. and Rauh, A. 'SoundCloud and Bandcamp as Alternative Music Platforms'. *Social Media + Society*, Vol. *5*, No. 4, (2019), 2056305119883429.
13. Schwetter, H. 'From Record Contract to Artrepreneur? Musicians' Self-Management and the Changing Illusio in the Music Market'. *Kritika Kultura, No. 32*, (2018), pp. 183–207.
14. O'Grady, P. 'The Master of Mystery: Technology, Legitimacy and Status in Audio Mastering'. *Journal of Popular Music Studies*, Vol. *31*, No. 2, (2019), pp. 147–164.
15. Wierzbicki, J. 'Max Weber and Musicology: Dancing on Shaky Foundations'. *The Musical Quarterly*. Vol. *93*, No. 2, (2010), pp. 262–296.

16. Denning, M. *Noise Uprising: The Audiopolitics of a World Musical Revolution*. Verso, London, (2016).
17. Roessner, J. Radio in Transit: Satellite Technology, Cars, and the Evolution of Musical Genres. In: Purcell, R. and Randall, R., editors. *21st Century Perspectives on Music, Technology, and Culture: Listening Spaces*. Palgrave Macmillan, London, (2016), pp. 55–71.
18. Jones, R. 'Technology and the Cultural Appropriation of Music, International Review of Law'. *International Review of Law, Computers & Technology*, Vol. *23*, No. 1–2, (2009), pp. 109–122.
19. Sinnreich, A. Music, Copyright, and Technology: A Dialectic in Five Moments. *International Journal of Communication*, Vol. *13*, (2019), pp. 422–439.
20. Théberge, P. 'The Network Studio: Historical and Technological Paths to a New Ideal in Music Making'. *Social Studies of Science*, Vol. *34*, No. 5, (2004), pp. 759–781.
21. Novak, D. 'The Sublime Frequencies of New Old Media'. *Public Culture*, Vol. *23*, No. 3 (2011), pp. 603–634.
22. Navas, E. and Gallagher, O., editors. *The Routledge Companion to Remix Studies*, Routledge, London, (2015).
23. Cavicchi, D. Foundational Discourses of Fandom. In: Booth, P., editor. *A Companion to Media Fandom and Fan Studies*. Wiley, New York, (2018), pp. 27–46.
24. Liebowitz, S. J. 'Research Note—Testing File Sharing's Impact on Music Album Sales in Cities'. *Management Science*, Vol. *54*, No. 4, (2008), pp. 852–859.
25. Bottomley, A. J. 'Home Taping is Killing Music': The Recording Industries' 1980s Anti-home Taping Campaigns and Struggles Over Production, Labor and Creativity'. *Creative Industries Journal*, Vol. *8*, No. 2, (2015), pp. 123–145.
26. Hesmondhalgh, D. and Baker, S. *Creative Labour: Media Work in Three Cultural Industries*. Routledge, London, (2013).
27. Harrison, N. Reflections on the Amen Break: A Continued History, an Unsettled Ethics. In: Eduardo Navas, O. G., editor. *The Routledge Companion to Remix Studies*, Taylor & Francis, London, (2014), pp. 444–452.
28. Frane, A. V. 'Swing Rhythm in Classic Drum Breaks From Hip-Hop's Breakbeat Canon'. *Music Perception: An Interdisciplinary Journal.*, Vol. *34*, No. 3, (2017), pp. 291–302.
29. Otzen, E. *'Six Seconds that Shaped 1,500 Songs'*. BBC News, London, (2015), Accessed May 2020 from https://www.bbc.com/news/magazine-32087287.
30. Allington, D., Dueck, B. and Jordanous, A. 'Networks of Value in Electronic Music: SoundCloud, London, and the Importance of Place'. *Cultural Trends*, Vol. *24*, No. 3, (2015), pp. 211–222.
31. Stevenson, D. and Magee, L. 'Art and Space: Creative Infrastructure and Cultural Capital in Sydney, Australia'. *Journal of Sociology*, Vol. *53*, No. 4, (2017), pp. 839–861.

32. Lena, J. C. *Banding Together: How Communities Create Genres in Popular Music*. Princeton University Press, Princeton, (2012).
33. Gallan, B. and Gibson, C. 'Mild-mannered Bistro by Day, Eclectic Freak-land at Night: Memories of an Australian Music Venue'. *Journal of Australian Studies*, Vol. *37*, No. 2, (2013), pp. 174–193.
34. Gibson, C., Brennan-Horley, C., Laurenson, B., Riggs, N., Warren, A., Gallan, B., et al. 'Cool Places, Creative Places? Community Perceptions of Cultural Vitality in the Suburbs'. *International Journal of Cultural Studies*, Vol. *15*, No. 3, (2012), pp. 287–302.
35. Michailidis, T., Dooley, J., Granieri, N. and Di Donato, B. 'Improvising Through the Senses: A Performance Approach with the Indirect Use of Technology'. *Digital Creativity*, Vol. *29*, No. 2–3, (2018), pp. 149–164.

15

Transformations and continuities in the mastering sector

Steve Collins, Adrian Renzo, Sarah Keith, and Alex Mesker

INTRODUCTION

The histories of music production and consumption are inscribed with disruptions facilitated by new technologies and practices. Until recently, the mastering sector has largely avoided the challenges faced elsewhere in the music industries. This has started to change over the past decade, with new online services such as LANDR offering a 'one-click-fits-all' mastering solution. This chapter charts such disruptions in the mastering sector in the context of broader historical changes in the music industries. Our argument has three parts. First, the disruption currently faced by the mastering sector is part of a long tradition stretching back at least to the early twentieth century. The music industry has always been subject to some type of disruption and we 'set the scene' by noting a series of disruptive events that significantly reconfigured various aspects of the music industries. Second, those disruptions have often involved non-music-focused corporations encroaching on territory that was formerly seen as properly belonging to 'the music industry' (variously defined). In the most recent iteration of this pattern, information technology (IT) companies have usurped some of the space that was once occupied by music corporations. Third, the disruption currently taking place in the mastering sector is often couched in terms of 'quality' (that is, as a contest between mastering professionals and online automated systems), but is actually more about convenience and price points. Amateur musicians who may not have previously considered having their work mastered are being targeted by companies that promise a low-price alternative to professional mastering, achievable via a 'one-click' solution. Continuing a previous study [1] this research involves surveying professional mastering engineers to develop a picture of how the sector is responding to interventions from cloud-based automated mastering services such as LANDR and CloudBounce.

TRANSFORMATIONS AND CONTINUITIES

Discussions of disruption to the recording industry more often than not refer to the *distribution* of recorded works – the radical decentralisation that occurred at the turn of the millennium. Although both music production and music distribution have been disrupted by digital and networked technologies since the late 1990s [2], distribution is more often the topic under examination (for example, see Alderman [3,4], David [5], and Kusek and Leonhard [6,7]). In the dominant narrative, the advent of file-sharing and peer-to-peer software such as Napster caused major record labels to flounder until the Apple iPod and iTunes Store demonstrated there was a viable market for digital music files. New economic configurations stemming from legal downloads were further solidified by streaming platforms such as Spotify, Google Play, and YouTube. Within two decades, the recording industry had shifted its emphasis from sales towards the 'heavenly jukebox' [8]. The Internet reconfigured the distribution of music away from established arrangements involving record labels, distributors, and retail outlets, and towards anyone with Internet access and a new host of intermediaries – some legal, some not – and file-sharers could (freely) distribute the copyrighted works of record labels using software like Napster or Kazaa, and websites like The Pirate Bay or Oink.

Disruption is by no means limited to distribution. Every point in the music value chain has experienced disruption [9, p. 1327], including music production. The personal computer, combined with affordable Digital Audio Workstations (DAWs) and plugins, created new sites for music production away from traditional studios. Attics, basements, spare rooms, and bedrooms became spaces for music production. While home studio arrangements certainly lend themselves more readily towards some genres than others (it is far more difficult to accommodate a symphony orchestra than a Push controller in the spare room), they can produce professional results [10, pp. 127–128]. As early as October 2000, Moby recounted that he wrote, recorded, and mixed the entirety of *Play* in his bedroom before sending the album off to a mastering studio for final treatment [11, p. 344]. Similarly, Flume's eponymous 2012 album was produced on his laptop, 'either in his bedroom at home or while backpacking around Europe, layering beats and samples while sitting in German pubs, Spanish cafes and Amsterdam backpacker joints' [12].

The radical decentralisation of production and distribution extends to other areas of the industries. At the very earliest stage of (pre-)production, crowdfunding has provided an alternative to the traditional advance from a record deal, which once served as the mark of a musician having 'made it'. Amanda Palmer notably raised over $1 million on Kickstarter to fund an album and tour [13, p. 127] and has continued to explore 'micro-patronage' on Patreon [14, p. 283] which offers a subscription-based model for fans to financially support artists on a monthly basis. At the other end of the production chain, podcasting and internet radio stations can serve niche genres and their audiences as well as unsigned, independent artists that would struggle to reach wider audiences through mainstream broadcast services. These outlets provide a diversity that mainstream broadcasters are often unable to offer [15, p. 25]. Such disruptions provide opportunities for musicians and do not necessarily replace other, more established processes. Record labels still exist, as do commercial radio stations, managers, tour promoters, studio engineers, record producers, and session musicians. Many aspiring artists still vie for record deals. These disruptions are perhaps better understood in the context of 'transformations and continuities' [16, pp. 7–11] where new practices sit alongside traditional arrangements.

While much of the commentary in this field emphasises recent instances of disruption (for example, see Hughes et al. [17]), we note that disruption has been a defining characteristic of the music industries for over a hundred years. In the latter stages of the 1800s, for example, the music industry consisted of publishers selling sheet music in response to the increasing popularity of the piano. In the US, Tin Pan Alley publishers successfully lobbied for a copyright in the performance of the music they published and secured 'control over the two main modes by which the public consumed music' – the production of sheet music and its public performance [18, p. 414]. The introduction of the pianola further disrupted the economy established by sheet music publishers, as manufacturers of piano rolls had no obligation to compensate music publishers for use of their songs, until legislated to do so in the Copyright Act 1909. Shortly thereafter, the music industry was again disrupted by recording technologies. Recordings rather than performances (mechanically or by live musicians) soon became the dominant means by which audiences consumed music, and this shift in turn affected how music itself was created. Of course, music was still performed publicly but 'its organization and profits were increasingly dependent on the exigencies of record making. The most important way of publicizing pop – the way most people heard most music – was on the radio, and records were made with radio formats and radio audiences in mind' [19, p. 236].

The music industry as we know it today bears scant resemblance to the sheet publication trade of 1887. Over the course of time, and with some regularity, new formats have emerged accompanied by new business practices and economies. The current (and now dominant) streaming model led by Spotify marks the latest point in a continuing series of technological, economic, cultural, and legislative changes. A notable and recurring feature of these changes – encompassing both distribution and production – is that change is often driven from *outside* the music industries, as explored below.

TECH COLONIES

The notion of technology companies encroaching on the music industries is a well-established pattern stretching back at least as far as the early twentieth century. Frith [20, p. 57] notes that electronics manufacturers often drove changes in the music industries. For example, manufacturers often need to become involved in the production of 'software' (such as vinyl records or compact discs) in order to have a product with which to promote their new hardware (record players, compact disc players). As stockbroker Edward Lewis put it, 'a company manufacturing gramophones but not records was rather like making razors but not the consumable blades' (cited in Frith [20, p. 57]). This principle was echoed many years later when Apple released the iPod closely followed by the iTunes Music Store. Apple's entry into the music distribution industry is attributed to Universal Music Group's then-CEO Doug Morris being swayed by the tech company's limited market share ([21, 158]). In other words, major record companies saw digital sales as a potential *parallel* revenue stream to the sales of tangible media such as CDs, rather than as a replacement of tangible media. The alliance between the major labels and Apple was, however, somewhat uneasy [22, p. 370]. In 2007, Rio Caraeff, Universal's executive vice-president in charge of digital strategy, remarked that Apple's entry into the music industry had placed 'golden handcuffs' on record labels: iTunes may have been profitable for labels, but the iTunes Store monolith also prevented the labels from launching their own stores [23]. From iTunes onwards, the prominent shifts in the narrative of music distribution have been made by IT companies – Spotify, Amazon, Google, etc. Indeed, Hesmondhalgh and Meier [24, p. 1556] observe that 'the IT industries are now the primary sector determining change' in the music industries.

Of course, interventions from the IT sector are not specific to the distribution aspect of music. Music production has a long and closely entwined relationship with IT: after all, digital audio workstations (DAWs) and plugins are software packages. DAWs introduced affordable and industry-grade technologies into computer-based home studios [10, pp. 122–125]. Software such as Ableton Live, Cubase, Logic, and an abundance of plugin effects, synths, and sample libraries have afforded amateur musicians access to music technologies identical to those used by their professional peers (a potential previously only hinted at by the cassette-based Portastudio and its ilk). Although traditional recording and production spaces still exist, 'in-the-box' and networked technologies have displaced the loci of many activities related to the production of music [10, p. 125]. Prior [25, pp. 398–400] writes of visiting a recording studio and noting how 'globalized networks of communication [...] and the internet' will 'soften' and even threaten the traditional sites of music production as processes move into a 'global space of flows'. The widespread availability of technologies previously only accessible to professionals, coupled with the Internet's distributive capabilities, allows contemporary musicians (professional and amateur alike) to compose, record, mix, produce, and distribute music from home or mobile studios.

Until recently, the mastering process has proved remarkably resilient in the face of such technological developments and the decentralisation of many sites of music production. As recently as the early 2000s, mastering was still predominantly seen as a job that needed to be handled by professionals, with all that that entailed: dedicated, treated studio spaces, dedicated monitors, and a particular (sometimes jealously guarded) signal processing chain. In the past decade, however, the disruptions and disintermediations present in many other points of the music value chain have begun to appear in the professional mastering sector. Here, we focus on the emergence of a number of cloud-based automated mastering services (CAMS).

Mastering has traditionally been regarded as a craft requiring significant expertise and experience. As an industry, mastering to some extent trades on its enigmatic status as a 'dark art' [26, p. 183; 27, p. 241; 28, p. 1]. O'Grady [29, pp. 147–148] writes of his own experience as a musician sitting in on a mastering session:

> *In 2009, before attending a mastering session for a project I was working on, a colleague told me that the mastering engineer kept a black box underneath the mastering console. While the box was probably a customized signal processor, the engineer would not disclose what it was or how it affected the sound produced. The black box ostensibly played a significant role in the alleged distinct sound of the work produced [...] the contents of the black box intrigued me–and still does.*

The operational specifics of CAMS are likewise opaque, but the generic principle of CAMS is very simple: upload a file, select some options, and wait for a few minutes while the mastered file is prepared for download. LANDR allows its users to select the desired characteristics of their mastered audio from a limited set of options; there are three 'intensities' with no control over the variables involved. In contrast, CloudBounce offers several genre-based settings. CAMS typically offer a number of subscription packages, with LANDR's options ranging from $5 per month to $25 per month including a variety of output options. For example, the 'Basic' package gives subscribers an unlimited number of low quality MP3s but charges $23.99 for a high quality WAV file, whereas the 'Pro' package offers unlimited renderings of every format. LANDR's subscription packages include distribution to 'iTunes, Spotify, Google Play + all other major stores' marking LANDR's move into the aggregation/distribution sector. CloudBounce offers a similar array of packages, including pay-as-you-go at $4.90 per track, an arrangement that yields a 320-kbps MP3 and a 24-bit WAV file. Other platforms offer additional features incorporating a mixture of traditional analogue hardware and digital workflows. Aria, for example, provides an automated mastering service using proprietary analogue mastering equipment assisted by a robot arm that works the pots.

The narrative of mystery [29, pp. 153–160] that surrounds the mastering process has a great deal of purchase among musicians, and the processes underpinning CAMS are no less mysterious. As mentioned previously, LANDR offers three 'intensities' but its operational sequences are not transparent [30, p. 11]. The opacity of the inner workings of each CAMS, invoking expertise, specialist equipment, machine learning, or other proprietary technology, help to create its value.

'REPLACING' THE ENGINEER

At first glance, CAMS appear to replace the need for a human mastering engineer. As Birtchnell and Elliott write, 'progress in machine learning, big data analytics and algorithms has inspired efforts to scientifically and methodically plot the skills and practices involved [in mastering] and replicate them in software' [31, p. 80]. Inherent in these systems lies the provocation that mastering can be distilled to an automated process that adjusts levels, frequencies, and stereo balance to parameters required by a musical genre. This view is evident in LANDR's publicly available 2014 patent application, 'System and method for performing automatic audio production using semantic data' [32]. In short, 'chromosomal' identifiers are extracted from the uploaded audio file, which indicate the file's genre. Based on the identified genre, a common configuration of effects are applied, such as certain filters, spatial processors, and multi-band compression.

In our research, mastering engineers have been wary of the idea that mastering can be reduced to a set of algorithms. To some, this idea was seen as an assault on the professionalism of mastering engineers. In several cases, the mere mention of CAMS such as LANDR was enough to provoke discussion of whether an algorithm can ever master a track to the same standard as an experienced human mastering engineer. Many of our respondents emphasised the importance of human involvement in mastering because algorithms lack creative judgement and the capacity for problem-solving in individual cases: 'Software has yet to replace room acoustics and experience'; 'These [CAMS] just set the level properly which is fine. But they can't make artistic judgments'; 'Mastering requires the input of someone who has reached a master skill level of dealing with audio'. Another respondent enumerated a number of specific decisions that a mastering engineer might make during the mastering process – decisions that an automated process might overlook or mishandle:

> *Can automated services get every track "right" on a project? What are the chances? If you send in a fantastic sounding finished product, will it come back from LANDR untouched? How does it know if an intro/verse/chorus needs different treatment to the rest of a song? Or treating a radio edit or vocal up or instrumental version with subtly different processing decisions?*

The concerns expressed by our respondents are commonplace in the mastering sector and are reflected in dozens of articles found online, in trade publications, and spirited forum discussions. For example, Geoff Pesche makes the case that an algorithm cannot exercise creative decision making (for example, crossfading two tracks or 'heavy sonic resculpting of tracks') [33]. Similarly, Vlado Meller argues that there is no algorithm capable of creating a cohesive-sounding album, pointing out that the 'professional human touch in music production is vital', as is 'human decision in sculpting sound' [33]. A common theme for justifying the importance of the mastering engineer is the cohesion of tracks in an album format. However, Justin Evans, co-founder of LANDR, counters that his service's focus on individual tracks is a reflection of how music is predominantly consumed today:

> *We live in, and love, the world of SoundCloud and streaming services. Today, most songs are listened to alone, or as part of a streaming playlist instead of a coherent album. We think this raises many interesting questions about what mastering means. In the '40s, mastering was part of the job of a transferring engineer. Then vinyl and radio changed that. Now we are living in a streaming world. What should mastering be now? That question shaped our original product vision* [33].

Evans' suggestion that the history of mastering is marked by disruption and change fits within the discourse of wider changes to the music industry discussed above. Whether CAMS can ever be as good as a human mastering engineer is something of a strawman. The viability of CAMS is less a question of quality and more about business models.

'ONE CLICK FITS ALL': TARGETING THE NEW AMATEURS

Prior [34–36, p. 89] identifies a sector of amateur music production populated by what he labels 'new amateurs'. These musicians are not 'particularly interested in "making it"–although some do–but pursue meaningful, self-sustained music-making. The 'new amateurs' are mobile, dislocated individuals, who are engaged in global and social musical networks for content and distribution. While music *making* has become increasingly accessible to amateurs over the last 20–30 years, facilities allowing amateur producers to *distribute* and (potentially) *capitalise* on their music are a much more recent development. Certainly, earlier forms of democratised music distribution existed [13, pp. 50–52], but websites such as mp3.com or peoplesound.com predominantly catered for the amateur sector. The emergence of new intermediaries such as TuneCore and Distrokid allow amateurs to distribute their music on the same platforms (Spotify, iTunes, Apple Music, etc.) as professional artists. Distributing music on Spotify, for example, is now a simple and affordable process for the amateur musician. The ability for the amateur to release music using the same services as professionals reflects what Prior [34, p. 85] identifies as the 'marketisation of leisure time'. Today, music-making by amateur producers can yield capital for both the producer (through distribution and streaming of recordings) and for digital music services (through the user-pays subscription model). There is a growing recognition of casual, low-budget and amateur music producers as a distinct market sector, and arguably, this is what CAMS like LANDR are targeting as their core user base.

While it is possible that a few musicians might switch from professional mastering engineers to CAMS, the stated intention of services like LANDR is to fill (or marketise) a gap. Pascal Pilon, CEO of MixGenius (the company behind LANDR) states, 'We're targeting musicians [...] Over 95% of songs that are recorded by amateur musicians never get mastered [...] People don't have the money to do that. Nowadays mastering is an elitist process that only a few can manage to pay for. And it's way too complicated for most people as well' [40]. By focusing on discussions of quality rather than more pragmatic decision-driving factors like price point and turn-around times, mastering engineers overlook the potential threats and opportunities presented by CAMS. Most CAMS do not aim to displace institutions such as Abbey Road: they simply capitalise on the widespread notion that, as a contemporary musician, 'you can do everything yourself'.

The fact that CAMS target new amateurs is made particularly clear by the fact that many promise a 'one-click' solution – or something very close to it – in their marketing material. When first visiting the LANDR website, the user's eye is immediately drawn to the prominent tagline, 'Create, we'll do the rest'. Reflecting a growing trend for the simplification of complex musical, computational, and technical processes, such statements have become a well-established selling point for software developers targeting the amateur market. For example, ToonTrack's popular EZmix plugin – a combined guitar amp, mixing, and mastering tool – promises 'One Click [...] Great Sound' and offers 'professionally designed audio processing effect chains' in 'one convenient package'. Similarly, iZotope's Neutron 3 offers 'assistive audio technology' that 'can automatically sculpt your tracks and balance your levels, so you can get right into the creative flow'. In the world of CAMS, it is not just LANDR that has adopted such an approach: eMastered offers to 'Master Your Track, Instantly', while CloudBounce calls for users to 'Have your music mastered immediately during & after your creative process'. These services clearly position mastering as a process that is both easily automated *and* distinct from the creative process of music-making. Moreover, these services are aimed squarely at musicians who are increasingly the target of software designed to ease the cognitive and labour burdens of achieving professional results [38, pp. 52–53]. Technological advancements promise professional results without having to invest the time and effort in acquiring knowledge and developing skills in music production.

There are also signs that major record companies are now seeking to capitalise on amateur producers. In 2015, LANDR raised $6.2 million from various sources including musicians, producers, private equity firms and Warner Music Group [39]. As with previous changes to the music industry, what was once seen as a threat gradually comes to be seen as an opportunity for the major record labels. Warner Music Group's diversification and investment in LANDR is a step towards accessing the 'long tail' of independent, non-label content, which in 2018 comprised 8% of the entire recording business [40]).

THE REAL AND PERCEIVED THREATS

CAMS appear to be targeting 'new amateurs' who are attracted by the fast-turnaround and low costs involved. Given the inherent de-personalisation and de-professionalisation of tasks once handled by specialist engineers, it comes as no surprise that some of our respondents were quick to defend their role in music production. The mastering engineers we surveyed expressed a range of views on the real or perceived threat of CAMS. Some respondents emphasised that currently, the human engineer can produce far superior results to the algorithm. However, some of the same respondents also conceded that this will change in the coming years as the technology advances. The 'Bob Ludwig client' and the 'LANDR user' currently represent predominantly separate markets, but some of our respondents predict that more of the former will eventually 'cross the floor' if/as the technology improves.

Respondents to our survey frequently underscored the role of mastering engineers as a form of quality assurance. For example, one respondent described mastering as 'quality check control', while another noted that mastering can compensate for the 'mistakes' made by musicians. He continued:

> *Some of the key elements for the mastering that are not available to the general public is acoustically [treated] rooms, properly chosen speakers, properly positioned speakers, plus many years of experience and training. It took me 15 years to say that 'now I understand the mastering process' (maybe:). It happened after I built my 4th studio, learned a bit of acoustics, got proper monitors with proper DA controller, and produced hundreds of songs.*

Similarly, another respondent suggested that 'high level mastering' depends on '[m]onitors and [r]oom acoustics' and therefore 'provides a different, faster and more consistently accurate' result.

As noted above, many of these comments are logical given the substantial time and money that many mastering engineers have invested in their profession. However, we would suggest that for many listeners and producers, *convenience* is often more important than 'quality'. Just as many listeners embraced CDs for their convenience rather than for any purported leap in sound quality (Frith 1996, p. 25), CAMS have the potential to disrupt mastering engineers' business model regardless of whether they deliver an objectively 'superior' (or even adequate) product. It is worth noting, then, that relatively few of our respondents perceived services such as LANDR as a threat to their current business model. However, they were only able to maintain such a position by distinguishing between 'high level mastering' and 'other' types of mastering.

Birtchnell [41, p. 14] suggests that mastering engineers have little to fear from LANDR as its 'results are not yet on a par with "professional" level mastering'. This conclusion, however, is determined by a claimed absence of LANDR-mastered tracks featured in popular music charts and is perhaps a little spurious – it is, after all, well established that Top 40 charts are not necessarily an accurate measure of 'popular' taste (Frith 1996, p. 15). Although somewhat cagey about its revenue, LANDR ('Who Uses LANDR?') itself claims in excess of 1.8 million users from over 100 countries who have mastered over ten million tracks. One such user, an independent artist called ViN, has over 1.4 million plays of his LANDR-mastered track 'About You' on Spotify; this is not an insignificant achievement for an independent artist. LANDR also mastered Gwen Stefani's single 'Make Me Like You' and remixes of Lady Gaga's 'Til It Happens to You' [42]. Perhaps the viability of CAMS should not be determined by chart success, but by their ability to monetise a previously untapped market sector and attract new users.

Sterne and Razlogova [30] question whether the service that CAMS provide is, technically, mastering. They argue that in order for an automated system to be successful, its creators must necessarily define the terms for success; 'a "finished" recording is an aesthetic judgment and a moving target', and the emergence of CAMS 'marks a moment of social and definitional contest' [30, p. 4]. This contest was reflected by several of our respondents who distanced themselves from the perceived target market for automated marketing services: the assumption here appeared to be that 'they' might make use of such services, but 'we' will remain safe from economic disruption because 'we' cater for a particular, high-end market. One respondent linked this type of distinction with specific genres of music: 'These services are great for those who make music for games, DIY movies, college concerts, electronic (high compressed) styles and similar'.

Some engineers did, however, see such services as a threat to the mastering sector more broadly. These respondents admitted that the technology is still in its infancy, but they recognised that in years to come, mastering-via-algorithm could well become more widespread and have an impact on the market for engineers:

> *When I survey the mastering sector as a whole these services are absolutely a threat to many mastering businesses that exist currently. Much of what mastering engineers produce is about consistency of sound based on the commercial demands of particular musical genres. The automated services [...] can achieve the same outcome at lower cost so by my estimation they will consume 80% of most mastering studios client base within the next 5–10 years.*

Another respondent suggested that mastering would achieve the same status as 'craft beer': relevant to 'high end projects where it provides that extra quality control [...] But long term I think it will evaporate'. Several respondents were quite pessimistic: 'The big mastering houses like Gateway will survive', but the middle tier of mastering engineers 'will get smaller and smaller due to more and more musicians using the automated mastering sites [...] I do not predict a healthy climate for middle tier mastering studios in the years to come'; 'the call or need or desire for dedicated pros will continue to shrink'; 'automated mast[ering] will halve the number of mastered tracks, to say the least. Having mastering as [a] living would be harder for non major-affiliated rigs'; 'The big mastering studios will probably always be around. The low end and DIY studios will also be around. The middle tier mastering studios will all probably be closed within the next 10 years for lack of business'.

CONCLUSION

As Sterne and Razlogova [30, p. 15] note, '[t]he relationship between mastering engineers and LANDR is not a John Henry-like battle between man and machine'. Indeed, we are not suggesting that mastering carried out by a human engineer will disappear, but the landscape has certainly changed and will continue to do so. The emergence of CAMS in some ways mirrors the data-driven 'one-click' solutions that have become prevalent in other areas of production and distribution (such as the instant recommendations offered by Spotify). Current music producers are now accustomed to these 'one-click' efficiencies and may actually demand more of them as part of their workflow. For example, LANDR now has DAW integration in Bandlab's (formerly Cakewalk's) Sonar, and LANDR's distribution features mean that an artist can compose, record, mix, master, and distribute content all within the confines of a single DAW.

Critics and industry practitioners primarily concerned about CAMS' quality are overlooking the larger significance of such systems. Whether CAMS' primary user base will be those who previously would not have mastered their music or whether we will see increasing numbers of professionals eschewing human mastering engineers is debatable, but some of our respondents certainly predict the latter happening. Arguably, the new 'democratisation' of music distribution sees companies such as LANDR targeting the 'new amateurs' who lack the funds to pay for professional mastering services. Human engineering will not disappear (at least, not entirely) in the near future. For now, it appears that parallel markets for different 'tiers' of mastering will and can coexist, but disruption to the mastering sector will continue to occur.

REFERENCES

1. Collins, S., Renzo, A., Keith, S. and Mesker, A. *'Mastering 2.0: The Real or Perceived Threat of DIY Mastering and Automated Mastering Systems'. Journal of Popular Music and Society*, DOI: 10.1080/03007766.2019.1699339 (2019).
2. Williamson, J. and Cloonan, M. 'Rethinking the Music Industry'. *Popular Music*, Vol. *26*, No. 2, (2007), pp. 305–322.
3. Alderman, J. *Sonic Boom: Napster, Mp3, and the New Pioneers of Music*. Perseus Books, Cambridge, MA, (2001).
4. Anderton, C. 'Audio Mastering in Your Computer'. *Sound on Sound*, (2004). Available at: https://www.soundonsound.com/techniques/audio-mastering-in-your-computer [Accessed 20 July 2019].
5. David, M. *Peer to Peer and the Music Industry: The Criminalisation of Sharing*. Sage, London, (2010).
6. Kusek, D. and Leonhard, G. *The Future of Music: Manifesto for the Digital Music Revolution*. Berklee Press, Boston, (2005).
7. LANDR. Who Uses LANDR? *LANDR*, (n.d.). Available at: https://support.landr.com/hc/en-us/articles/360004943034-Who-uses-LANDR- [Accessed 29 June 2019].
8. Mann, C. The Heavenly Jukebox. *Atlantic Online*, (2000). Available at: http://www.theatlantic.com/past/docs/issues/2000/09/mann.htm [Accessed 25 March 2019].
9. Leyshon, A. 'The Software Slump?: Digital Music, the Democratisation of Technology, and the Decline of the Recording Studio Sector within

the Musical Economy'. *Environment and Planning*, Vol. *41*, No. 6, (2009), pp. 1309–1331.

10. Goold, L. and Graham, P. The Uncertain Future of the Large-Format Recording Studio. In: Gullö, J. O., editor. *Proceedings of the 12th Art of Record Production Conference Mono: Stereo: Multi*. Royal College of Music (KMH) & Art of Record Production, Stockholm, (2019).
11. Young, S. and Collins, S. 'A View from the Trenches of Music 2.0'. *Popular Music and Society*, Vol. *33*, No. 3, (2010), pp. 339–355.
12. Guilliatt, R. Electric Dreams. *The Australian*, (2018). Available at: https://www.theaustralian.com.au/weekend-australian-magazine/harley-streten-aka-flume-from-bedroom-dj-to-global-star/news-story/49e8befc583ea7be5e76cee8d8635430 [Accessed 2April 2019].
13. Collins, S. and Young, S. *Beyond 2.0: The Future of Music*. Equinox, Sheffield, (2014).
14. Landström, H., Parhankangas, A. and Mason, C. *Handbook of Research on Crowdfunding*. Edward Elgar, Cheltenham, (2019).
15. Markman, K. M. and Caroline S. E. 'Why Pod? Further Explorations of the Motivations for Independent Podcasting'. *Journal of Radio & Audio Media*, Vol. *21*, No. 1, (2014), pp. 20–35.
16. Meikle, G., and Young, S. *Media Convergence: Networked Digital Media in Everyday Life*. Palgrave Macmillan, Basingstoke, (2012).
17. Hughes, D., Evans, M., Morrow, G. and Keith, S. *The New Music Industries: Disruption and Discovery*. Springer, Springer Nature, Cham, Switzerland, (2016).
18. Keyes, M. J. 'Musical Musings: The Case for Rethinking Music Copyright Protection'. *Michigan Telecommunications and Technology Law Review*, Vol. *10*, No. 2, (2004), pp. 407–433.
19. ———. The Industrialization of Music. In: Bennett, A., Shank, A. and Toynbee, J., editors. *The Popular Music Studies Reader*. Routledge, Oxford, New York, (2006).
20. Frith, S. The Industrialization of Popular Music. In: Lull, J., editor. *Popular Music and Communication*. SAGE, Newbury Park, (1992).
21. Levy, S. *The Perfect Thing: How the iPod Shuffles Commerce, Culture and Coolness*. Simon & Schuster, New York, (2006).
22. Negus, K. 'From Creator to Data: The Post-Record Music Industry and the Digital Conglomerates'. *Media, Culture & Society*, Vol. *41*, No. 3, (2019), pp. 367–384.
23. Mnookin, S. Universal's CEO Once Called iPod Users Thieves. Now He's Giving Songs Away. *Wired.com*, (2007). Available at: https://www.wired.com/2007/11/mf-morris/ [Accessed 24 March 2018].
24. Hesmondhalgh, D. and Meier, L. 'What the Digitalisation of Music Tells us About Capitalism, Culture and the Power of the Information Technology Sector'. *Information, Communication & Society*, Vol. *21*, No. 11, (2018), pp. 1555–1570.
25. Prior, N. The Rise of the New Amateurs: Popular Music, Digital Technology and the Fate of Cultural Production. In: Hall, J. R., Grindstaff, L. and Lo, M., editors. *Handbook of Cultural Sociology*. Routledge, New York, (2010).

26. Bregitzer, L. *Secrets of Recording: Professional Tips, Tools & Techniques*. Focal Press, Oxford, (2009).
27. Hepworth-Sawyer, R. and Golding, C. *What Is Music Production?* Focal Press, London, (2011).
28. Cousins, M. and Hepworth-Sawyer, R. *Practical Mastering: A Guide to Mastering in the Modern Studio*. Focal Press, Burlington, MA, (2013).
29. O'Grady, P. 'The Master of Mystery: Technology, Legitimacy and Status in Audio Mastering'. *Journal of Popular Music Studies*, Vol. *31*, No. 2, (2019), pp. 147–164.
30. Sterne, J. and Razlogova, E. 'Machine Learning in Context, or Learning from LANDR: Artificial Intelligence and the Platformization of Music Mastering'. *Social Media + Society*, Vol. *5*, No. 2, (2019), [online].
31. Birtchnell, T. and Elliott, A. 'Automating the Black Art: Creative Places for Artificial Intelligence in Audio Mastering'. *Geoforum*, Vol. *96*, (2018), pp. 77–86.
32. Terrell, M. *et al* United States Patent No. 9,304,988. *United States Patent and Trademark Office*, (2016). Available at: http://patft.uspto.gov/netacgi/nph-Parser?Sect1=PTO1&Sect2=HITOFF&d=PALL&p=1&u=%2Fnetahtml%2FPTO%2Fsrchnum.htm&r=1&f=G&l=50&s1=9,304,988.PN.&OS=PN/9,304,988&RS=PN/9,304,988 [Accessed 19 June 2019].
33. Inglis, S. 'LANDR, CloudBounce & The Future of Mastering'. *Sound on Sound*, (2016). Available at: https://www.soundonsound.com/techniques/landr-cloudbounce-future-mastering [Accessed 19 June 2019].
34. Prior, N. *Popular Music, Digital Technology and Society*. SAGE Publications, (2018).
35. Rogerson, B. MixGenius Discusses the LANDR Automated Online Mastering Service. *Music Radar*, (2014). Available at: https://www.musicradar.com/news/tech/mixgenius-discusses-the-landr-automated-online-mastering-service-603304 [Accessed 15 June 2019].
36. Rumsey, F. 'Mastering in an Ever-Expanding Universe'. *Journal of the Audio Engineering Society*, Vol. *58*, No. 1, (2010), pp. 65–71.
37. Strauss, K. 'Making Money on the Music Mix: Meet MixGenius'. *Forbes*, (2014). Available at: https://www.forbes.com/sites/karstenstrauss/2014/06/16/making-money-on-the-music-mix-meet-mixgenius/ [Accessed 18 June 2019].
38. Mueller, G. *Media Piracy in the Cultural Economy: Intellectual Property and Labor under Neoliberal Restructuring*. Routledge, New York; Oxford, (2019).
39. Falconer, K. Landr Audio Raises $6.2 mln from Warner Music Group, VCs and Artists. *The PE Hub Network*, (2015). Available at: https://www.pehub.com/canada/2015/07/landr-audio-raises-6-2-mln-from-warner-music-group-vcs-and-artists/ [Accessed 30 March 2019].
40. Mulligan, M. The Frank Ocean Days May Be Gone, but Streaming Disintermediation Is Just Getting Going. *MIDIA: Music Industry Blog*, (2010). Available at: https://musicindustryblog.wordpress.com/tag/spotify/ [Accessed 24 July 2019].

41. Birtchnell, T. (2018). 'Listening Without Ears: Artificial Intelligence in Audio Mastering'. *Big Data & Society*, Vol. 5, No. 2, pp. 1–16.
42. Brown, H. LANDR Brings Music Mastering to the Cloud, Helping Major Labels & Bedroom Acts Alike *Billboard.* (2019). Available at: https://www.billboard.com/articles/business/8499104/landr-music-mastering-online-cloud-major-labels-bedroom-artists [Accessed 11 July 2019].

16

Mastering in music

A CONFERENCE, A DISCUSSION, AND A 'WHAT-IF' – A MANIFESTO FOR AN ENCAPSULATED AUDIO DELIVERY FORMAT

J.P. Braddock and Russ Hepworth-Sawyer

CAVEAT

Before we lay out our stall, so to speak, what must be placed here is a caveat. We describe the need for a format shift and tools. To date, we have been unable to find details of any such product in development, but given market sensitivities surrounding intellectual property, it is quite possible we're not permitted to know. Therefore, we apologise if we've not mentioned someone who is already pioneering this, and we applaud them!

A CONFERENCE

What follows is formed from informal discussion between two mastering engineers and colleagues around the activities of the Audio Engineering Society's first UK Mastering Conference in 2018 at the University of Westminster. The conversation was unstructured and held in the spirit of a 'what if' frame of mind. Of the many topics discussed, one 'what if' dominated our discussions: what if there was a way to save us all the new, additional work, we're not really getting paid value for. We have tried to capture and relay our thinking below.

A DISCUSSION

Since the 1990s, there has been some form of stability for the mastering engineer when it comes to file formats for replication or digital delivery. Typically, it has been based around the creation of a Compact Disc (CD) ready Disc Description Protocol image file (DDPi), or some basic.wav or.mp3 exports for digital distribution or promotion. Even with the advent of iTunes at the turn of the century,

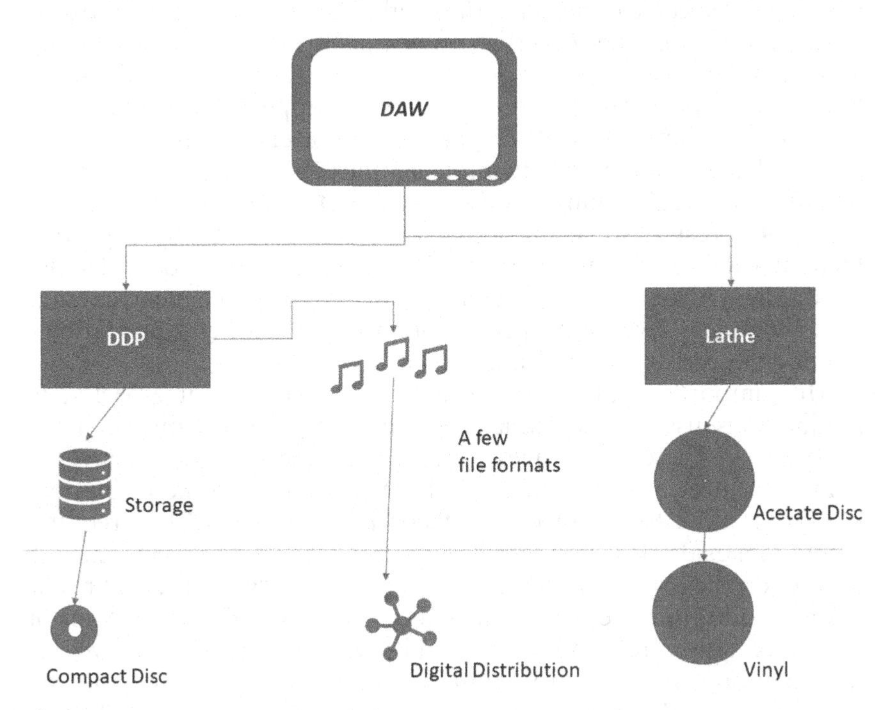

Figure 16.1 Typical early twenty-first century format creations from a mastering engineer's studio for later replication and distribution (below line).

the delivery was still largely a PCM export from the DDP or a manufactured CD rip. Those cutting vinyl would naturally prepare an acetate for pressing.

Despite growing interest and the much-publicised increase in vinyl record consumption, most consumer music is actually digital and distributed primarily using several differing digital platforms. This is confirmed by the 2019 IFPI Global Music Report which reveals that in 2018 physical format revenues dipped by 10.1%. Interestingly, sold music as download revenues dropped by a staggering 21.2%. This definitely shows a direction of travel as the report reports an overall global growth in music consumption of 9.7%! This growth indicates a significant shift away from the ownership of music to a licencing model of music provision which has led to the ultimate growth in the streaming services such as Tidal and Spotify amongst others.

Despite no longer having to store bulky CD jewel cases within our rather small living rooms, a fashionable outcome of consumers digesting streamed audio is the curation of playlists by oneself or by others. Streaming services, such as Spotify, present certain playlists for moods, activities or genres, but the humble playlist can also be curated by users or organisations for promotional activity. Due to the streaming services' library being totally open access to all users, it is possible for specific curation with the ability to mix and match tracks from infinite numbers of artists. With such diversity on offer, not only in audio tone and sonic quality, but mostly the perceived level of the material, providers have seized on what appears to be the fastest way of providing a smooth experience for users: loudness normalisation. The seemingly speedy implementation has led to the adoption of differing equal loudness normalisation standards between providers. As a result, some mastering engineers respond by exporting their mastering output accordingly to reflect each provider individually. Others will create a streaming master with a mean average lower loudness level appropriate for most normalised applications. Other engineers argue that the master is the master.

Most importantly, these significant market changes, irrespective of loudness discussions, continue to present new and ongoing delivery challenges. These challenges present themselves with deliveries aiming to meet a myriad of possible outputs in both file resolution, file types, True Peak (TP) with deciBel Full Scale (dBFS) requirements alongside differing loudness normalisation standards that all depend on the aforementioned consumer format the music will reside upon. In addition, the continuing expectation of multiple revisions or even mix source revisions after completion of a master, as well as multiple variants of a mix from ‘clean’ versions to ‘instrumental’ versions, have all added to the time required to complete a job. It is worth acknowledging that few mastering engineers are able to charge for the additional time expended engaging in these different deliveries, or certainly not at their full rate in any event.

At the AES Mastering Conference, one engineer expressed that this could be due to a partial lack of understanding of the role of mastering. Many clients are now very computer literate, and have access to extremely powerful and inexpensive Digital Audio Workstations (DAWs) such as Cockos’ Reaper (www.reaper.fm). Most clients will know how simple it is to export audio in a different file format. It is therefore possibly easy for clients to (mis)conceive that a mastering engineer might simply tweak and ‘print’ a new offline mix in the needed file format without listening to the resultant export, and quality controlling all studio outputs.

Related to this misunderstanding, all mastering engineers are incorrectly thought to be working within comparable digital workstation-based systems, and most ‘in-the-box’, meaning the engineer is using plugin technology rather than bespoke analogue processing units. For a large proportion of mastering engineers this assumption could not be further from reality, where high end analogue processing is employed to play the audio through the device(s) in real time. What appears to be a simple File > Export As File command on a home-based DAW, might actually involve real-time playback through several analogue processors. The client is paying for a mastering engineer to verify that the chosen audio we’ve outputted is indeed correct and is the best it can be. In the case of

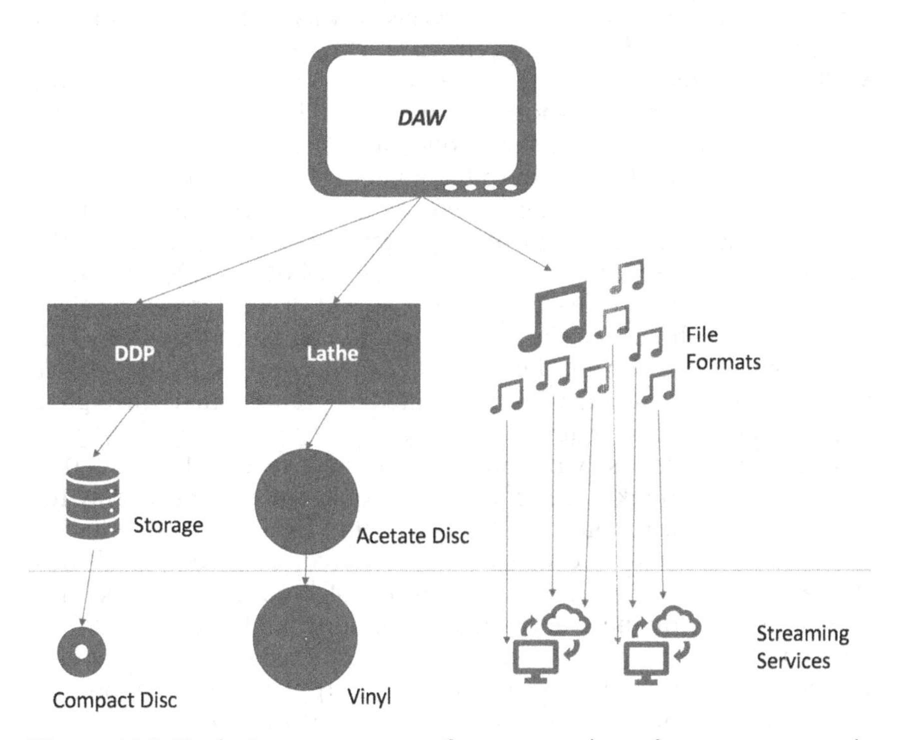

Figure 16.2 Typical contemporary format creations from a one song's master (radio edit for example) for later replication, distribution and streaming services (below line).

recalling a mastering session, time will need to be taken to ensure that the equipment is recalled to where it was at its last position before processing and exporting the audio real time and again ensure the necessary quality control.

The notion that an engineer would choose to listen all the way through an album DDP bounce is lost on some in the digital era, wherein there lies an assumption that all exports work correctly every single time. All mastering engineers have encountered anomalies in the digital realm at points and is a reason to maintain quality control mechanisms. Quality Control (QC) is a function that is not often covered in the popular discourse of audio mastering. Several books such as those by the author [1,2] and Bob Katz do indeed discuss the importance of QC, but since those books were released, the burden has increased across many formats, versions and platform-ready encodes.

LOUDNESS ASIDE

The AES UK Mastering Conference offered valuable discussion around loudness levels, the delivery of these levels to clients and the streaming services. Whilst, a consensus was neither sought nor agreed, two predominant views emerged. One view provided by Mike Wells (Mike Wells Mastering) was that he was already, as a standard offering, providing two exports for his clients [3]. First were Compact Disc (CD) ready, loud, masters, following what we'd probably consider a form of loudness akin to the much-discussed Loudness Wars. The second was what he considered *streaming-ready* masters, with a mean loudness in the region of the streaming services, which at the time of the conference typically aimed for −15 dB LUFS. The other emerging view from other engineers was that 'a master is a master', and therefore should be the best the audio can be represented. This view suggests a given master should remain consistently *acceptable* (by the client's standards) if translated or represented through a loudness normalised system.

Delivering masters with different loudness levels also, as other chapters in this book have reflected, not only causes controversy for mastering engineers, but also for the clients themselves. Clients are not always sure why two masters have arrived, one that sounds just the same as what they sent you (in terms of loudness), intended for streaming and loudness normalised situations; and another that offers a representation more like they expect themselves to sound – exciting and powerful. Historically, this output was the epitome of the process. To the artist, and of course the record label, this was the best version of the product that ever existed. This is the copy from which all other copies would be made, the copy from which fans would listen and admire their artist, and allow the music to enter into the annals of history, forever.

During our 'what if' discussion, and subsequent blue-sky thinking, *loudness* was not the prominent driver, but, naturally, could be addressed by the solutions dreamt during our conversation, which are further considered below.

STRESSES AND STRAINS

In contemporary mastering, as expressed in other chapters of this book, one might argue that the artform is under stress. Artificial Intelligence in the form of online mastering services appeal to a sector of the original mastering clientele who want fast sweetening of their demos, or productions. In many cases the cost of these services, such as market leaders www.landr.com and www.cloudbounce.com are inexpensive, or free for some export formats.

Another stress results from an increased awareness, interest and revelation of the mastering process in recent years. The mantle of the *dark art* of mastering is now becoming eroded as the Internet advertises *the engineer's* processes and approaches openly [4–5]. As a result, significant numbers of new (human-based) mastering services have suddenly appeared and now compete in terms of price. Even established mastering houses, such as Abbey Road and Metropolis in London, both offer online mastering services at competitive prices, presumably allowing them to completely fill the diaries of their mastering engineers.

All of the factors mentioned above place pressure on the mastering engineer and the process. Time is of the essence if the studio is to earn money as it once did. Add to this backdrop the aforementioned matter that mastering engineers are not only expected export for different formats, but potentially numerous versions at differing loudness levels as well. Each of these different exports require adequate quality control, thus potentially duplicating or even quadrupling the amount of time spent on performing QC measure for an album, for example. If we lived in a world where the mastering engineer charged clients for time spent on a given job, we suspect that clients would be less inclined to request so many additional versions. And, an added strain is that in many cases, artist and label mastering budgets have reduced. The engineer is doing more for less essentially. Can the aforementioned boom in machine learning and artificial intelligence help us? Time to dream…

A DREAM

Let's consider the historical backdrop of a tremendous industry collaboration such as the 100 plus companies that worked on the Digital Audio Tape (DAT), and the altruistic collaboration of companies such as Roland, Korg, Prophet, EMI, and others that created MIDI 1.0. This collaboration spirit appears to be evident also in the recent ratification of MIDI 2.0. Is it therefore not possible to draw both physical replication companies of CDs and Vinyl, the streaming services all together with the mastering community to develop a robust delivery mechanism that will not only solve an income balance problem within audio mastering, but reduce quality control issues in the factory and the Internet?

We dare to dream to propose the creation of a universal and agreed delivery format, taking cues from the enormously successful Disc Description Protocol (DDP) from Doug Carson and Associates and in part the principle behind Mastering for iTunes (MFiT) as was, now Apple Digital Masters.

With careful design and implementation, a new delivery format, player, and extractor could be created to solve some, if not all, of the problems outlined in the last section of this chapter. It must be stressed that this proposal does not intend to stipulate or stagnate artistic expression through strict levels or formatting restrictions, but instead leave room for however an artist wishes to be represented now and in the future. By this logic, *Death Magnetic* by Metallica, a prototypical example of an extraordinarily loud album, could still exist a possible aesthetic decision – but, an approved one!

The dream proposed here has been dubbed the Universal Description Protocol image (UDPi). In principle UDPi would present the possibility to convert a single print of high-resolution audio and metadata from the mastering studio into several predetermined potential exports. We acknowledge the idea will not be new to the community. There are other digital ingest formats and principles such as the original *Mastered for iTunes* droplet, Fraunhofer's work with Sonnox on the *Pro-Codec* for example. There are other examples such as downmixes from surround to stereo. All of these pre-existing examples could be drawn upon here.

Labels and artists would also use the UDPi file set as a permanent archive, future proofing their assets for the next generation of audio specification requirements. It is anticipated that future users could export multiple requirements from the one file on the UDPi platform, safe in the knowledge that quality control was once 'taken care of' at creation by the mastering engineer, and approved by the artist and label (in fact that approval could be recorded within the file as metadata!). Additionally, providing a high-resolution stereo (or surround) audio master is provided to make the UDPi master, then it is possible, with new UDPi iterations to deliver the same audio to future file formats, resolutions, and versions.

DELIVERING THE DREAM

In general, we mastering engineers are a fickle bunch by the nature of the job. To gain confidence in such a proposed system, the mastering engineer would need to have options to control each stage of the delivery. The proposed protocol should include the ability to output format, level and set other important parameters, such as dither application preference for all formats in the 'delivery portfolio'. These could include outcomes for: Compact Disc (CD) master quality print 44.1 kHz 16 bit wav file; a Disc Description Protocol (DDP) file set; current Apple Digital Masters (ADM) master prints format; as well as a lossy resolution stream with a True Peak (TP) at −1 dbFS or −2 dbFS; a 'hot' master; universal LUFS loudness calculations; pre file export for album as a whole and singles splits; High definition (HD) masters, dependant transfer path could be 24 bit and 44.1 kHz to 192 kHz or even 32 bit float; plus high resolution split A side and B side master files for vinyl cutting; Lossy file set for promo (promotional releases); in addition to a video-ingestion *ready* file at 48 kHz, 24 bit.

In a world with multiple commercial avenues for music consumption and numerous differing format specifications, we can only expect that the future of mastering will become even more sophisticated. As a recent example, one might consider the once dominance of iTunes and its associated music store, which, at one time, sold compressed mp4s; before later aiming for a streaming-only format

Table 16.1 Proposed delivery outcomes from UDPi

Proposed Delivery Formats	Notes
PCM High Definition outputs	Set to the agreed maximum ingest of UDPi (192kHz) plus any outputs dependent on the user output chosen. Could include 192/32float, 96/24 etc.
DSD High Definition outputs	Possible DXD conversion output to PCM.
DDP image file set	with MD5 checksum and associated PQ TOC tables.
Apple Digital Master (ADM) print	Built in afclip error reporting.
User definable true peak adjustment	Including lossy encoding error checking. Lossy file export for promo or user testing.
User definable multiple loudness normalized outputs	Can be unpacked to different formats or levels dependent on needs/services.
Vinyl Ready AB Split's	High definition output and associated TOC tables.
Metadata entry over and above the usual norms	Inclusion of credit based information, could be added to the UDPi after completion by label or artist.
Video ingest ready file	48kHz/24bit, EBU128 levelling options
Flexibility to output future unknown file sets.	All parameters user definable to accommodate new end user requirements.

which requires an understanding of the Mastered For iTunes (MFiT) paradigm. Spotify, at the time of writing, is the market leader. Even Spotify is changing its goalposts from ReplayGain (RG) to the more standard ITU1770 loudness standards, another shift that could mean potentially different mastering processes and procedures to some in the studio (https://artists.spotify.com/faq/mastering-and-loudness).

What if the mastering engineer, had the power to make concrete the ideal outcome in terms of the aesthetic in tone, dynamic range and resolution with any given piece of audio. Surely, it would be very helpful not to have to export and manage multiple formatted outcomes?

On the face of it, this would seem to be an easy aspect to deliver in the digital world with the assistance of artificial intelligence. However, there are varied conversion issues that must be considered to 'fit' all requirements and new possible standards; as any proposal must future proof itself as much as is possible.

SPECIFICATIONS

Based on the aforementioned requirements and potential outcomes, we recommend the following specifications for the UDPi protocol. The UDPi protocol will:

1. Ingest the highest possible PCM resolution *at source.*
2. Act as an archive to future proof outcomes for unknown new delivery (HD vinyl being one such potential new requirement).
3. Feature fully selectable parameters for the mastering engineer to set the type of sample rate conversion tool and dithering outcomes. The latter being quite specific for engineers to choose different algorithms such as POW-r or Apogee, for example.
4. Output to DDP 2.0
5. Include all current metadata tagging.
6. Include additional metadata information for archiving potential, and label sign off
7. Set file naming strings for differing file types.
8. Feature the ability to set TP and LUFS outcomes upon export. These may be adjustable to match any given standard or requirement, with alert warnings issued for outcomes that do not meet restrictions.
9. Feature the ability to add future formatting outcomes without revisiting the original masters.

If the above features were included in the UDPi export, mastering engineers would only have to generate one master file set rather than several. The task of exporting could be delegated to the end user utilising an export tool like Sonoris software's DDP Player which many mastering engineers are currently familiar with.

Generally, current consumer playback systems are lower resolution than most mastering engineers transfer paths. It's not unusual to have a 32 bit Float, 192 kHz or 96 kHz final resolution print in the context of the digital realm. The restriction of analog-to-digital (AD) conversion from an analogue transfer path might limit this to 24 bit in the capture. This final 'print' without conversion would be the input format for the UDP. To make this function, an auditioning tool would be required. This, in one part, would permit real-time feedback into the mastering process but also for end users to be able to audition and sign off their individual required outcomes. Again, audition systems from Sonnox or Nugen Audio are functionally close to meeting the required type of tool, and Sonoris's DDP Player already has a user feedback system integrated (www.sonoris.com).

BENEFIT ANALYSIS

What would be the benefit of such a system in both the short and long term? If the UDPi protocol could be agreed and implemented, there are several positive outcomes for all involved.

Firstly, UDPi would provide a robust, yet flexible, format for archival purposes, therefore *future proofing* the audio for the generations ahead. The next obvious benefit would be the reduction in time

Table 16.2 Benefits chart for both mastering engineer and the artist label paradigm

Mastering Engineer Benefit	Label/Artist benefit
Reduces time by automating file version exports	Reduces time by automating file version exports
More robust quality control	More robust quality control
Quality archiving	Quality archiving
Robust metadata collection	Artist & Label data more widely and accurately portrayed
Enables more accurate pricing for clients	Clearer pricing structure – anticipated budgets
Additional tools that could be written into the UDPi encoder such as DSP, level checking etc.	Filesets can be drawn from the one encapsulated file
Retains best source quality for future reuse.	Future proofing the investment

it would take for the mastering engineer to generate multiple file formats, and to various loudness or level formats. UDPi could also save time through integral analysis of audio and whatever artificial intelligence and machine learning necessary to export audio based on settings chosen. This could potentially prevent export errors in file type conversion for different platforms.

For several years there have been numerous attempts to promote a widespread system to ensure correct and detailed metadata follows its audio. The Music Producers Guild's (MPG) *Credit Where Credit Is Due* campaign and Barry Grint and Ray Staff's *Broadcast WAV* (bwav) promotion both assist in this matter [6], but widespread management of metadata could still benefit from what UDPi has to offer. UDPi provides such an opportunity, which then can be distilled to every exported format (presuming the formats ultimately carry the supplied information).

Most importantly, as a high definition archive, the audio could be extracted for new file formats in years to come as they come along. Naturally an updated UDPi player or exporter algorithm would be necessary.

All of the above would serve to save the client costs that are usually unseen, or largely unexpected at the engagement stage. Equally time and accuracy would be of benefit to the artist and the label alike.

Whilst all of the above benefit the mastering engineer, it also demonstrates clear benefit to all involved. One major advantage for all would be the ability to produce clear pricing policies as each master and each release variant, such as 'clean', 'sync', and 'instrumental' versions would be contained within the new UDPi fileset.

SLEEPWALKING

It is possible that our proposed dream could fall foul of many obstacles. We've purposefully not discussed, in any great detail, the loudness of files in any great depth here, as we feel that this would cloud the issue, and therefore the benefit of UDPi. The loudness debate, if this book is anything to go by, will continue, long after we've awoken from this dream. That said, it is therefore not worth avoiding the possibility to encourage a new, flexible and future-proof, format for file delivery in mastering.

As with many formats, in fact too many to mention, the benefits we conceive of are nothing but undeniable for the intended market place, but due to other unforeseen factors, whether that be marketing, or poor public relations, the format has either shifted or gone by the wayside completely. The former, could in many ways, relate to the Digital Audio Tape, originally a consumer format, but ultimately blighted by poor PR around the USA phonographic industry's desire to suppress its success. Naturally, DAT sidestepped from a consumer format to that of a professional one for use as a master recorder. Other formats come and go and indeed UDPi could succumb to the same fate before it hits the inventor's workbench and has its name ratified.

Whilst we believe our dream is the right dream, another dream is perfectly possible. Perhaps a universal standard could be discussed which would see a unified file format be produced for all streaming services? Currently, their reasons for level differences, file format choice, and resolution depths are perplexing. We suspect that there is a desire to persist with these 'differences' in available platforms in a quest for market dominance. One streams from Tidal for the quality, from Spotify for the playlists and large library, and from Apple Music for the kudos perhaps. As Scott Harker explores in this book, there's an argument that each of these delivery platforms also have a different 'sound'.

For all of these suspicious 'differences' in listening platforms, the effect upon the mastering engineer is concerning in the long term. It is possible that mastering might, as the artform it is today, sleepwalk into oblivion as more and more musicians and producers engage in the well-publicised artificial 'assistance' on offer for mastering online. Instead of removing the human character a mastering engineer can provide with artificial intelligence, let's use the power of this technology to help us deliver the ever-increasing version and file types of the future.

CONCLUSION

With consideration from the music industry and the audio mastering community towards these requirements, the positive ramifications of UDPi's implementation in the main would permit additional time to be focused on musical outcomes and less around making *parts*. The mastering engineer could be sure the delivery would be to their specification from the UDPi exporter in addition to the fact that the client would be fully aware they had signed off all the audio outcomes from listening on the UDP player. Additionally, the client would be able to react to future format requirements for their original project quickly, without the need to approach the mastering engineer since they can simply export any new version directly from their original master UDP fileset.

REFERENCES

1. Cousins, M. and Hepworth-Sawyer, R. *Practical Mastering: A Guide to Mastering in the Modern Studio.* Focal Press (Routledge), Burlington, (2013).
2. Grint, B. 'Embedding International Standard Recording Code (ISRC) in Broadcast WAV files.' *EBU Technology & Inovation*, Vol. *23*, (2015), pp. 17–19.
3. Wells, M. *Dangerous Technical Presentation*, AES Mastering Conference, 23rd September, (2018).
4. Nardi, C. *Zen in the Art of Sound Engineering.* The Proceedings of the 2005 Art of Record Production Conference, (2005).
5. Sonoris. Mastering Software, (2019). Available at: https://www.sonorissoftware.com/index.php/product-category/mastering-software/. Last accessed 15 March 2020.
6. Toulson, R., Grint, B. and Staff, R. *Embedding ISRC Identifiers in Broadcast Wave Audio Files*, (2015).

17

Signature mastering

NEW SOUND AESTHETICS IN POST-PRODUCTION AND THE NEW ROLE OF MASTERING

Holger Lund

INTRODUCTION

This article approaches the field of mastering by taking two observations into consideration: firstly, sounds heard on vinyl are changing and, secondly, the self-conception of the mastering engineer is changing. These changes will be identified, described and subsequently analysed in relation to the economic, cultural, and aesthetic developments that they are part of.

By analysing the discourse on mastering and vinyl cutting mainly in music press and musicology, the article aims to give insight into recent developments in the field of high-end music production, connecting these to economic developments (the Western music market and its rules and restrictions) as well as cultural developments (the decline of club culture in some Western countries). It will further raise the question in which way and why sound aesthetics have changed and if the changes may foster a kind of 'museum of club sounds'.

Sounds heard on vinyl are changing

For some years now, new sounds have come up on vinyl records, especially in bass-related club music. Records from genres like (Dirty) Grime, (Post-)Dubstep, Trap, Juke, Blip Hop, Future Bass, and Drum & Bass as well as from Sound Art offer an entirely new kind of deepness, spatiality, and presence in sound.

The changing self-conception of the mastering engineer

The phenomenon of new sounds heard on vinyl records is closely linked to achievements in mastering and vinyl cutting. These have been accomplished by mastering engineers, who rather tend to be *mastering artists*, developing signature sounds and emphasising the creative part of mastering, to an extent that allows to talk about *signature mastering* [1]. Just as in the transition from anonymous medieval craftsmen to renaissance artists known by name, nowadays mastering engineers get more and more attention for their work by musicians, record labels, music press and the public alike. *Mastering artists* are thus increasingly appreciated for what they can do: create a specific sound in post-production. Historically, there have been precursors of this development, like Doug Sax, George Marino, Bernie Grundman, or Bob Ludwig, praised and known for their signature sound-like treatment of music, but they have been and are still labelled as mastering or audio engineers, not as mastering artists. The shift towards the new status is happening more or less right now.

The specific sounds themselves, in literature often perceived as kinetic soundscapes or sonic sculptures, point to the growing relevance of post-production in music. These sonic achievements lead to the questions where they have been gained and by whom.

Cultural production as a whole seems to be subject to a shift of focus from production to post-production. Film may be seen as one of the best examples for this development, as more and more filmic elements are pushed into post-production. A standard Hollywood production today is being made in post-production for about 80% at least. And computer-generated images (CGI) are even made 100% in post-production. However, they are made completely digitally. Like in film, post-production gains an increasing importance in vinyl production through sound operations realised in mastering and the vinyl cut – only these operations are mostly achieved in the realm of the analogue. Therefore, music production – at least for some genres – shifts attention from recording and mixing to post-production and, in parallel, especially in the field of high-end music production, from digital to post-digital processes, be they (retro-)analogue or analogital [2]. The post-digital mastering artists not only master the music that has already been recorded and mixed, but create new sounds with new aesthetics – and that is exactly what they are in demand for by musicians and praised for by listeners.

Case studies

What do these new sounds sound like? The works of three signature mastering engineers/artists are investigated in this text as case studies for the aforementioned developments and for a close analysis of the sound aesthetics developed through their specific approaches to mastering: Rashad Becker, Matt Colton, and Taylor Deupree. A very interesting development that goes along with the new conception of mastering engineers as mastering artists and the expanded role of post-production is the predilection for a certain aesthetics of dirty mastering, of a 'fucked up' sound, as mastering engineer Matt Colton puts it in radical words: 'Maybe it sounding wrong is better than it sounding right' [3]. This approach, unfolded in quite a number of interviews, favours distorted over clean sound (Becker), raw over polished sound (Colton) and coloured over transparent sound (Deupree). Thus, one could use the term 'dirty mastering' as opposed to the 'old-school' ideal of indispensable cleanness and 'neutral' mastering. This 'dirty mastering' serves certain aesthetic purposes, which will be explained later on, as well as it serves as a signature sound making the mastering artists (intentionally) audible as part of the music.

From the club to the 'museum of club sounds' on vinyl

The new sounds heard on vinyl records have been developed mainly in relation to club music genres. One could ask oneself, why this relation is so clearly experienced. As music clubs for example in Britain, France, and Germany are brushed away by neoliberal real-estate investors, who do not want to have their clients exposed to any kind of value-reducing noise, club music, one may suggest, is somehow looking for a new home – a home where it can be as visceral and vibrating as possible. It seems that the vinyl record, by means of post-production used by mastering artists, can offer such a new home, being a museum for club sounds, offering preservation and diffusion of a club music feeling.

SOUND RESEARCH – POINT OF VIEW AND STARTING POINTS

Having published academic essays on pop music [4] as well as more sound-oriented pieces [5], I have never explicitly done academic research on sound itself. The ideas for this article were formed by a new listening experience that started in the early 2010s, with new sounds and new sound spaces one could encounter coming from loudspeakers and based on vinyl records, both in private listening and at clubs. The phenomenon is related to vinyl records, especially to varieties of bass music, or different styles related to bass music, and records belonging to an avant-garde in sound, in club audiophilia. Examples include releases from recent years by acts such as Overlook, Mumdance, Logos, Klein, Beatrice Dillon, and Raime/Yally or labels such as Silent Season, Hemlock, Dom & Roland Productions, and Well Rounded Dubs [6].

It is a fringe sector: editions for record releases are often between 200 and 500 copies, the preferred format is the 12" single played at 45 rpm. It is a small sector also regarding technology and personnel: there are only about 100 relevant mastering and cutting engineers worldwide [7], apparently almost all of them male [8], and there are only ca. 300 vinyl cutting machines left worldwide [9], most of them still from the 1980s or earlier.

One can also notice a new understanding of music, as exemplified by the London duo Raime. Although they create avant-garde bass music for clubs, it might just as well be called sound art or sound design than conventional music, or maybe even be viewed as a club-audiophile exhibition and highlighting of sounds in the form of music. One of the relevant mastering engineers, Taylor Deupree from 12k Mastering, comments on this situation: 'The music my peers make is so much about sound and the quality of sound. The sounds themselves are the art, the focus; this isn't pop music' [10].

Amidst the central figures, apart from the musicians themselves, are the mastering engineers. Mastering is used to achieve precision of sound and sonic impact, and mastering engineers work at times with sounds that show very complex and dynamic treble and bass frequencies, so the sounds seem to stand directly in front of the listeners or draw them immersively. Both in rhetorics and in musicology, such effects of directness, presence, and impact are termed 'hypotyposis', and that is exactly what mastering can achieve. It brings the music very close to you or places you inside the music, when played on a proper system, of course.

It is not just in the sounds coming from the loudspeakers that one finds indicators of the growing importance of mastering: the auctorial formula 'mastered and cut by…' increasingly appears in paratexts on music, e.g. on the product pages of vinyl traders such as boomkat [11], or the mastering engineer even makes a claim to full artist status, for example on the product sticker on a record by Oye Records Berlin (2017), where the headline informs the customer: Blackbush Orchestra, 'Familia EP (incl. Kay Suzuki Remix & VAKULA Mastering)'.

RESEARCH ON MASTERING: NEW ATTENTION FOR A NEGLECTED DISCIPLINE

Parallel to these developments, recent years have seen the beginning of an academic discourse on mastering – a discourse which started with a lament on the absence of such a discourse. Jens Gerrit Papenburg remarked that mastering was 'heavily understudied' [12], discerning that 'there is still a gap in the scholarship regarding a cultural and historical analysis of mastering' [13]. Carlo Nardi diagnosed 'an apparent lack of scholarly attention' [14] in academic circles, seeing that 'audio mastering has remained out of the radar' [15]. But he also notes: 'Between 2008 and 2012, *MusicTech*, one of the most popular magazines for music production, dedicated four of its special issues *Focus* to audio mastering … witnessing to a sudden interest in a phase of record production that has been traditionally given scant attention' [16].

In October 2017, Vinyl Factory – who had already published the short video 'Sculpting Sound: The Art of Vinyl Mastering' in 2014 [17] – published a text on present and future developments in music, in which they interviewed three mastering engineers [18]. Such interviews would have been unthinkable just a few years earlier, simply because a mastering engineer would not have been regarded as a relevant voice on the future of music at all.

The new attention for mastering can also be investigated as a post-digital phenomenon [19]. Unlike in film, where post-production first rose in relevance, but where it is strictly digitally-based, in music the increased importance of post-production with mastering and vinyl cutting has to be defined as post-digital: the end product is analogue vinyl, and wherever the signal chain may have originated, it always ends up on an analogue medium. Most of the relevant mastering engineers either exclusively use analogue equipment for the signal chain, or work with analogital equipment, but never purely digital, which is typical for reanalogisation in post-digital tendencies.

Of course, mastering engineers are very aware of what is happening in the consumer market. Taylor Deupree puts it like this: 'Mastering engineers always talk about how ironic it is that we use all of this expensive analogue outboard equipment only to have songs listened to by people on their phone speakers as they walk down the streets' [20]. And yet: 'There are still a lot of people out there who care about great sounding music and those are the people for whom I work hard to please. Engineers can't just give up and let the cell phones win' [21]. Deupree here takes a post-digital viewpoint in opposition to the digitalisation of music. Bartmanski and Woodward amplify this view adding the idea of mastering and cutting as a post-digital soundcraft: 'The production of a vinyl remains a craft' [22]. Just like barbers or brewers, mastering engineers are beholden to their craft. What they lack in hipsterdom, they make up for in nerdiness, as Matt Shelvock notes: 'How does mastering factor into audiophile culture? Electro and dubstep fan subcultures exhibit elitist tendencies through prioritizing high resolution records with ultra-wide-band frequency content. Examples can be seen of fans comparing masters of their favourite artists on hobbyist forums, where spectral diagrams are posted to verify audio resolution' [23].

The increased attention to mastering can also be seen from a scientific background in the history of theory. With *Art Worlds* (1982), Howard S. Becker delivered a theory that explains all kinds of artistic production as a system of processes not involving the artists only – like in an aesthetics of genius – but many other people and their objectives in relation to the artistic processes. Becker speaks of 'networks of people cooperating' [24] as a prerequisite that allows the creation of artistic productions. Praxology as a sociology of practice, actor network theory and the theory of media environments all build on Becker's approach [25]. These theories share the basic premise of understanding artistic or artificial results as the outcome of systemically generated, all-involving production processes within networks; and then they proceed to test the relevance of the more inconspicuous or utterly disregarded links within the process. Such a test of relevancy now seems due for mastering. As Becker writes: 'The status of any particular activity, as a core activity which requires special artistic gifts or as mere support, can change' [26]. He offers the recording engineer and the sound mixer as examples for this situation, detailing their practices and histories as well as their processes of perception. Their status changed especially when rock musicians started working with elements of the recording studio (multichannel recording, effects placement) and started controlling their sound in the mixing process, which now required a 'special artistic talent' [27]: 'Sound mixing, once a mere technical specialty, had become integral to the art process and recognized as such' [28]. The transformation of Beatles' recording engineer Alan Parson to a musician and producer of Alan Parsons Project at the end of the 1970s is a prominent example for this change, approved by the record companies Charisma and Arista when trusting him to be able to carry out this change.

FIRST CONCLUSIONS

For now, the following can be summed up: In recent years, a development can be observed in bass-oriented musical styles such as (Dirty) Grime, (Post-)Dubstep, Trap, Future Bass, or Drum & Bass – but also in Sound Art [29] – where artists attempt to lend their music a more forceful presence via sound design, through the use of pitching, bass and sub-bass on the one hand, through new spatialities of sound on the other. The music becomes virtually (sound-) sculptural, gaining defined expansion and depth in space. Both the sculpturality and hypotyposis of sound are conflated by Andreas Lubich from Calyx Mastering in his historical view on mastering: 'Vinyl mastering was more or less being a sculptor, so you have a big block of sound and then you have to get the track out of it till everything is in front of you as it should be' [30]. Likewise, Rashad Becker metaphorically labels his activity as 'sculpting sound' [31], 'because it's all about the physical dimensions' [32].

In these views, slowly an insight begins to establish itself that Bartmanski and Woodward have described like this: 'The quality of what we eventually hear on vinyl depends not only on how a given track was produced by an artist or recorded in a studio, but how it is mastered, and how it is cut,... how the cut is calibrated' [33].

The question that follows is: who is responsible for which elements of the sound, or, framed differently, does one still hear music that is produced or is it already recognisably post-produced? A new appreciation of music and its sound characteristics becomes apparent where the temporally last steps of post-production suddenly move far ahead – in relevance. One could literally speak of *signature mastering*, as mastering engineers not only perceptibly sign off their work like auctorial artists, but also want their work understood as artistic work. Within the practice of mastering, Rashad Becker, Matt Colton and Taylor Deupree – mastering's masters, so to speak – are the forerunners in that direction, especially since each of them has developed a specific sound aesthetics. In the following, I want to explore and introduce their individual cases.

CASE 1: RASHAD BECKER, DUBPLATES & MASTERING, BERLIN & CLUNK, BERLIN

Resident Advisor describes the mastering engineer, vinyl cutter and musician Rashad Becker as much more than a mastering engineer, calling him a 'sound scientist' and certifying: 'In the last 15 years he's become something of a rock star in the field of mastering' [34]. Becker himself is more modest: 'Mastering is finishing a musical piece post mix towards the release of that piece. … Technically speaking it is a very little thing. Like with all audio treatment you refer to frequencies, to amplitudes, and to phase relations, and these are the three parameters you can work on. For that purpose, you utilise equalisers, filters, compressors … and that's it' [35].

And yet it does not stop there. Historically, with the spatio-temporal separation of recording and mixing engineers on the one hand and mastering engineers on the other, the first independent mastering studios opened at the end of the 1960s [36]. Papenburg writes: 'The master engineer mediates between technology and aesthetics; according to mastering guru Bob Ludwig, mastering is "the last creative step and the first manufacturing step in the record-making process"' [37]. In this constellation, the mastering engineer can become an author, and subsequently has been given a credit in musical productions since the early 1970s: 'As early as the 1970s, disco mastering became increasingly personalized by specifying the master engineer on the record sleeve or by "vinyl graffiti" [carrying the name of the engineer or the studio] on the record's dead wax' [38].

Not only does everybody involved get a text credit, but in rare cases they are also presented visually. A nice example of complete appreciation for all contributors is offered on the backside of the LP cover for Eduardo and Silvia Araujo's *Rebu Geral*, released in Brazil in 1981 [39]. Half of the images show the musicians, the other half everyone else who helped realise the music on the vinyl record – pictured in the exact same dimensions. There is the recording engineer and his assistant, the soundmixer and his assistant, the studio owner, plus images of the mastering engineer and vinyl cutter. Visually the arrangement of the personnel is almost paratactical; it shows who carries which responsibilities – and all are equally recognised, not even the assistants are hierarchically devalued. A very democratic tableau – and one that along with the tasks performed also ties aspects of authorship to specific names.

Staying with the question of authorship the focus is set again on mastering. Authorship always is based on distinction. This is where taking a closer look at Becker's specific distortion aesthetics becomes interesting. Distortion usually is very unpopular in mastering circles, but he has put it to special use, regarding it not as an evil to avoid but as a quality of sound that can be exploited: 'The other type of tool I came to like and use very much in a mastering application… is tube distortion.… Sometimes the distortion is rather brutal but mixed in at very very low levels' [40].

Becker is conscious about what he is doing: 'The term distortion is of course *stigmatised* […]. I mean distortion is any deviation from the original waveform, adding harmonics is (non-linear) distortion and as soon as you apply any kind of filter you have (linear) distortion. But you can apply the parameter of sound quality to distortion as well. There is low quality distortion and there is high quality distortion and what I can provide here is rather high quality distortion. Here I can add harmonics, in a way that is impossible to achieve with any kind of equaliser or filter alone, because they can only enhance what is already there' [41].

At this point it becomes obvious that Becker interferes with the music, adding new sound qualities that were not present before. How and why does he do that? When Robert Henke asks him in an interview whether 'the distortion adds harmonics which are not present in the original signal but which are related to the original signal and which are applied carefully to make the overall sound more rich', Becker answers: 'Richer and with more dynamic. I can apply harmonics, I can apply subharmonics, I can emphasize the fundamentals – it depends on the application. I always use the distortion in parallel. The original signal does not get replaced by the distorted signal but gets mixed with the distorted signal, so I have a very precise control over the amount and colour of the distortions' [42]. Is there anything else distortion contributes to the sound? 'Adding distortion also can help in definition of spatial parameters, helps broaden the stereo image in a subtle way' [43]. Becker concludes: 'I did not expect that distortion could be so useful, I thought is was more a production tool but is has become a very central piece for mastering for me' [44, sic!].

However, an aesthetics of distortion can be traced back to the early 1950s, as Papenburg notes: 'As early as the 1950s, a lot of engineers who were not impressed by the standardization of the recording curve [RIAA standard, no treble above 20,000 Hz, no bass below 50 Hz] were primarily interested in making "hot" records, which were supposed to "stand out when played on jukeboxes". Sun Records owner Sam Phillips points out that he had added "intended distortions" during mastering to achieve "proper loudness". … Indie labels like Specialty [with artists including Little Richard] "allowed more noise into a record's groove than the majors would tolerate". Little Richard's "Long Tall Sally" 1956 sounds very loud because of being cut at the margins of distortion' [45].

To what extent these aesthetics of distortion differ, more spatially-oriented in Becker versus more volume-oriented in the 1950s, would still require a more exact study.

CASE 2: MATT COLTON, METROPOLIS MASTERING, LONDON

Juno Plus magazine says about Matt Colton formerly of Alchemy Mastering in London: 'For electronic music, Matt Colton has become one of the most sought after mastering engineers working today' [46]. In 2013, he received the award of the Music Producers Guild as Mastering Engineer of the Year [47].

At first impression, he also appears quite modest, describing himself both as a technician (creating a master) as well as a creative person (presenting the sound to the listener) [48]. 'There's the creative side of mastering and there's the technical side that you need to know and understand,... but in terms of the actual signal processing, sonic bending stuff, it's just about trying to make the sonic resonate with the recording' [49].

If he had not become a mastering engineer, Colton says, he would like to be a sculptor, 'hands-on-mechanical, hand-craft' [50]. This possibly influences his idea of mastering, where the cut determines his approach – 'understanding vinyl engineering informs his mastering' [51], as it was put in a *Resident Advisor* podcast.

And yet Colton takes up his position closer to the aesthetic than to the technical side, which he somewhat contemptuously labels as 'button-pressing'. He says: 'It's mostly about the listening', and after that about making aesthetic decisions [52]. Here he has developed an aesthetics of the 'fucked up', as he calls it, an aesthetics of mistakes, of craziness: 'I think, working with Rephlex [Records], from such an early age, really shaped my view on what is right and wrong in a recording in terms of... they bring me stuff, and it'll be like – is it supposed to sound like that? And they would say: "Yeah, absolutely, that's exactly how it should sound." And after a couple of sessions: "Hey, cool, I get it!" Things don't have to sound... it's not necessary that we're trying to make things sound as polished and as smooth and as good and as right as possible, because that's not necessarily right, you know, sometimes things can be fucked up, and maybe it sounding wrong is better than it sounding right' [53].

Colton elaborates: 'When you got something, you know, it's fucked up, it's distorted, it's like a hiss, you know, it's all over the place – I can go through and I can de-hiss it. Give me a day and I'll have it sounding fairly polished, but that's not the intention, that's not... the artist, the label, they don't want the listener to be hearing a piece of polished music, they want them to be racked into something that's the opposite of that. So, that's my brief, let's not tidy this up, let's keep it raw, do we make it rawer, do you know what I mean? Maybe we do, maybe we get more crazy with it!' [54].

In the podcast, he uses strong words such as 'distortion', 'inserting dirt', 'keep it raw', 'make it rawer', 'get more crazy with it'. Similar to Becker, he aims for a deliberate impurification of the sound, not for a transparent cleanness in mastering.

Again, coming back to the vinyl cut. Colton describes the cutting machines as having a life of their own; they are wilful, 'esoteric, temperamental', 'horrific when they start working' yet at the same time 'magical' [55]. This evaluation chimes in with the story of a studio owner in Jamaica who was called the 'master of the electric amphibians', since he was the only one 'who knew how to tame the electric amphibians' hidden in the confusing cable system of his studio. And Jackson Bailey aka Tapes speaks of the 'ghosts in the cables', when he says: 'Always, always, ALWAYS record everything. Because there are ghosts in the cables, in the electricity; spooky shit that pops out at a complete random moment that ends up being the focal point of an entire work and you're like: hey man, eureka!' [56].

It is obvious that Colton knows about the aesthetic resistance of the machines – and knows how to utilise it instead of suffering from it.

CASE 3: TAYLOR DEUPREE, 12K MASTERING, NEW YORK

Similar to Colton, Taylor Deupree from 12k Mastering focuses on choices: 'Mastering is all about decision-making' [57], he says – including creative, aesthetic and technical decisions. For himself, he clearly postulates the creative end of things: 'I love the creative side of mastering' [58].

Deupree has developed an aesthetics of colouring, which he pits against a prevalent aesthetics of transparency: 'There are sort of two categories of analog outboard equipment: "transparent" and "colored". Mastering engineers often lean towards the transparent gear just because it can make corrections without being heard. I have a couple of pieces that would fall into this category, but I also really love the tubey and saturated gear that impart some sort of vibe or "color" to a recording... and I'm not shy about using it. It's important for me to make this claim because if someone is looking for a really transparent mastering engineer they may not want to use me' [59].

Just like Colton, he stands for an aesthetics of mistakes: 'We can try things that may seem "wrong" to other people and they really work' [60]. He sees himself as part of a more general development, of which he is the frontrunner, both in his (self-)conception and his ambition: 'Traditionally, the role of a mastering engineer is a non-creative one. Technical-only, I would say. This can still be true today and many engineers strive for a transparency of sound. However, because I come from an arts background my mastering tends to take on a more creative roll. I always call myself a very non-transparent engineer, unafraid to inject a bit of my own sound' [61].

In conclusion: 'I think I often have a sonic stamp to my mastering, which would probably be an idea criticized in many circles, but I think people come to me because of that sound. For me, sound design and songwriting are nearly one and the same' [62].

The term 'sonic stamp' refers to a signature sound and thus to the idea of signature mastering. His key statement that 'sound design and songwriting are nearly one and the same' points to an equal valuation of the songwriter's and the mastering engineer's creative achievements, or at least it claims this equality. None of the other mastering engineers taken into consideration have gone that far.

With all the discussion about mastering, vinyl cut should not be forgotten. Even if Nardi has described mastering as the 'final step of sound manipulation' [63], there is in fact another step of sound manipulation after mastering, and that is the vinyl cut. According to Robert Henke: 'It's a process which always changes the result ... like developing a colour photo by yourself. It's a process which, if you do it five times, leads to five different results of brightness, colours, and so on' [64]. Andreas Lubich describes the way that the sound is shaped during the cutting: 'Then I have to decide if it's better to tame the highs a bit for particular tracks because otherwise I would get too much distortion, or even emphasize the highs in a certain way' [65].

Lubich addresses the specific challenges of electronic dance records, which are the focus of our observations. These challenges include regulating 'the amount of bass and excessive highs' at the 'maximum possible level'. This will lead to 'high amplitudes for the groove within the same timeframe'. The result: the process 'heats up the cutterhead up to 200 degrees Celsius then security circuit disconnects the cutter head to prevent it from damage'. Here appear the 'hot cuts' mentioned above, also in the literal sense.

The format of the 12" record offers a way out of this problem. Papenburg writes: 'Its physics allows for the cutting of records with a wide cut. The design of the bass frequencies benefits from this additional space in particular. Moreover, the twelve-inch single's higher speed makes a sophisticated resolution of high frequencies possible' [66]. Which is why the initial sound phenomenon that started the research for this article can be found mostly on 12" discs arranged for 45 rpm.

For the record: while by now a discourse on mastering has been established, hardly any publication dedicated to vinyl cut can be found. This part of very last sound manipulation is at best mentioned in the context of mastering, especially since the relevant mastering engineers usually also take care of the cutting, or it is totally neglected.

A HISTORY OF POP MUSIC AS SOUND HISTORY?

Amongst the academic historians of pop music, Diedrich Diederichsen ventures farthest into the topic of mastering. Where Nardi sees mastering as the 'final step of sound manipulation' [67], which allows him to call it at least 'a kind of musical practice' [68], Diederichsen thinks in larger terms. He sees pop as something based in sound, based in sound effects: 'Melody lines are short and quickly comprehended .... It appears they only exist to carry a sound, mostly a broader coating of sound, so to speak, and a few more sound effects in more selected spaces' [69]. He elaborates on 'what pop music really is about: totem sounds ... and signature sounds ... and shadings of sound. These attractions depend on the decisions and ideas of producers and engineers, and there is no musical tradition that would ascribe authorship to these tasks' [70] – since that would mean rewriting the history of pop music into a sound history of pop music. And this is exactly what Diederichsen starts pondering: 'Pop music ... has made its huge developments not on the level of musical notation; the developments in sound aesthetics especially over recent years have not happened on the level of a musical architecture of handling time [song, melody, harmony, rhythm], but within the quasi-sculptural aesthetics of exhibiting sounds. This goes for so-called electronic music after techno as much as for the drones in doom metal' [71].

In 2017, Immanuel Brockhaus published such a sound history of pop music with *Kultsounds: Die prägendsten Klänge der Popmusik 1960–2014* [72]. Unfortunately, he does not discuss issues of post-production in any depth – though admittedly, as one could have seen, this discourse has become noticeable only since ca. 2013.

MORE CONCLUSIONS

If you thought that everything the vinyl medium has to offer had already been done, you could reach a different conclusion after listening to some of the earlier mentioned recent vinyl releases. These records take new paths towards quasi-haptic, bodily sonic sculptures or into carefully sound-designed kinetic soundscapes by using hypotyposis, the effects of presence. Parallel to the increased importance of post-production in film, a marked revaluation and new estimation of post-production in music can be witnessed. Most prominently, mastering has achieved a place in music discourse, also the non-professional one with music magazines and music blogs. That means it has to be freshly determined which parts of the music come from whom: from the musicians, from the technical stages (production, mixing) or from mastering and vinyl cutting, the post-production. By now, 80% or more of an average Hollywood blockbuster is created in post-production. What happens in music? Is a new post-productive music on vinyl emerging? Evidently it seems so [73]. If Becker explicitly advises that 'people should not make music for post-production' [74] – he says so because that is exactly what is happening. And because it is happening through the masters of mastering, including himself. Different from Deupree, Becker is simply still too modest to see himself in the role of the mastering musician.

So, what do mastering and vinyl cutting achieve in post-productive music? The aesthetics of mastering are determined between two poles: Becker distinguishes between distorted versus clean, Colton between raw versus polished and Deupree between coloured and transparent. And all of them position themselves as colourists against the clarifiers. They put the case for a dirty mastering to advance hypotyposis, and even mention their respect for the inner life of the machines. This goes against all the efforts of the clarifiers who want to control everything and sometimes even plug into an own independent power source to take care of the treacherous electric amphibians: 'We also run our studio off batteries and make our own 60 Hz here, so we have very clean power. They're like the size of a refrigerator' [75].

FINAL CONCLUSIONS – THE VINYL RECORD AS A MUSEUM OF CLUB SOUNDS

In the end, the why has to be determined: Why can a post-productive music executed by mastering artists be observed in the making right now? The idea of signature mastering discussed above mainly applies to electronic club music. In fact, exactly in the present era of dying clubs and the decline of party culture caused by investor-friendly regulations and more general political decisions on the one hand and young people's changing job and leisure behaviours on the other. Great Britain – where in large part signature mastering originated – is hit especially hard, and recent times have seen massive closures of clubs [76].

An ensuing theory might suggest that one could see club music leaping onto the vinyl record, as it were, to survive in another medium than the music-medial dispositive of the club. Of course, the tendency of shutdowns does not erase all of the clubs, but possibly enough to create a demand for a new solution. The medium of the vinyl record thus might even serve as a sound museum of the club, which can be privately reactivated playing and listening to the music at home – but this theory would still have to be investigated.

Recently Diederichsen lamented – related to the era of a dying club culture – the loss of a visceral impacted musical vibration as a 'collective resonance that went through many bodies both metaphorically and literally' [77]. Perhaps the new post-productive, corporeally immersive and forcefully present music can offer a new private vibration aesthetics in compensation. The old vibration aesthetics of listening to loud music together, where the volume directly addresses the body, is vanishing today. In almost every club automatic sound limiters have been installed for permanent noise control. In Portugal, for example, these limiters were prescribed by law in 2016. To protect the audience, a maximum of only 80 decibels is still allowed in almost all locations [78]. The visceral, abdominal beating experience of music, where 'architecture and the body [are set to move] by vibration' [79] as well as the socially inclusive 'spatio-corporeal expansion of music' [80], are thus being prevented by limiters. And yet the music still needs to be visceral, even if the clubs are dying. And here is where post-productive music on vinyl records may offer a way out of the dilemma.

REFERENCES

1. On sonic signatures in music production cf. Brøvig-Hanssen, R. and Danielsen, A. *Digital Signatures: The Impact of Digitization on Popular Music Sound*, The MIT Press, Cambridge, MA, (2016) and Zagorski-Thomas, S. *The Musicology of Record Production*, Cambridge University Press, Cambridge (2014).
2. On post-digitality cf. Kulle, D., Lund, C., Schmidt, O. and Ziegenhagen, D., editors. *Post-digital Culture*, (2017), accessed September2019 from http://post-digital-culture.org/.
3. Rothlein, J. EX. 195 Matt Colton, (2014), accessed September2019 from https://www.residentadvisor.net/podcast-episode.aspx?exchange=195, ca. 28:00–29:00 min.
4. Cf. Lund, C. and Lund, H. 'No Bass: Roland TB 303 Bass Line – der kreative Missbrauch eines Musikinstrumentes und seine innovativen musikhistorischen Folgen.' In: Keller, D. and Dillschnitter, M., editors. *Zweckentfremdung: 'Unsachgemäßer' Gebrauch als kulturelle Praxis*, Wilhelm Fink, Paderborn, (2016), pp. 123–132. Cf. also Lund, H. Den Sorte Skole: Musical Expression as a Black Market Practice and the Search for a Grey Area, Akademie Schloss Solitude, Stuttgart (2015), accessed December2017 from http://www.quotes-and-appropriation.de/speaker/holger-lund.
5. Cf. Lund, H. The Aesthetics of Imperfection and Hybridization – What is so Interesting About Turkish Funk and Pop Music of the 1960s and 1970s? (2011), accessed September2019 from http://www.fluctuating-images.de/the-aesthetics-of-imperfection-and-hybridization/. Lund, H. Anatolian Rock: Phenomena of Hybridization (2013), accessed September2019 from http://norient.com/academic/anatolian-rock/. Lund, C. and Lund, H. 'Style and Society – Istanbul's Music Scene in the 1960s and 1970s: Musical Hybridism, the Gazino and Social Tolerance', in: Helms, D. and Phleps, T. (eds.). *Speaking in Tongues: Pop lokal global*, Transcript, Bielefeld (2015), pp. 177–198, accessed September2019 from http://geb.uni-giessen.de/geb/volltexte/2017/12973/pdf/Popularmusikforschung42_11_Lund.pdf.
6. Some recommended listening examples: https://www.discogs.com/TAR-VGB-Loonies-Rough-Surface/release/10492683 VGB, 'Rough Surface', released on Well Rounded Dubs, 2017; https://www.discogs.com/NUMBer-2-Headland-%C3%86ther-Headland-Remix-Local-Tetrad-Remix666/release/11504191 Unknown Artists, 'RMXZ 1', released on Well Rounded Dubs, 2018, https://www.discogs.com/Logos-Glass-Boylan-Devils-Mix/release/9843807 https://www.discogs.com/NUMBer-2-Headland-%C3%86ther-

Headland-Remix-Local-Tetrad-Remix666/release/11504191 Logos, 'Glass', Boylan Devils Mix, released on Devils, 2017; https://www.discogs.com/Unknown-Artist-Dreadz/release/10774887 S.P.Y, 'Termination', released on Hospital Records, 2017; https://www.discogs.com/SPY-Alone-In-The-Dark-EP-2/release/10703223 Unknown Artist, 'Dreadz', released on Dreadz, 2017; https://www.discogs.com/Unknown-Artist-Dreadz/release/10774887 Asda, 'Three Tracks [especially Track B1]', released on Fuckpunk, 2014/2018 (repress); https://www.discogs.com/Asda-Three-Tracks/release/12185465 Siskiyou, 'Swerve', released on chestplate, 2018; https://www.discogs.com/Siskiyou-Mirrors/release/11909435 Pinch & Mumdance, 'Strobe Light', released on Tectonic Recordings, 2017 https://www.discogs.com/Pinch-2-Mumdance-Control-Strobe-Light/release/10885327.

7. According to Anton Spice: 'We asked three of the world's premier mastering and cutting engineers (of which not many more than a hundred exist)', In: Spice, A. What's Actually Going on When People Talk About Digital vs. Analogue Masters, (2017), accessed September2019 from https://thevinylfactory.com/features/analogue-digital-vinyl-mastering-interviews/.
8. Amongst the in 2019 like "woman in 2019 to win a grammy" very few female exceptions one can find Mandy Parnell (http://www.blacksaloonstudios.com/), Emily B. Lazar, who was the first woman to win a Grammy in the 'best engineered album' category after 60 years, and Katie Tavini (https://www.katietavini.co.uk/). Thanks to Katia Isakoff for pointing these out to me. See also recently Wolfe, P.: *Women in the Studio. Creativity, Control and Gender in Popular Music Sound Production*, Routlegde, Oxon, New York (2019).
9. Cf. Spice, A. Sculpting Sound: The Art of Vinyl Mastering (2014), accessed September2019 from https://thevinylfactory.com/films/watch-our-new-short-film-on-the-art-of-vinyl-mastering/. Yet, with the vinyl revival and the vinyl market constantly growing new cutting machines and new cutting technologies like Rebeat's HD-Vinyl (2020) are developed in recent times. Even semi-professional cutting machines like the PhonoCut (2019) are developed, a consumer friendly at-home vinyl record cutting machine, which allows for DIY lathe cutting.
10. Deupree, T. In: Sage, C. Deupree Pt. Two (2011). Accessed December2017 from http://www.fluid-radio.co.uk/2011/02/12k-live-in-tokyo-taylor-deupree-part-two-%e2%80%93-the-arms-race-primum-non-nocere/.
11. Cf. e.g. YALLY (Raime), 'Burnt/Sudo' (2017), boomkat product page accessed September2019 from https://boomkat.com/products/burnt-sudo.
12. ———., '(Re-)Mastering Sonic Media History', in: Papenburg, J. G. and Schulze, H., editors. *Sound as Popular Culture: A Research Companion*, MIT Press, Cambridge, MA, (2016), p. 374.
13. Papenburg, J. G., '(Re-)Mastering Sonic Media History', in: Papenburg, J. G. and Shulze, H., editors. *Sound as Popular Culture: A Research Companion*, MIT Press, Cambridge, MA, (2016), p. 380.
14. ———. 'Gateway of Sound: Reassessing the Role of Audio Mastering in the Art of a Record Production', *Dancecult: Journal of Electronic*

Dance Music Culture, Vol. 6, No. 1, (2014), p. 8. Accessed September2019 from https://openmusiclibrary.org/article/541300/.

15. ———. 'Gateway of Sound: Reassessing the Role of Audio Mastering in the Art of a Record Production', *Dancecult: Journal of Electronic Dance Music Culture*, Vol. 6, No. 1, (2014), p. 10. Accessed September2019 from https://openmusiclibrary.org/article/541300/.
16. Nardi, C. 'Gateway of Sound: Reassessing the Role of Audio Mastering in the Art of a Record Production', *Dancecult: Journal of Electronic Dance Music Culture*, Vol. 6, No. 1, (2014), p. 9. Accessed September2019 from https://openmusiclibrary.org/article/541300/.
17. ———. Sculpting Sound: The Art of Vinyl Mastering (2014), accessed September 2019 from https://thevinylfactory.com/films/watchour-new-short-film-on-the-art-of-vinyl-mastering/.
18. Spice, A. What's Actually Going on When People Talk About Digital vs. Analogue Masters, (2017), accessed September 2019 from https://thevinylfactory.com/features/analogue-digital-vinyl-mastering-interviews/.
19. Cf. on post-digitality, see: Lund, H. 'Make It Real and Get Dirty! On the Development of Post-digital Aesthetics in Music Video', English version of: 'Make It Real & Get Dirty! Zur Entwicklung postdigitaler Ästhetiken im Musikvideo' (2015). In: Kulle, D., Lund, C., Schmidt, O. and Ziegenhagen, D., editors. *Post-digital Culture* (2017). Accessed September2019 from http://post-digital-culture.org/.
20. Fifteen Questions. Fifteen Questions with Taylor Deupree. Master of the microcosm. The irony of analogue (n.d.). Accessed September 2019 from https://15questions.net/interview/fifteen-questions-taylor-deupree/page-3/.
21. Fifteen Questions. Fifteen Questions with Taylor Deupree. Master of the microcosm. The irony of analogue (n.d.). Accessed September 2019 from https://15questions.net/interview/fifteen-questions-taylor-deupree/page-3/.
22. Bartmanski, D. and Woodward, I. *Vinyl: The Analog Record in the Digital Age*, Bloomsbury Academic, London, Oxford, New York, (2015), p. 81.
23. Shelvock, M. *Audio Mastering as Musical Practice, Master's Thesis (Popular Music and Culture)*, University of Western Ontario, London, ON, (2012), p. 59. Accessed September2019 from http://ir.lib.uwo.ca/cgi/viewcontent.cgi?article=1709&context=etd.
24. Becker, H. S. *Art Worlds*, University of California Press, Berkeley, Los Angeles, London, (1982), p. 35.
25. Cf. Weber, T. 'Vorwort des Herausgebers', In: ———. *Kunstwelten, Avinus*, Hamburg, (2017), p. 13.
26. ———. *Art Worlds*, University of California Press, Berkeley, Los Angeles, London, (1982), p. 17.
27. ———. *Art Worlds*, University of California Press, Berkeley, Los Angeles, London, (1982), p. 18.
28. Becker, H. S. *Art Worlds*, University of California Press, Berkeley, Los Angeles, London, (1982), p. 18.

29. For sound art cf. Sharma, G. K. *Komponieren mit skulpturalen Klangphänomenen in der Computermusik*, Diss. Universität für Musik und Darstellende Kunst Graz, (2016). Accessed September2019 from https://iem.kug.ac.at/fileadmin/media/osil/gksh_dissertation.pdf.
30. Lubich, A., in: Spice, A. Sculpting Sound: The Art of Vinyl Mastering, (2014) accessed September 2019 from https://thevinylfactory.com/fi lms/watchour-new-short-fi lm-on-the-art-of-vinyl-mastering/.
31. ———., in: Rothlein, J. Rashad Becker: Non-traditional Music, (2013). Accessed September2019 from https://www.residentadvisor.net/features/1941.
32. Becker, R., In: Hall, J. Revolutionary Intent: Mastering Legend Rashad Becker on Lenin, Pig Bladders and Killing the Author, (2013). Accessed September 2019 from http://www.factmag.com/2013/09/10/revolutionary-intent-mastering-legend-rashad-becker-on-lenin-pig-bladders-and-killing-the-author/2/.
33. Bartmanski, D. and Woodward, I. *Vinyl: The Analog Record in the Digital Age*, Bloomsbury Academic, London, Oxford, New York, (2015), p. 75.
34. Rothlein, J. Rashad Becker: Non-traditional Music (2013). Accessed September 2019 from https://www.residentadvisor.net/features/1941.
35. Becker, R., In: Henke, R. Mastering (2008). Accessed September2019 from http://roberthenke.com/interviews/mastering.html.
36. Papenburg, J. G. '(Re-)Mastering Sonic Media History', in: Papenburg, J. G. and Shulze, H., editors. *Sound as Popular Culture: A Research Companion*, MIT Press, Cambridge, MA, (2016), p. 374.
37. Papenburg, J. G. '(Re-)Mastering Sonic Media History', in: Papenburg, J. G. and Shulze, H., editors. *Sound as Popular Culture: A Research Companion*, MIT Press, Cambridge, MA, (2016), p. 375.
38. Papenburg, J. G. '(Re-)Mastering Sonic Media History', in: Papenburg, J. G. and Shulze, H., editors. *Sound as Popular Culture: A Research Companion*, MIT Press, Cambridge, MA, (2016), p. 379.
39. Eduardo und Silvia Araujo, *Rebu Geral*, FIF, Brazil 1981, https://www.discogs.com/de/Eduardo-E-Silvia-Araujo-Rebu-Geral/release/1876570. For the back cover see https://www.discogs.com/de/Eduardo-Araujo-Sylvia-Araujo-Rebu-Geral-/release/11644330, where images on the CD reissue show the original LP back cover. Accessed September2019.
40. Becker, R., in: Henke, R. Mastering, (2008). Accessed September 2019 from http://roberthenke.com/interviews/mastering.html.
41. Diederichsen, D., Über Pop-Musik, Kiepenheuer and Witsch, Cologne, (2014), p. 327.
42. Diederichsen, D., Über Pop-Musik, Kiepenheuer and Witsch, Cologne, (2014), p. 327.
43. Diederichsen, D., Über Pop-Musik, Kiepenheuer and Witsch, Cologne, (2014), p. 327.
44. Diederichsen, D., Über Pop-Musik, Kiepenheuer and Witsch, Cologne, (2014), p. 327.
45. Papenburg, J. G. Papenburg, J. G. '(Re-)Mastering Sonic Media History', in: Papenburg, J. G. and Shulze, H., editors. *Sound as Popular*

Culture: A Research Companion, MIT Press, Cambridge, MA, (2016), p. 376f.
46. Manning, J. Hit Factory: An Interview with Matt Colton, (2013). Accessed September2019 from https://www.juno.co.uk/reviews/2013/09/10/signal-path-hit-factory-an-interview-with-matt-colton/.
47. Manning, J. Hit Factory: An Interview with Matt Colton, (2013). Accessed September 2019 from https://www.juno.co.uk/reviews/2013/09/10/signal-path-hit-factory-an-interview-with-matt-colton/.
48. Cf. dBs Music. Guest Lecture: Matt Colton Mastering Engineer, (2016). Accessed September2019 from https://www.youtube.com/watch?v=wzaCZGNRY4s.
49. Manning, J. Hit Factory: An Interview with Matt Colton, (2013). Accessed September 2019 from https://www.juno.co.uk/reviews/2013/09/10/signal-path-hit-factory-an-interview-with-matt-colton/.
50. Ivey, J. Interview – Matt Colton of Alchemy Mastering, (2017). Accessed September2019 from https://www.youtube.com/watch?v=OSd6fFREqts, ca. 13:30 min.
51. Rothlein, J. X. 195 Matt Colton (2014). Accessed September2019 from https://www.residentadvisor.net/podcast-episode.aspx?exchange=195.
52. Colton, M., in: Rothlein, J. EX. 195 Matt Colton (2014). Accessed September 2019 from https://www.residentadvisor.net/podcast-episode.aspx?exchange=195.
53. Colton, M. in: Rothlein, J. EX. 195 Matt Colton (2014). Accessed September 2019 from https://www.residentadvisor.net/podcast-episode.aspx?exchange=195.
54. Colton, M., in: Rothlein, J. EX. 195 Matt Colton (2014). Accessed September 2019 from https://www.residentadvisor.net/podcast-episode.aspx?exchange=195.
55. Colton, M., in: Rothlein, J. EX. 195 Matt Colton (2014). Accessed September 2019 from https://www.residentadvisor.net/podcast-episode.aspx?exchange=195.
56. Jackson Bailey aka Tapes, In: Hurkx, M. 'Ghosts in the Cables', in: *House of Music, No. 5*, Rushhour, Amsterdam, (2017), p. 29. For 'ghosts in the cables' also cf.: ARK, 'Arkstrated Rhythmachine Komplexities: Machinic Geisterstunde and Post-Soul Persistences', in: Rohlf, J., Gärlid, A., Replansky, T. et.al. (eds.). Persistence: *CTM-Festival 2019. Adventurous Music & Art* (2019), pp. 60–65.
57. Deupree, T. In: Sage, C. Deupree Pt. Two (2011). Accessed December 2017 from http://www.fluid-radio.co.uk/2011/02/12k-live-in-tokyo-taylordeupree- part-two-%e2%80%93-the-arms-race-primum-non-nocere/.
58. Deupree T. In: Fifteen Questions. Fifteen Questions with Taylor Deupree. Master of the microcosm. The irony of analogue (n.d.). Accessed September 2019 from https://15questions.net/interview/fi fteen-questions-taylor-deupree/page-3/.
59. Deupree, T. In: Sage, C. Deupree Pt. Two (2011). Accessed December 2017 from http://www.fluid-radio.co.uk/2011/02/12k-live-in-tokyo-taylordeupree-part-two-%e2%80%93-the-arms-race-primum-non-nocere/.

60. Deupree, T. In: Sage, C. Deupree Pt. Two (2011). Accessed December 2017 from http://www.fluid-radio.co.uk/2011/02/12k-live-in-tokyo-taylordeupree-part-two-%e2%80%93-the-arms-race-primum-non-nocere/.
61. Deupree T. In: Fifteen Questions. Fifteen Questions with Taylor Deupree. Master of the microcosm. The irony of analogue (n.d.). Accessed September 2019 from https://15questions.net/interview/fi fteen-questions-taylor-deupree/page-3/.
62. Deupree, T. In: Fifteen Questions. Fifteen Questions with Taylor Deupree. Master of the microcosm. The irony of analogue (n.d.). Accessed September 2019 from https://15questions.net/interview/fi fteen-questions-taylor-deupree/page-3/.
63. Nardi, C. 'Gateway of Sound: Reassessing the Role of Audio Mastering in the Art of a Record Production', Dancecult: Journal of Electronic Dance Music Culture, Vol. 6, No. 1, (2014), p. 8. Accessed September 2019 from https://openmusiclibrary.org/article/541300/.
64. Robert Henke, in: Bartmanski, D. and Woodward, I. Vinyl: The Analog Record in the Digital Age, Bloomsbury Academic, London, Oxford, New York, (2015), p. 75.
65. Lubich, A., Spice, A. Sculpting Sound: The Art of Vinyl Mastering (2014), accessed September 2019 from https://thevinylfactory.com/fi lms/watchour- new-short-fi lm-on-the-art-of-vinyl-mastering/. Yet, with the vinyl revival and the vinyl market constantly growing new
66. Papenburg, J. G. '(Re-)Mastering Sonic Media History', in: Papenburg, J. G. and Shulze, H., editors. Sound as Popular Culture: A Research Companion, MIT Press, Cambridge, MA, (2016), p. 378f.
67. Nardi, C. 'Gateway of Sound: Reassessing the Role of Audio Mastering in the Art of a Record Production', Dancecult: Journal of Electronic Dance Music Culture, Vol. 6, No. 1, (2014), p. 8. Accessed September 2019 from https://openmusiclibrary.org/article/541300/.
68. Nardi, C. 'Gateway of Sound: Reassessing the Role of Audio Mastering in the Art of a Record Production', Dancecult: Journal of Electronic Dance Music Culture, Vol. 6, No. 1, (2014), p. 13. Accessed September 2019 from https://openmusiclibrary.org/article/541300/.
69. Diederichsen, D. *Über Pop-Musik* Kiepenheuer and Witsch, Cologne (2014), p. 78.
70. Diederichsen, D., *Über Pop-Musik*, Kiepenheuer and Witsch, Cologne, (2014), p. 52.
71. Diederichsen, D., *Über Pop-Musik*, Kiepenheuer and Witsch, Cologne, (2014), p. 263. Diederichsen also arrives at an early critique of this development: 'The sonic means to make something sound bigger, louder, more dramatic, fatter or more voluminous tend to separate from that which they were supposed to magnify, especially in recent times. Pedestals become minimalist sculptures' (ibid., p. 316).
72. Brockhaus, I. *Kultsounds: Die prägendsten Klänge der Popmusik 1960–2014*, Transcript, Bielefeld, (2017).
73. Rainer Maillard, recording engineer and producer for the Emil Berliner Studios, observes a similar development: 'I have worked digitally for more than 20 years and I've noticed that slowly productions are shifting

more and more into post-production because of the huge number of possibilities.' Quoted In: Thomas, D. 'Platte zum Aufnehmen', in: *Mint*, No. 21, (2018), p. 16.

74. Becker, R., in: Henke, R. Mastering (2008). Accessed September 2019 from http://roberthenke.com/interviews/mastering.html.
75. Ludwig B. In: Owsinski, B. *The Mastering Engineer's Handbook*, Cengage Learning, Boston, (2015), p. 148.
76. Cafe, R. 'Last call: What's happened to London's nightlife?', BBC News, 7 October 2016. Accessed September2019 from http://www.bbc.com/news/uk-england-london-37546558. For Berlin see also: Perdoni, S. 'Klubszene: Kampf ums Überleben in der Innenstadt', Berliner Zeitung, 9 November 2017, accessed September2019 from https://www.berliner-zeitung.de/berlin/klubszene-kampf-ums-ueberleben-in-der-innenstadt-28819888.
77. Diederichsen, D. *Körpertreffer: Zur Ästhetik der nachpopulären Künste*, Suhrkamp, Berlin (2017), p. 122.
78. Although clubs can buy themselves out of these regulations at a steep price. 'Protection of the population' here also offers a very profitable business model for state or council. Source: personal encounters with club owners in Porto, Portugal, December 2016.
79. Diederichsen, D., Über Pop-Musik, Kiepenheuer and Witsch, Cologne, (2014), p. 321.
80. Diederichsen, D., Über Pop-Musik, Kiepenheuer and Witsch, Cologne, (2014), p. 327.

Index

Printed in the United States
By Bookmasters